T0181892

Communications
in Computer and Information Science   1181

Commenced Publication in 2007
Founding and Former Series Editors:
Phoebe Chen, Alfredo Cuzzocrea, Xiaoyong Du, Orhun Kara, Ting Liu,
Krishna M. Sivalingam, Dominik Ślęzak, Takashi Washio, Xiaokang Yang,
and Junsong Yuan

More information about this series at http://www.springer.com/series/7899

Guangtao Zhai · Jun Zhou · Hua Yang ·
Ping An · Xiaokang Yang (Eds.)

Digital TV and Wireless Multimedia Communication

16th International Forum, IFTC 2019
Shanghai, China, September 19–20, 2019
Revised Selected Papers

 Springer

Editors
Guangtao Zhai 🆔
Shanghai Jiao Tong University
Shanghai, China

Hua Yang
Shanghai Jiao Tong University
Shanghai, China

Xiaokang Yang
Shanghai Jiao Tong University
Shanghai, China

Jun Zhou 🆔
Shanghai Jiao Tong University
Shanghai, China

Ping An
Shanghai University
Shanghai, China

ISSN 1865-0929   ISSN 1865-0937   (electronic)
Communications in Computer and Information Science
ISBN 978-981-15-3340-2   ISBN 978-981-15-3341-9   (eBook)
https://doi.org/10.1007/978-981-15-3341-9

This Springer imprint is published by the registered company Springer Nature Singapore Pte Ltd.
The registered company address is: 152 Beach Road, #21-01/04 Gateway East, Singapore 189721, Singapore

Preface

The present book includes extended and revised versions of papers selected from the 16th International Forum of Digital Media Communication (IFTC 2019), held in Shanghai, China, during September 19–20, 2019.

IFTC is a summit forum in the field of digital media communication. The 2019 forum was co-hosted by SIGA, the China International Industry Fair (CIIF 2019), and Shanghai Association for Science and Technology, and co-sponsored by Shanghai Jiao Tong University (SJTU), China Telecom Group, Shanghai Telecom Company, IEEE BTS Chapter of Shanghai Section, Shanghai University of Engineering Science, and Shanghai Key Laboratory of Digital Media Processing and Transform. The 16th IFTC serves as an international bridge for extensively exchanging the latest research advances of digital media communication around the world as well as the relevant policies of industry authorities. The forum also aims to promote the technology, equipment, and application in the field of digital media by comparing the characteristics, framework, significant techniques and their maturity, analyzing the performance of various applications in terms of scalability, manageability, and portability, and discussing the interfaces among varieties of networks and platforms. Also, a "TianYi" Cup AI competition was held, sponsored by Shanghai Telecom Company.

The conference program included invited talks delivered by 4 distinguished speakers from Macau, China, as well as an oral session of 6 papers and a poster session of 26 papers. Another 3 papers were accepted from the AI competition. The topics of these papers range from audio/image processing to image and video compression as well as telecommunications. This book contains 35 papers selected from IFTC 2019.

The proceeding editors wish to thank the authors for contributing their novel ideas and visions that are recorded in this book, and all reviewers for their contributions. We also thank Springer for their trust and for publishing the proceedings of IFTC 2019.

November 2019

Guangtao Zhai
Jun Zhou
Hua Yang
Ping An
Xiaokang Yang

Organization

General Chairs

Xiaokang Yang   Shanghai Jiao Tong University, China
Ping An   Shanghai University, China
Guangtao Zhai   Shanghai Jiao Tong University, China

Program Committee

Jun Zhou   Shanghai Jiao Tong University, China
Yue Lu   East China Normal University, China
Hua Yang   Shanghai Jiao Tong University, China
Jiantao Zhou   University of Macau, China

Competition Chairs

Yiyi Lu   China Telecom Shanghai Company, China
Zhijun Fang   Shanghai University of Engineering Science, China
Hanlong Guo   China Telecom Shanghai Company, China

International Liaisons

Weisi Lin   Nanyang Technological University, Singapore
Patrick Le Callet   Nantes University, France
Lu Zhang   INSA de Rennes, France

Finance Chairs

Yi Xu   Shanghai Jiao Tong University, China
Lianghui Ding   Shanghai Jiao Tong University, China

Publications Chairs

Xianming Liu   Harbin Institute of Technology, China
Qiudong Sun   Shanghai Second Polytechnic University, China
Liquan Shen   Shanghai University, China

Award Chairs

Changwen Chen   SUNY Buffalo, USA
Wenjun Zhang   Shanghai Jiao Tong University, China

Publicity Chairs

Xiangyang Xue   Fudan University, China
Yuming Fang   Jiangxi University of Finance and Economics, China

Industrial Program Chairs

Zhenning Zhang   Think Force Inc., China
Guozhong Wang   Shanghai University, China

Organizing Committee

Cheng Zhi   SIGA, China
  (Secretary-General)

Contents

Quality Assessment

Telecommunication

Video Surveillance

Virtual Reality

Image Processing

Fast Traffic Sign Detection Using Color-Specific Quaternion Gabor Filters

Shiqi Yin🆔 and Yi Xu^(✉)

Institute of Image Communication and Network Engineering,
Shanghai Jiao Tong University, 800 Dongchuan Road, Minhang District,
Shanghai, China
{yinshiqi,xuyi}@sjtu.edu.cn

Abstract. A novel and fast traffic sign detection method is proposed based on color-specific quaternion Gabor filtering (CS-QGF). The proposed method is based on the fact that traffic signs are usually specialized in color and shape. Accordingly, we apply a quaternion Gabor transformation to extract the color and shape features of traffic signs simultaneously. Statistical color distribution of traffic sign is analyzed to optimize the construction of quaternion Gabor filters. The feature extracted via CS-QGF is robust to the distortion of color, the change of image resolution, and the change of lighting and shading conditions, which helps the following traffic sign detector reduce the search range of proposal regions. Experiments on GTSDB and TT100K datasets demonstrate that the proposed method helps to localize traffic signs in images with high efficiency, which outperforms state-of-the-art methods on both detection speed and final recognition accuracy.

Keywords: Traffic sign detection · Traffic sign recognition ·
Color-specific quaternion Gabor filtering

1 Introduction

Traffic sign recognition (TSR) is an indispensable part in many applications, such as advanced driver assistance system, highway maintenance and intelligent traffic enforcement, which requires us to detect small traffic signs quickly from the large-size images captured in different lighting and shading conditions and recognize their semantic meanings according to the color and shape information of the detected objects. In the detection phase of a TSR system, the main challenge is that the extraction of region proposal is often negatively influenced by the variations of color, contrast, and resolution in different scenarios, as illustrated in Fig. 1. How to extract features that are robust to these changes and keep a fast detection speed at the same time is the main task of this work.

To achieve real-time processing, image processing-based methods are widely used to generate proposal regions of traffic signs, *e.g.*, color thresholding [4,10,12,14,18] or shape analysis [1]. In addition, Shao [17] utilized simplified

© Springer Nature Singapore Pte Ltd. 2020
G. Zhai et al. (Eds.): IFTC 2019, CCIS 1181, pp. 3–12, 2020.
https://doi.org/10.1007/978-981-15-3341-9_1

(a) Template sign and its over-lighting and low-resolution case

(b) Template sign and its dark shading case

Fig. 1. An illustration of the traffic signs with different shapes and color distortions.

Gabor wavelets (SGW) to extract features through monochromatic traffic signs. However, these methods are sensitive to the change of scenarios, which leads to high detection and recognition errors. On the contrary, the recent learning-based methods utilize neural networks (NNs) [2,6,15] and support vector machine (SVM) [3,19] to detect and recognize signs with high accuracy, but their huge computational cost makes them difficult to process in real-time. To achieve a balance between performance and speed, the hybrid systems are proposed, combining image processing-based features for detection and learning-based models for recognition. The typical features include the descriptors of MSER (maximally stable extremal region) [19,20] and Histogram of Oriented Gradient (HOG) [1,13,19,20], while the recognition models can be sparse representations [9], Capsule Networks [7] and Scale-aware CNN [21], etc. In the framework of hybrid system, we propose a traffic sign detection method based on color-specific quaternion Gabor filtering (CS-QGF), which is robust to color distortions, low contrast and low resolution. In particular, considering the fact that traffic signs are characterized by their colors and shapes, we formulate RGB signals as quaternions and apply Gabor filtering in the quaternion space. The quaternion Gabor filters inherit the high-pass property of traditional real-valued Gabor filters, and extend this property to color space via setting their rotation axes according to the specific colors of the traffic signs. Additionally, a multi-scale mechanism is introduced to enhance the robustness of the filtering results to the change of scale. Accordingly, we extract color-edge regions as the sign proposals, which reduces the search range of proposals and accelerates the following detector greatly. Given the detected regions, we modify a CNN traffic-signs classifier [22] to achieve a trade-off between classification/recognition accuracy and recall rate.

2   Proposed Method

2.1   Proposed Framework

As illustrated in Fig. 2, our traffic sign recognition system includes a detection stage and a recognition stage. In the detection stage, based on the fact that traffic signs can be divided into three categories according to background or outline color (*i.e.*, red, yellow/orange and blue), we establish a set of multi-scale quaternion Gabor filters to extract color edges for these three particular colors and locate the traffic sign boundaries. Then, the shape constraints are imposed on these color edge maps, which consists of shape detectors like circle, rectangle, triangle etc., and we obtain the marks of the traffic signs. In the recognition stage, these masked region proposals will be transported to traffic sign recognition network.

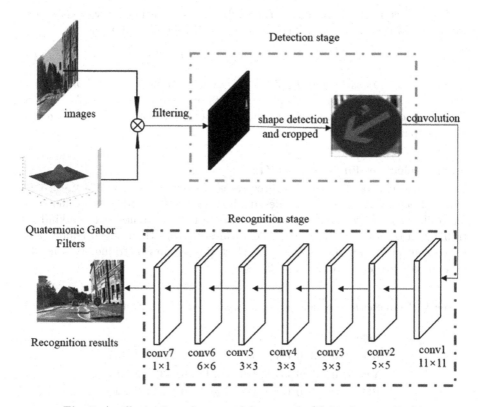

Fig. 2. An illustration of proposed framework. (Color figure online)

2.2 Constructions of CS-QGFs

2.2.1 Fundamentals of QGFs

Gabor filters are sensitive to edges in images, which show the similar properties like human vision system. The complex Gabor filters could only process monochromatic channels, in this paper, we adopt multi-scale Quaternion Gabor Filters to capture edges around particular colors. The QGFs can be build as follows:

$$g(x, y) = \frac{1}{2\pi\sigma_x\sigma_y} \exp\left[-\frac{1}{2}\left(\frac{x^2}{\sigma_x^2} + \frac{y^2}{\sigma_y^2}\right) + 2\mu\pi\left(\boldsymbol{W}x + \boldsymbol{V}y\right)\right], \qquad (1)$$

where σ_x and σ_y are space constant along the X-axis and Y-axis of Gaussian Envelope Function, μ denotes the pure imaginary unit, which represents a particular color of traffic signs and its modulus is equal to 1. $\boldsymbol{W}$ and $\boldsymbol{V}$ in the QGFs are the center frequency of the filter along the X-axis and Y-axis respectively.

Correspondingly, its Quaternion Fourier Transformation (QFT) is:

$$QFT(g) = \exp\left\{-2\pi^2 \left[\sigma_x^2 (u - \boldsymbol{W})^2 + \sigma_y^2 (v - \boldsymbol{V})^2\right]\right\}. \tag{2}$$

To extract image information from different orientations at various scales, we can build the multi-scale Quaternion Gabor Filter (CS-QGF) as:

$$g_{sn} = a^{-2s}g(x', y'), a > 1, \tag{3}$$

where $x' = a^{-s}(x \cos\theta + y \sin\theta)$, $y' = a^{-s}(-x \sin\theta + y \cos\theta)$, and $\theta = n\pi/N, n = 0, 1, \ldots N-1$, the orientation number of the filters, while $s = 0, 1, \ldots M-1$, CS-QGF extracts M-scale image information through the scale factor a^{-2s} ($a = 2$ in the experiments).

2.2.2   Selection for μ of CS-QGFs

Traffic signs can be divided into three categories based on color features: blue, red, and yellow. Accordingly we design three sets of CS-QGFs with different μ for traffic signs with particular colors. To obtain RGB values of three kinds of traffic sign colors, we cluster the color values of more than 30000 traffic signs using K-means method ($k = 3$). The clustering results of traffic signs in two datasets are given in Tables 1 and 2 respectively.

Table 1. Parameter μ_{gsn} of quaternion Gabor filters to detect traffic signs in GTSDB.

Color	c_i	c_j	c_k	μ_{gsn}
Red	0.823	0.434	0.354	$0.823i + 0.434j + 0.354k$
Yellow	0.731	0.533	0.426	$0.731i + 0.533j + 0.426k$
Blue	0.261	0.475	0.841	$0.261i + 0.475j + 0.841k$

2.3   Fast Detection for Traffic Sign Proposals

Color image could be represented as a pure quaternion matrix. In this case, using the imaginary part of CS-QGF to extract image color information will be

Table 2. Parameter μ_{gsn} of quaternion Gabor filters to detect traffic signs in TT100K.

Color	c_i	c_j	c_k	μ_{gsn}
Red	0.963	0.212	0.169	$0.963i + 0.212j + 0.169k$
Orange	0.708	0.639	0.300	$0.708i + 0.639j + 0.300k$
Blue	0.273	0.497	0.826	$0.272i + 0.497j + 0.826k$

more effective. According to derivation process in [11] , we can divided QGF into 2 parts: $g = g_R + \mu g_I$, the scalar part g_R and imaginary part g_I. Functions are shown in the followings:

$$g_R(x,y) = \frac{1}{2\pi\sigma_x\sigma_y} \exp\left[-\frac{1}{2}\left(\frac{x^2}{\sigma_x^2} + \frac{y^2}{\sigma_y^2}\right) + \cos 2\pi\left(\boldsymbol{W}x + \boldsymbol{V}y\right)\right], \quad (4)$$

and

$$g_I(x,y) = \frac{1}{2\pi\sigma_x\sigma_y} \exp\left[-\frac{1}{2}\left(\frac{x^2}{\sigma_x^2} + \frac{y^2}{\sigma_y^2}\right) + \sin 2\pi\left(\boldsymbol{W}x + \boldsymbol{V}y\right)\right]. \quad (5)$$

Each color pixel can be regarded as $\mu_0 = (r_0, g_0, b_0)$. The QGF can be expressed as $\mu_{gsn} = \mu_g g_I$ (where $\mu_g = (c_i, c_j, c_k)$), denoting as the basic color of traffic signs. The selection of μ_{gsn} is illustrated in Table 1. We can obtain the filtering results through multiplication of quaternion Gabor filters μ_{gsn} and quaternion pixel μ_0. The derivation process is as follows:

$$I = \mu_0 \times \mu_{gsn} = g_I\left(\mu_0 \times \mu_g\right) = g_I \begin{vmatrix} i & j & k \\ r_0 & g_0 & b_0 \\ c_i & c_j & c_k \end{vmatrix}, \quad (6)$$

where $|\cdot|$ is the determinant of a matrix. Ideally, for the region has similar color to the traffic signs, $i.e.$, $(r_0, g_0, b_0) \approx c\,(c_i, c_j, c_k)$, c is a constant. Accordingly, we can rewrite (6) as:

$$I = cg_I\left(\mu_0 \times \mu_g\right) = g_I \begin{vmatrix} i & j & k \\ r_0 & g_0 & b_0 \\ c_i & c_j & c_k \end{vmatrix} \approx 0. \quad (7)$$

It means that the smaller the difference in color between a color image region and the basic traffic sign color, the filtering response is closer to zero, which means the region can be regarded as traffic sign proposals.

2.4   Traffic Sign Recognition

In this paper, we integrate our detection method with the modified CNN [22] classifier for traffic sign recognition, which is a modified robust end-to-end CNN model based on *Overfeat* [16] and can detect and classify objects simultaneously. Compared with existing traffic sign algorithms, the modified CNN based on *Overfeat* is more robustness to the small objects.

For the first stage of the traffic sign detection task, we adopt a set of multi-scale CS-QGFs to filtering out traffic signs which can be classified into three different categories in particular colors. However, a lot of objects with similar background colors may increase the detection error in the reality and wild scenes. Thus, in this paper we add further shape descriptors to the color-filtered binary feature map through *Hough Transformation* in order to classify traffic signs more effectively and quickly. It is known that traffic signs are generally designed in particular shapes like circle, triangle and rectangle.

As for the second stage, we modified the-state-of-art method of [22] as our baseline to classify and recognize traffic signs, which demonstrates the superiority of our proposed CS-QGFs. In [22], Zhu et al. evaluate performance of CNNs on traffic sign detection in the wild scenes, which is a network based on *Overfeat* framework with a bounding box regression step. Their network is fully convolutional, and the last layer is branched into three streams: a pixel layer, a bounding box layer and a label layer. The output of the pixel layer represents the probability of a certain 4×4 pixel region in the input image containing a target object. For the bounding box layer, the result represents the distance between that region and the four sides of the predicted bounding box of the target. While for the function of label layer, it outputs a classification vector with n elements, where each element is the probability it belongs to a specific class.

In general, the TSR method we proposed can not only extract the color regions of traffic signs with higher efficiency, but also validate the high robustness of the classifier network to the small signs. In the experiment, we cut the pixel layer to adapt to the datasets. The details of modified CNN for recognition stage are illustrated in blue dotted box in Fig. 2, and the experimental results in Sect. 3 also proved promising performance on traffic sign proposals using CS-QGFs.

3   Experimental Results

We evaluate our approach on two traffic sign datasets: GTSDB [5] and TT100K [22]. Detection procedure of approaches is conducted on MATLAB platform and classification process is implemented by Caffe . The approach is running on Geforce GTX TITAN with image size of 800×1360 and 2048×2048 pixels. Our experiment is designed as two parts. First, Considering the characteristics of different public datasets, we evaluate our traffic sign detection approach with the state-of-the-art methods on GTSDB. Then we integrate the state-of-the-art classification networks [22] with our detection algorithm on both two datasets to validate our method's effectiveness.

The GTSDB dataset contains 900 images and more than 1200 traffic signs, which can be mainly classified into three categories: prohibitory (circular, white ground with red border), mandatory (circular, blue ground) and danger (triangular, white ground with red border). TT100K is a Chinese traffic sign dataset which follows international patterns, and can also be divided into three categories: warnings (yellow triangles with black boundary and information), prohibitions (white surrounded by a red circle) and mandatory (blue circles with white information). Figure 3 shows three kinds of traffic signs in two datasets.

(a) prohibitory signs

(b) danger signs & warning signs

(c) mandatory signs

Fig. 3. Major classes in two datasets, the left column are signs in GTSDB, and the ones on the right are signs in TT100K. (Color figure online)

3.1 Experimental Results of Sign Detection

First we now considering how well our detection algorithm performs by comparing some existing state-of-the-art approaches [1,8,17,20]. During the detection stage, a predicted bounding box is considered as a positive example if it has an Intersection-over-Union (IoU) ratio greater than 0.75 with the ground truth box, and otherwise negative. Our detection results are shown in Fig. 4. Our approach achieves over 0.96 accuracy. (In this paper, we compare with the state-of-the-art algorithms.) Although the detection rate of our method is slightly lower than

Fig. 4. Detection results of the proposed approach. The left column lists the original images, the middle one lists the ground truth with green bounding boxes and the right column shows our results after binarization; and the first two rows are images from GTSDB and the bottom row is image from TT100K. (Color figure online)

Table 3. Detection results between the state-of-the-art and our approach. DR: detection rate; DS: detection speed.

Design	DR (%)	DS (ms/image)
HOG [1]	91.69	13–17
CPM [20]	97.72	67
SFC-trees [8]	95.37	187
SGW [17]	91.24	70
Ours	**96.86**	**11.9**

that of [20], our method is 6 times faster. The detection results are shown in Table 3. Generally, our algorithm can achieve the great trade-off between the detection rate and detection speed.

3.2   Experimental Results of Sign Detection and Recognition

We finally realize unified traffic detection and recognition by integrating the CS-QGFs with the modified CNN classifier. The modified CNN is a classification network [22] which based on the *Overfeat* [16]. It has been proved that the modified CNN could achieve higher recognition precision than the existing TSR algorithms especially for small traffic signs. Therefore, we select the modified CNN as the classifier to recognize the cropped signs filtered by CS-QGFs, where the sign detection module effectively improves the performance of following sign recognition due to reduced detection errors and faster processing speed. Figure 5 shows the results on both GTSDB and TT100K datasets. It indicates that our algorithm outperforms the-state-of-the-art methods. The comparison between our method and state-of-the-art methods on the two datasets is shown in Table 4.

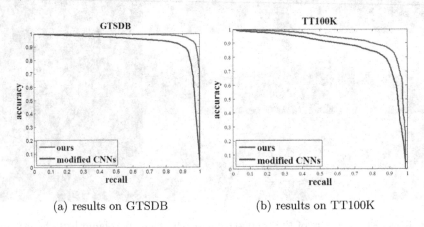

(a) results on GTSDB   (b) results on TT100K

Fig. 5. Comparison results of our approach and the-state-of-the-art [22] on GTSDB and TT100K datasets.

Table 4. Performance comparison between the state-of-art and our approach. OA: our method's accuracy, SA: the-state-of-art's accuracy [22]; OR: our method's recall, SR: the-state-of-art's recall [22]; OM: our mAP, SM: the state-of-art's mAP.

Datasets	OA	SA [22]	OR	SR [22]	OM	SM
GTSDB	**0.96**	0.93	**0.94**	0.92	**0.98**	0.93
TT100K	**0.96**	0.88	**0.92**	0.91	**0.92**	0.87

In both datasets, we obtained an average mAP of 0.98 and 0.92 respectively, which is superior to that of the state-of-the-art method.

4   Conclusion

In this paper, we present a new approach based on CS-QGF for traffic sign detection, which can accelerate the speed of proposal extraction by considering color and shape characteristics simultaneously. In addition, we integrate our method with CNN classifier to further improve the recognition accuracy. The experiments validate high robustness and efficiency of our approach in traffic sign detection and recognition tasks.

Acknowledgements. This work was supported in part by National Natural Science Foundation of China 61671298, STCSM (17511105401,18DZ2270700) and 111 project B07022.

References

1. Chen, Z., Huang, X., Ni, Z., He, H.: A GPU-based real-time traffic sign detection and recognition system (2014)
2. Ciresan, D., Meier,U., Masci, J., Schmidhuber, J.: A committee of neural networks for traffic sign classification. In: International Joint Conference on Neural Networks (2011)
3. Gonzalez, Á., et al.: Automatic traffic signs and panels inspection system using computer vision. IEEE Trans. Intell. Transp. Syst. **12**(2), 485–499 (2011)
4. Hechri, A., Mtibaa, A.: Automatic detection and recognition of road sign for driver assistance system. In: Electrotechnical Conference (2012)
5. Houben, S., Stallkamp, J., Salmen, J., Schlipsing, M., Igel, C.: Detection of traffic signs in real-world images: the German traffic sign detection benchmark. In: International Joint Conference on Neural Networks (2013)
6. Jin, J., Fu, K., Zhang, C.: Traffic sign recognition with hinge loss trained convolutional neural networks. IEEE Trans. Intell. Transp. Syst. **15**(5), 1991–2000 (2014)
7. Kumar, A.D.: Novel deep learning model for traffic sign detection using capsule networks. Int. J. Pure Appl. Math. **118**(20) (2018)
8. Liu, C., Chang, F., Chen, Z.: Rapid multiclass traffic sign detection in high-resolution images. IEEE Trans. Intell. Transp. Syst. **15**(6), 2394–2403 (2014)

9. Lu, K., Ding, Z., Ge, S.: Sparse-representation-based graph embedding for traffic sign recognition. IEEE Trans. Intell. Transp. Syst. **13**(4), 1515–1524 (2012)

10. Malik, R., Khurshid, J., Ahmad, S.N.: Road sign detection and recognition using colour segmentation, shape analysis and template matching. In: International Conference on Machine Learning and Cybernetics (2007)

11. Manjunath, B.S., Ma, W.Y.: Texture features for browsing and retrieval of image data. IEEE Trans. Pattern Anal. Mach. Intell. **18**(8), 837–842 (1996)

12. Nguwi, Y.Y., Kouzani, A.Z.: Detection and classification of road signs in natural environments. Neural Comput. Appl. **17**(3), 265–289 (2008)

13. Overett, G., Tychsen-Smith, L., Petersson, L., Pettersson, N., Andersson, L.: Creating robust high-throughput traffic sign detectors using centre-surround hog statistics. Mach. Vis. Appl. **25**(3), 713–726 (2014)

14. Ruta, A., Li, Y., Liu, X.: Real-time traffic sign recognition from video by class-specific discriminative features. Pattern Recogn. **43**(1), 416–430 (2010)

15. Sermanet, P., Lecun, Y.: Traffic sign recognition with multi-scale convolutional networks. In: International Joint Conference on Neural Networks (2011)

16. Sermanet, P., David, E., Zhang, X., Mathieu, M., Fergus, R., Lecun, Y.: OverFeat: Integrated recognition, localization and detection using convolutional networks. Eprint (2013)

17. Shao, F., Wang, X., Meng, F., Rui, T., Tang, J.: Real-time traffic sign detection and recognition method based on simplified gabor wavelets and cnns. Sensors **18**(10), 3192 (2018)

18. Wanitchai, P., Phiphobmongkol, S.: Traffic warning signs detection and recognition based on fuzzy logic and chain code analysis (2008)

19. Xue, Y., Hao, X., Chen, H., Wei, X.: Robust traffic sign recognition based on color global and local oriented edge magnitude patterns. IEEE Trans. Intell. Transp. Syst. **15**(4), 1466–1477 (2014)

20. Yang, Y., Luo, H., Huarong, X., Fuchao, W.: Towards real-time traffic sign detection and classification. IEEE Trans. Intell. Transp. Syst. **17**(7), 2022–2031 (2016)

21. Yang, Y., Liu, S., Ma, W., Wang, Q., Liu, Z.: Efficient traffic-sign recognition with scale-aware CNN (2018)

22. Zhu, Z., Liang, D., Zhang, S., Huang, X., Hu, S.: Traffic-sign detection and classification in the wild. In: IEEE Conference on Computer Vision and Pattern Recognition (2016)

Smoke Detection Based on Image Analysis Technology

Huiqing Zhang[1,2], Jiaxu Chen[1,2(✉)], Shuo Li[1,2], Ke Gu[1], and Li Wu[1]

[1] Faculty of Information Technology, Beijing University of Technology,
Beijing 100124, China
chen_jia_xu@126.com
[2] Engineering Research Center of Digital Community, Ministry of Education,
Beijing 100124, China

Abstract. Ecological problems and pollution problems must be faced and solved in the sustainable development of a country. With the continuous development of image analysis technology, it is a good choice to use machine to automatically judge the external environment. In order to solve the problem of smoke extraction and exhaust monitoring, we need the applicable database. Considering the number of databases that can be used to detect smoke is small and these databases have fewer types of pictures, we subdivide the smoke detection database and get a new database for smoke and smoke color detection. The main purpose is to preliminarily identify pollutants in smoke and further develop smoke image detection technology. We discuss eight kinds of convolutional neural network, they can be used to classify smoke images. Testing different convolutional neural networks on this database, the accuracy of several existing networks is analyzed and compared, and the reliability of the database is also verified. Finally, the possible development direction of smoke detection is summarized.

Keywords: Smoke detection · Image analysis technology · Image database

1 Introduction

With the development of three industrial revolutions, the quality of our life has been continuously improved. At the same time, the earth on which we live has been continuously polluted [1]. In the era of steam engines, haze weather first appeared in London, so London is called the fog capital. The industrial center of Germany at the turn of the 19th and 20th centuries was covered by gray and yellow smoke for a long time. Acid rain in Europe in the 1960s. In recent years, fog and haze weather has appeared in northern China. There are signs that the global environment is deteriorating. Therefore, more and more

This work is supported by the Major Science and Technology Program for Water Pollution Control and Treatment of China (2018ZX07111005).

© Springer Nature Singapore Pte Ltd. 2020
G. Zhai et al. (Eds.): IFTC 2019, CCIS 1181, pp. 13–22, 2020.
https://doi.org/10.1007/978-981-15-3341-9_2

countries pay more attention to environmental protection [2]. Flue gas and smoke dust has the greatest impact on environment. In addition to its toxicity and irritation, the flue gas emitted by various factories or mechanical equipment will combine or react with other substances to produce new harmful pollutants, these new pollutants are also called secondary pollutants [3]. These pollutants causing serious damage to atmosphere, water and soil, and endanger the health of humans and other organisms [4–9]. Steel plants and heat supply factories contain a large number of harmful pollutants such as smoke, dust, sulfur dioxide, nitrogen oxides, carbon dioxide, carbon monoxide, heavy metal mercury, arsenic, etc. The smoke pollutants of steel and cement plants also increase particulate pollutants in manufacturing process. The smoke from burning straw and automobile exhaust will increase more organic pollutants [10].

Different fumes have different colors. Such as black smoke, white smoke, yellow smoke, blue smoke, red smoke and green smoke. They come from different sources and have different hazards to human health. Some black fumes are produced by burning fossil energy sources such as oil and coal mines, which contain a large number of dust particles and sulfur dioxide, this situation is mainly due to incomplete combustion or unburned fine particles and larger particles of dust. Large areas of white smoke emitted from some factory chimneys are water vapor produced by cooling towers or desulfurization towers [11]. There are two kinds of yellow smoke, yellow-red smoke mainly contains iron, mainly from iron and steel plants, yellow-green smoke contains harmful gases such as chlorine and nitrogen oxides, mainly from related chlor-alkali chemical plants. Light blue smoke is called photochemical smog [12,13], it's caused by secondary pollutants, hydrocarbons will react with nitrogen oxides in the air under the action of sunlight (ultraviolet light). Blue smoke may also be automotive exhaust, it is caused by the engine oil entering the combustion chamber and burns when the internal fault occurs in the engine. In addition, organic exhaust is generally colorless. Based on the above characteristics, we can preliminarily judge the possible pollutants in the flue gas according to the color.

Real-time monitoring of flue gas is an important part of pollution control. At present, flue gas detection requires a large number of Sensor equipment, and usually requires manual participation, time-consuming and labor-intensive. Using computer image recognition technology to detect smoke is a good choice, more and more researchers pay attention to it. As an important part of machine learning, deep learning has made continuous progress. Convolutional Neural Networks (CNN) is one of the representative algorithms of deep learning. The basic structure of CNN includes two layers, feature extraction layer and feature mapping layer. This net is based on imitating the visual perception mechanism of organisms, it has a stable effect on the learning of pixels. If you want computer to learn how to analyze images, you need to use CNN structure, which is one of the core algorithms of image recognition. Air quality detection based on image recognition has been developed continuously, such as PM2.5, PM10 and so on [14–17]. Our goal is to make the computer recognize whether there is smoke or

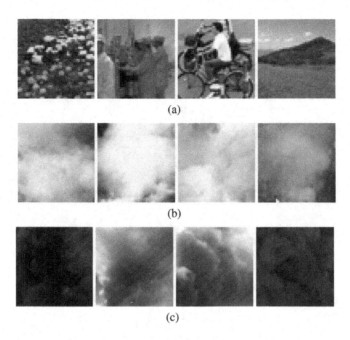

Fig. 1. Three categories of pictures. (a) smokeless; (b) white smoke; (c) smoke of other colors. (Color figure online)

not, and feedback the smoke color in the case of smoke. In order to get good recognition ability, we need CNN structure to learn enough image data.

The rest of this article is organized as follows. The second section introduces the database used for training and testing, introduce some commonly used algorithms. In the third part, we compare and analyse the performance of these algorithms. Finally, we draw conclusions in the fourth part and summarize the possible development direction of smoke detection.

2   Image Database and Convolutional Neural Network

In this part, we mainly introduce the reconstructed database for smoke detection and convolutional neural networks for smoke detection and classifying smoke images. As the learning data of model training, database is the basis of acquiring high performance models. We use this database to analyze and compare the performance of different deep convolutional networks. We have used eight networks, namely VGGNet [18], ResNet [19], GoogLeNet [20], Xception [21], DenseNet [22], DNCNN [23], DCNN [24] and MobileNets [25], different networks have their own characteristics.

2.1   Image Database

Yin et al's smoke detection database includes four sub-datasets, two larger sub-datasets are used for training and validation, and two smaller data sets are used

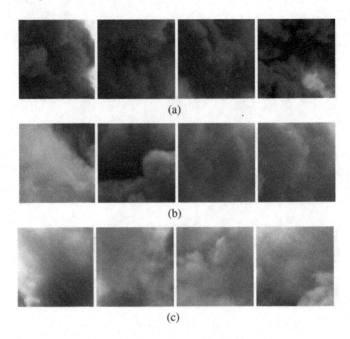

Fig. 2. Smoke in other colors. (a) black smoke; (b) blue smoke; (c) yellow smoke. (Color figure online)

for precision testing. All the pictures are classified as smoke and smokeless, two types of pictures are in PNG format with 48×48 pixels. A class of pictures with smoke are part of the scene where smoke exists, mainly including black smoke, white smoke, blue smoke and yellow smoke. Each smokeless picture is a part of a random scene, including characters, indoor environment, buildings, plants, mountains, rivers, skies, roads, vehicles, etc.

On the basis of the above image database, we subdivide it again. The first method is to divide them into three categories, they are smokeless, white smoke and smoke of other colors. Figure 1 shows examples of three categories of pictures. The second method is to divide them into five categories, they are smokeless, white smoke, black smoke, blue smoke and yellow smoke.

Smoke color, contour and detail are changeable [26,27], using image recognition to classify them is a challenging attempt. Through thousands of images, the convolutional neural network can learn enough image data and obtain stable performance.

We subdivide smoke images into different sub-categories. Namely black smoke, white smoke, blue smoke and yellow smoke, Fig. 2 shows examples of smoke pictures of different colors. Thousands of images can make CNN model learn enough image data and get stable performance.

2.2 Convolutional Neural Networks

With the development of deep learning technology, convolutional neural networks have become increasingly deeper during the past several years. Machine vision has made great achievements in deep learning contests, such as ILSVRC (ImageNet Large Scale Visual Recognition Challenge), the error rate of various networks is lower than that of human vision. So deep learning is the best choice for smoke detection. In this part, we will introduce the widely used convolutional neural networks, two networks specially used for smoke detection and one network specially used for mobile and embedded terminals.

VGG represents Oxford Visual Geometry Group of Oxford University, VGG is a deep CNN developed by Oxford Visual Geometry Group and Google DeepMind researchers. VGG has two structures, VGG16 and VGG19. There is no essential difference between VGG16 and VGG19, but the depth of the network is different(16 and 19 represent the number of weight layers in the network respectively). VGG explores the depth of CNN, it increases the depth of the network by directly superimposing convolution layers. The structure of VGG is very simple, the whole network uses the same size convolution core size (3×3) and maximum pooling size (2×2). VGG has been widely used in image feature extraction.

ResNet (Residual Neural Network), also known as Deep Residual Network, was proposed by Kaiming He of Microsoft Research Institute. With the increase of layers, the effect of CNN will not increase or even decrease, ResNet can solve the above problems and help us train deeper network. The structure of ResNet can accelerate the training of convolutional neural network. Compared with VGG, its parameters are much less. In addition, residual learning is used to improve the training efficiency, it is also a widely used CNN feature extraction network.

GoogLeNet is a deep learning structure proposed by Christian Szegedy. Compared with VGG, the common feature is deeper level, but it's much smaller than VGG. GoogLeNet based on inception architecture can make more efficient use of computing resources and extract more features under the same amount of computation. GoogLeNet has 5 million parameters, VGG has 36 times as many parameters as GoogLeNet, but the performance of GoogLeNet is better. It is a good choice in the case of limited memory and resources.

Xception is another improvement of Inception-v3 proposed by Google after Inception. It considers that the correlation between channels and spatial correlation should be dealt with separately. Xception uses a new concept called depthwise separable convolution to replace the convolution operation in inception-v3 [28], Although it has the same number of parameters as Inception-v3, it can use model parameters more efficiently.

DenseNet (Dense Convolutional Network) has fewer parameters than ResNet, another obvious difference is as follows, ResNet is summation, while DenseNet is splicing. The input of each layer is the output of all the previous layers, which has a very high parameter utilization. Its advantage is that information extraction is more thorough and abundant, and there is no problem of over-fitting [29] or accuracy decline when the number of layers is increased. Its disadvantage is high memory occupation and long training time.

DNCNN (Deep Normalization and Convolutional Neural Network) is a smoke detection network proposed by Yin et al. They specially construct a deep neural network to learn features directly from raw pixels of smoke pictures, this network completes both feature extraction and smoke recognition at the same time. They put forward a new network has 14 layers to implement automatic feature extraction and classification. The traditional convolution layers are replaced by normalization layers and convolutional layers, which speeds up the training process and improves the performance of smoke detection.

DCNN (Dual-Channel Neural Network) is a smoke detection network proposed by Gu et al. The proposed end-to-end network is mainly composed of two channels of deep sub-network. Its characteristic is that the end-to-end network is mainly composed of dual channels of deep subnetworks. The advantages of the sub-networks are as follows, the first sub-network can extract details information well, the second sub-network can capture the base information. Finally, two deep sub-networks are combined, they complement each other. DCNN uses fewer parameters and achieves very satisfactory results.

MobileNets is a network designed for deep learning applications on both mobile and embedded terminals, it was released in CPVR (Computer Vision and Pattern Recognition) in 2017. This network divides convolution operations into deep convolution and point convolution. Width multiplier and resolution multiplier are introduced, width multiplier changes the channel number of input and output to reduce the number of feature maps, resolution multiplier reduces the size of input images. This network reduces a lot of computation, and the accuracy decreases very little.

3 Experimental

In this section, we will analyze the performance of related networks based on the reconstructed smoke image recognition database. This part is composed of experimental settings and accuracy evaluation.

3.1 Experimental Settings

We use the reclassified database in our experiment. The reconstructed database has two main classification methods. The first method is to divide them into three categories, they are smokeless, white smoke and other colors. The second is divided into smokeless, white smoke, black smoke, blue smoke and yellow smoke. We use two sub-datasets with a large number of pictures to train and validate, while two smaller sub-datasets are used for accuracy testing.

In order to quantify the performance of these networks, we use the accuracy rate of typical evaluation indicators. For two classifications, AR is defined as the following formula:

$$AR = \frac{L_t + W_t + O_t}{Q_l + Q_w + Q_o} \times 100\% \tag{1}$$

Table 1. AR of each network (three classifications).

Accuracy Rate	VGG	ResNet	GoogLeNet	Xception	DenseNet	DNCNN	DCNN	MobileNets
AR_1	0.9378	0.9436	0.9342	0.9414	0.9472	0.9573	0.9646	0.9393
AR_2	0.9349	0.9355	0.9229	0.9534	0.9449	0.9542	0.9681	0.9435
$\overline{AR}$	0.9363	0.9394	0.9283	0.9477	0.9460	0.9557	0.9664	0.9415

$$AR = \frac{L_t + W_t + A_t + U_t + Y_t}{Q_l + Q_w + Q_a + Q_u + Q_y} \times 100\% \tag{2}$$

Equation 1 is suitable for three classifications and Eq. 2 for five classifications. Q_l, Q_w, Q_o, Q_a, Q_u, Q_y represent the total number of smokeless, white, other colors, black, blue and yellow smoke in the test data, respectively. Where $Q_a + Q_u + Q_y = Q_o$. L_t, W_t, O_t, A_t, U_t and Y_t represent the correct classification number of smokeless, white, other colors, black, blue and yellow smoke, respectively. The higher the AR value, the higher the accuracy of the convolutional neural network for smoke image classification.

Because there are two sub-datasets used for accuracy testing, we use the total number of images from two sub-datasets, weighting two accuracy rate to get the average value. The specific methods are as Eq. 3:

$$\overline{AR} = \frac{AR_1 Q_1 + AR_2 Q_2}{Q_1 + Q_2} \times 100\% \tag{3}$$

where Q is the total number of pictures used for accuracy testing, the total number of the first sub-database and the second sub-database are named Q_1 and Q_2 respectively. AR_1 and AR_2 are the accuracy rate obtained on two sub-databases by the same network, and $\overline{AR}$ is the weighted average accuracy rate of this network.

3.2   Accuracy Evaluation

The convolutional neural networks we use are VGGNet, ResNet, GoogLeNet, Xception, DenseNet, DNCNN, DCNN and MobileNets. After learning and training from the same database, each network is used to classify the images of two sub-datasets. According to the formula of AR, two accuracy rates can be obtained. Finally, calculating the value of $\overline{AR}$ for each network, we compare the performance of each network through $\overline{AR}$. The specific results are as follows.

In the case of three classifications, the AR of each network is listed in Table 1. The experimental results show that the $\overline{AR}$ of each Network is ranked as DCNN > DNCNN > Xception > DenseNet > MobileNets > ResNet > VGG > GoogLeNet. It can be seen that the performance of DCNN is significantly higher than other convolutional neural networks, because it pays more attention to smoke feature extraction.

In the case of five classifications, we list the AR of each network in Table 2. From the experimental results. When we add the classification of images, the AR

Table 2. AR of each network (five classifications).

Accuracy Rate	VGG	ResNet	GoogLeNet	Xception	DenseNet	DNCNN	DCNN	MobileNets
AR_1	0.8185	0.7975	0.8279	0.8332	0.7924	0.8628	0.8691	0.7866
AR_2	0.7588	0.7747	0.7707	0.8172	0.7720	0.8537	0.8564	0.7534
$\overline{AR}$	0.7874	0.7856	0.7981	0.8249	0.7818	0.8581	0.8625	0.7693

of each network has declined. The most serious is that the $\overline{AR}$ of MobileNets decreased by 17.2% points, and the $\overline{AR}$ of DenseNet decreased by 16.4% points. Smoke detection in multi-classification situation still has a lot of room for improvement. The experimental results show that the $\overline{AR}$ of each Network is ranked as DCNN > DNCNN > Xception > GoogLeNet > VGG > ResNet > DenseNet > MobileNets, top two performance networks are still DCNN and DNCNN.

4   Conclusion

Aiming at the problem of smoke detection or other exhaust monitoring, it is not advisable to use manual observation records. In order to simplify the use of equipment, it is a good choice to use image recognition to judge the smoke situation. This paper introduces the reconstruction of image database for smoke detection, and compare several kinds of networks, they can be used to extract smoke image features. After learning enough data, these networks have good accuracy in image classification, especially for smoke detection networks. But with the increase of image classification, AR decreases gradually, so it is necessary to continue to explore the network suitable for smoke detection.

Compared with the previous smoke detection, we not only use these networks to detect whether there is smoke or not, but also subdivide the smoke color when there is smoke. This can help us to judge the possible pollution components in the smoke, which has more practical value.

From the research point of view, we can continue to subdivide on the basis of current classification. The depth of color represents the different concentration of smoke, such as subdividing the blackness level in the case of black smoke. In addition, we can continue to study in time domain, such as changes in smoke area and smoke concentration, which can be used to estimate the amount of smoke exhaust.

From a practical point of view, smoke detection based on image recognition has not been widely used. Image recognition has many advantages and has developed rapidly, it is the future research direction to further transform theoretical research into practical application. We will continue to explore better network for smoke detection, so as to lay a good theoretical foundation for practical application.

References

1. Mgbemene, C.A., Nnaji, C.C., Nwozor, C.: Industrialization and its backlash: focus on climate change and its consequences. J. Environ. Sci. Technol. **9**(1), 301–316 (2016)
2. Tollefson, J., Weiss, K.R.: Nations approve historic global climate accord. Nature **528**, 315–316 (2015)
3. Wang, T., et al.: Air quality during the 2008 Beijing Olympics : secondary pollutants and regional impact. Atmos. Chem. Phys. **10**(16), 7603–7615 (2010)
4. Guan, W., Zheng, X., Chung, K., Zhong, N.: Impact of air pollution on the burden of chronic respiratory diseases in China: time for urgent action. Lancet **388**(10054), 1939–1951 (2016)
5. Yauk, C., et al.: Germ-line mutations, DNA damage, and global hypermethylation in mice exposed to particulate air pollution in an urban/industrial location. Proc. Natl. Acad. Sci. **105**(2), 605–610 (2008)
6. Voulvoulis, N., Georges, K.: Industrial and agricultural sources and pathways of aquatic pollution. In: Impact of Water Pollution on Human Health and Environmental Sustainability, pp. 29–54 (2016)
7. Saha, N., Rahman, M.S., Ahmed, M.B., Zhou, J.L., Ngo, H.H., Guo, W.: Industrial metal pollution in water and probabilistic assessment of human health risk. J. Environ. Manage. **185**, 70–78 (2017)
8. Landrigan, P.J.: Air pollution and health. Lancet Public Health **2**(1), e4–e5 (2017)
9. Orru, H., et al.: Residents' self-reported health effects and annoyance in relation to air pollution exposure in an industrial area in Eastern-Estonia. Int. J. Environ. Res. Public Health **15**(2), 252 (2018)
10. Sagna, K., Amou, K.A., Boroze, T.T.E., Kassegne, D., Almeida, A., Napo, K.: Environmental pollution due to the operation of gasoline engines: exhaust gas law. Int. J. Oil Gas Coal Eng. **5**(4), 39–43 (2017)
11. Ma, S., Jin, C., Chen, G., Yu, W., Zhu, S.: Research on desulfurization wastewater evaporation: present and future perspectives. Renew. Sustain. Energy Rev. **100**(58), 1143–1151 (2016)
12. Hallquist, M., et al.: Photochemical smog in China: scientific challenges and implications for air-quality policies. Natl. Sci. Rev. **3**(4), 401–403 (2016)
13. Aidaoui, L., Triantafyllou, A.G., Azzi, A., Garas, S.K., Matthaios, V.N.: Elevated stacks' pollutants' dispersion and its contributions to photochemical smog formation in a heavily industrialized area. Air Qual. Atmos. Health **8**(2), 213–227 (2015)
14. Yue, G., Gu, K., Qiao, J.: Effective and effifficient photo-based PM2.5 concentration estimation. IEEE Trans. Instrum. Meas. **68**(10), 3962–3971 (2019)
15. Gu, K., Xia, Z., Qiao, J., Lin, W.: Recurrent air quality predictor based on meteorology-and pollution-related factors. IEEE Trans. Multimedia **14**(9), 3946–3955 (2018)
16. Gu, K., Xia, Z., Qiao, J.: Stacked selective ensemble for PM2.5 forecast. IEEE Trans. Instrum. Meas. (2019)
17. Gu, K., Qiao, J., Li, X.: Highly efficient picture-based prediction of PM2.5 concentration. IEEE Trans. Industr. Electron. **66**(4), 3176–3184 (2019)
18. Simonyan, K., Zisserman, A.: Very deep convolutional networks for large-scale image recognition. arXiv preprint arXiv:1409.1556, September 2014
19. He, K., Zhang, X., Ren, S., Sun, J.: Deep residual learning for image recognition. In: Proceedings of the IEEE Conference on Computer Vision and Pattern Recognition, pp. 770–778, June 2016

20. Szegedy, C., et al.: Going deeper with convolutions. In: Proceedings of the IEEE Conference on Computer Vision and Pattern Recognition, pp. 1–9, June 2015
21. Chollet, F.: Xception: deep learning with depthwise separable convolutions. In: Proceedings of the IEEE Conference on Computer Vision and Pattern Recognition, pp. 1251–1258, July 2017
22. Huang, G., Liu, Z., Weinberger, K. Q., Maaten, L.: Densely connected convolutional networks. In: Proceedings of the IEEE Conference on Computer Vision and Pattern Recognition, pp. 4700–4708, August 2016
23. Yin, Z., Wan, B., Yuan, F., Xia, X., Shi, J.: A deep normalization and convolutional neural network for image smoke detection. IEEE Access **5**, 18429–18438 (2017)
24. Gu, K., Xia, Z., Qiao, J., Lin, W.: Deep dual-channel neural network for image-based smoke detection. IEEE Trans. Multimedia (2019)
25. Howard, A.G., et al.: Mobilenets: efficient convolutional neural networks for mobile vision applications. arXiv preprint arXiv:1704.04861, April 2017
26. Yuan, F., Shi, J., Xia, X., Fang, Y., Fang, Z., Mei, T.: High-order local ternary patterns with locality preserving projection for smoke detection and image classifification. Inf. Sci. **372**, 225–240 (2016)
27. Lin, G., Zhang, Y., Zhang, Q., Jia, Y., Xu, G., Wang, J.: Smoke detection in video sequences based on dynamic texture using volume local binary patterns. TIIS **11**(11), 5522–5536 (2016)
28. Szegedy, C., Vanhoucke, V., Ioffe, S., Shlens, J., Wojna, Z.: Rethinking the inception architecture for computer vision. In: Proceedings of the IEEE Conference on Computer Vision and Pattern Recognition, pp. 2818–2826, December 2016
29. Srivastava, N., Hinton, G., Krizhevsky, A., Sutskever, I., Salakhutdinov, R.: Dropout: a simple way to prevent neural networks from over-fifitting. J. Mach. Learn. Res. **15**(1), 1929–1958 (2014)

Non-local Recoloring Algorithm for Color Vision Deficiencies with Naturalness and Detail Preserving

Yunlu Wang[1], Duo Li[2], Menghan Hu[1(✉)], and Liming Cai[3(✉)]

[1] Shanghai Key Laboratory of Multidimensional Information Processing,
East China Normal University, Shanghai 200241, China
mhhu@ce.ecnu.edu.cn
[2] Hangzhou Hikvision Digital Technology Co., Ltd., Hangzhou 310051, China
[3] Shanghai Police College, Shanghai 200137, China
clm9978@163.com

Abstract. People with Color Vision Deficiencies (CVD) may have difficulty in recognizing and communicating color information, especially in the multimedia era. In this paper, we proposed a recoloring algorithm to enhance visual perception of people with CVD. In the algorithm, color modification for color blindness is conducted in HSV color space under three constraints: detail, naturalness and authenticity. A new non-local recoloring method is used for preserving details. Subjective experiments were conducted among normal vision subjects and color blind subjects. Experimental results show that our algorithm is robust, detail preserving and maintains naturalness. (Source codes are freely available to non-commercial users at the website (https://doi.org/10.6084/m9.figshare.9742337.v2)).

Keywords: Color blind · Recoloring · Color vision deficiency · Non-local algorithm

1 Introduction

Colors that are ubiquitous in the world bring lots of fun to us. Human beings can recognize plenty of colors due to the special structure of eyes. From Young-Helmholtz theory [1] and work of Svaetichin [2], we know that there are three types of retinal cone photoreceptors, different spectral sensitivities to the light. L (long wavelength) cone pigment is more sensitive to the light with long wavelength. M (middle wavelength) cone pigment is more sensitive to middle wavelength. S (short wavelength) cone pigment is only sensitive to short wavelength. Those cones make us accessible to color information.

According to the Cisco white paper released in 2019, the global IP video traffic will account for 82% of traffic by 2022 [3]. This suggests that images are the biggest form of information on the Internet. Color images are widely used in multimedia field such as websites, smartphones and televisions. Unfortunately, people with color vision

---

Y. Wang and D. Li—These authors contributed equally to this work.

© Springer Nature Singapore Pte Ltd. 2020
G. Zhai et al. (Eds.): IFTC 2019, CCIS 1181, pp. 23–34, 2020.
https://doi.org/10.1007/978-981-15-3341-9_3

deficiencies (CVD) may have difficulty in recognizing colors, which in turn affects their communication with others about color-related information. Dichromats, a kind of CVD, absent one of three cones in their eyes. Consequently, the original three-dimensional colors are only perceived as two-dimensional colors, resulting in a reduction in recognizable color space. Dichromats are divided into three kinds: pro-tanopia, deuteranopia, and tritanopia. Other types of color vision deficiencies and their prevalence is shown in Table 1. Deuteranomaly in Anomalous Trichromacy has a higher prevalence among all types of CVD.

Table 1. Classification of color vision deficiencies [4]

Type	Name	Cause	Prevalence
Anomalous Trichromacy	Protanomaly	L-cone defect	1.3%
	Deuteranomaly	M-cone defect	**4.9%**
	Tritanomaly	S-cone defect	0.01%
Dichromacy	Protanopia	L-cone absent	1%
	Deuteranopia	M-cone absent	1.1%
	Tritanopia	S-cone absent	0.002%
Monochromacy	Rod Monochromacy	No functioning cones	Rare

To help people with CVD solve this problem, one immediate idea is to simulate visual perception of CVD. Brettel [5] simulated the visual perception of dichromats under LMS color space, which was widely adopted. The LMS color space is a model to simulate the visual perception of human eyes, in which L, M, and S represent relative quantity of lights that human cone pigments receive. Recently, Yaguchi [6] proposed a method for simulating the color appearance for Anomalous Trichromats based on the spectral radiance of the stimulus. This method can predict the Rayleigh color matches for Anomalous Trichromats.

Unfortunately, the cure for Color Blindness is still unavailable [7]. That means people with CVD will have inconvenience in the whole life. In the past decades, many efforts have been done to develop information accessibility techniques for color blindness, and these techniques can be classified into two main categories: (1) tools that help designers in verifying colors and (2) recoloring the images for CVD viewers. For the first category, Jenny and Kelso [8] introduce Oracle Color, a software tool to assist designers in validating color schemes, which produces color maps that are easy to read for people with CVD. Recently, BLINDSCHEMES [9] was proposed to provide graph schemes sensitive to CVD. The schemes come with 21 new colors, of which seven colors are distinguishable for people suffering from color blindness. This method does help people with color blindness. However, selecting colors that are friendly for all types of CVD users will be a tough work. Moreover, these methods cannot be applied to existing natural images [10]. Thus, people with CVD can only obtain the 'designed' information instead of the colorful world.

By contrast, the recoloring method caught researchers' attention. One representative work conducted by Rasche [11] proposed a detail dimensional reduction algorithm to preserve more details in images for Monochromats and Dichromats to perceive.

Researches differ in varied color space. Doliotis [12] processed the images for people with CVD in LMS color space. Ruminski et al. [13] recolored images in HSB color space by scaling color difference, as well as scaling in brightness and saturation. Huang [14] recolored images aiming at preserving details as well as naturalness in CIELab color space. The conversion step is limited in a rotation in the a^*b^* plane. Huang et al. [10] also used a Gaussian Mixture Model to represent the color information and used EM algorithm to estimate the parameters. They weighted the 'key colors' by their importance in the optimization step. Meanwhile, some researchers focus on the naturalness of recoloring images [15] and the efficiency of the recoloring algorithm [16]. Recently, Zhu [17] summarized the shortcomings of existing recoloring algorithms and proposed a new recoloring approach for red–green dichromats without user-customized parameters. They transform the recoloring to an optimization problem that seeks a balance between contrast and naturalness.

We notice that most of the previous researches mainly aim at Dichromacy, whose one of three cones is absent. The total number of people with Dichromacy consists of about 2% of male. However, there are almost 5% of male are with Anomalous Trichromacy, whose cones work defectively [4, 18]. In this paper, we proposed a recoloring algorithm for Anomalous Trichromacy in HSV color space. Hue channel is a good representation for different colors. To recolor the images for people with CVD, unseen structural information of hue channel can be enhanced by the structural changes in saturation and value channel. There are particular colors that are not very distinguishable for color blind people, but they are as sensitive as normal vision people to the intensity of colors. This inspires us to convert the hue differences of these unrecognizable colors to the saturation and value differences.

To maintain the naturalness of images, the recoloring algorithm should be based on the principle of human visual system. Some previous works have used detail and naturalness as the criterion for recoloring algorithms [11, 14, 19]. In this study, three criteria based on human visual characteristics are taken into consideration:

- Detail: the model should convert the image while preserving detailed information.
- Naturalness: the distance between colors should be limited during color transformation.
- Authenticity: the color changes should not be too much in order not to confuse viewers. For instance, the red flower should not be converted into blue, and it should be converted into a color closely similar to the original color red.

There are two situations we are mainly concerned: (1) the recorded images for color blind people require to make CVD viewers see as many details as possible and preserve naturalness at the same time; and (2) the recorded images cannot contain unrecognizable shapes. To preserve details of the image, we used a new non-local recoloring method, which takes the global color information into consideration. To make the unrecognizable shapes in the image recognizable for CVD viewers, we figure out the structural information in the hue channel by analyzing non-local information. After getting a structural color contrast map, we convert the structural information into saturation channel and value channel. Thus, the unrecognizable information in hue channel becomes recognizable in the other two channels.

The proposed algorithm recolored images under three constraints: naturalness, detail preserving and authenticity. Shapes in recolored images should be recognized, while images should retain naturalness and more detail. To verify whether our algorithm can handle these two situations, we conducted subjective experiments for people with Deuteranomaly, a majority part of people with CVD. The performance of the proposed algorithm is evaluated by five metrics namely recognition accuracy, structural information, naturalness, detail preserving and authenticity. Subjects rated the recolored images by comparing them with original ones.

The organization of this paper is as follows: Sect. 2 introduces the non-local means algorithm and our proposed non-local based recoloring algorithm; Sect. 3 demonstrates the image processing step based on non-local color contrast map; In Sect. 4 we conducted subjective experiments and discussed results; and Sect. 5 is conclusion.

2 Non-local Algorithm

2.1 Non-local Means

Non-local means is a classical algorithm in the problem of image de-noising [20]. It takes not only local information but also global structural content into consideration. Non-local means filtering averages all pixels in the image, weighted by similarity of these pixels to the target pixel, weighted by similarity of these pixels to the target pixel.

In an image, suppose we have two points p and q in the area of Ω, the filtered value of point u(p) is:

$$u(p) = \frac{1}{N(P)} \int_\Omega v(q) f(p, q) dq \tag{1}$$

where $v(q)$ is the unfiltered value at point q; $f(p, q)$ is the weighting function designed to measure the difference and distance between point p and q; and $N(P)$ is the normalization factor.

2.2 Non-local Recoloring

As for the problem of recoloring for the color blind, our main concern is to separate unrecognizable colors from its spatial related area. One good way is to generate a color contrast map to indicate distinguishing degree. We afterwards adjust the color perception of the original image based on the generated color contrast map.

From the principle of color blindness, we know that even for a particular kind of color vision deficiency, the characteristics of different people varies. Prior to use, subjective experiments were carried out to determine the optimized parameters for the proposed algorithm, ensuing the better performance of algorithm.

The scheme of the algorithm is as follows: (1) obtain the Structural Color Contrast Map based on the non-local structural information and color contrast; (2) determine the scaling factor for the hue channel by generating Hue scaling Map; and (3) process the original image and transform it to the recolored image with Hue scaling Map. The whole scheme of our algorithm is shown in Fig. 1.

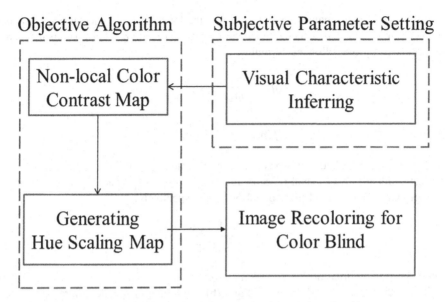

Fig. 1. Scheme of the proposed non-local recoloring algorithm.

2.2.1 Global Color Information

Before recoloring, global information is needed. Because this information helps us to have a better understanding of how 'unrecognizable' it is for the subjects. We convert the original image into HSV color space and pixel intensity of hue channel is extracted.

In the histogram of hue channel, there are usually several peaks of hue distribution. By analyzing the position of peaks and gaps between peaks, we get a knowledge of the hue spread of the whole image.

2.2.2 Color Contrast and Hue Scaling Map

Our purpose is to figure out the color distribution that are undistinguishable for the user. A simplest way is to label them pixel by pixel. But the drawback of this approach is that it doesn't consider the structural information of the whole image. If doing this, after hue scaling and naturalness compensation, the image losses its structural information too much. We should pay more attention to the non-local structural color contrast than only local color information.

To solve this problem, we utilize the modified non-local algorithm. In an area Ω of the image, p is the pixel to be calculated, and q is the pixel in Ω. The non-local Structural Color Contrast Map can be obtained via the formulas:

$$\mathrm{CC(p)} = \frac{1}{N(P)} \int_{\Omega} c(q) f(p, q) dq \tag{2}$$

$$N(P) = \int_{\Omega} f(p, q) dq \tag{3}$$

where $N(P)$ is the normalization factor. $c(q)$ is the local hue difference at pixel q, and the weighting factor $f(p, q)$ is the hue contrast between p and q. These parameters can be calculated by the following formulas:

$$c(q) = \frac{\alpha}{(h(q) - h_0)^2 + \delta} \tag{4}$$

$$f(p, q) = e^{-\frac{|B(q) - B(p)|^2}{\sigma^2}} \tag{5}$$

In equation, h_0 is hue of the mostly unrecognisable color for the user, resulting from the test to identify user visual perception characteristics. $B(p)$ and $B(q)$ are the local average hue values around pixel p and pixel q, respectively:

$$B(p) = \frac{1}{R(p)} \sum_{i \in \omega} h_i \tag{6}$$

where $R(p)$ is the total number of pixels in ω.

Using the color contrast map (See Fig. 2(b)), and the user specified unrecognizable hue, we can calculate the hue scaling factor for our algorithm, thus getting the Hue Scaling Map. The hue scaling factor is calculated as follows:

$$HS(p) = CC(p) \frac{B(p) - h_0}{|B(p) - h_0|} \tag{7}$$

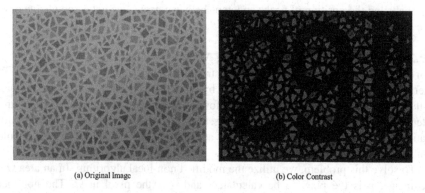

(a) Original Image   (b) Color Contrast

Fig. 2. One typical example showing original image (a) and color contrast (b).

3 Image Recoloring

After obtaining the Hue Scaling Map, the original image is required to be processed to get the recolored 'recognizable' image for CVD users. Noticing that people with CVD act to be sensitive to saturation and value change of colors that are unrecognizable in an image, the images in hue channel, saturation channel and value channel are respectively weighted by the Hue Scaling Map.

As a result of the above adjustment, the color contrast of recolored images will have a certain loss, resulting in the image losing details or looking unnatural. To minimize this side effect, we need to make reasonable and moderate channel adjustment, mainly based on the following three criteria:

3.1 Detail Preserving

The global hue change should be limited in case that the recolored image looks strange to people with CVD, especially for the area that is not unrecognizable. In the hue scaling step, the higher pixel value of the Color Contrast Map is, the larger structural contrast there will be.

$$\frac{\sum_{p \in \psi} CC(p)(h(p) - h'(p))}{\sum_{p \in \psi} CC(p)} \leq P \tag{8}$$

where ψ is the area of the image in which value of hue scaling map is less than 0.

3.2 Naturalness Preserving

To preserve naturalness, the total color change of the whole image should be as little as possible. We can compensate for the structural color contrast change by limiting the color change in local area.

$$\sum_{p \in \varsigma} \sum_{i \in \{H,S,V\}} |c_i(p) - c_i'(p)| \leq D \tag{9}$$

where $c_i(p)$ is the ith channel value of pixels in the image.

3.3 Authenticity

The color change in each channel should not be too high in order to reduce confusion with viewers. For instance, the red flower should not be converted into blue, and it should be converted into a color closely similar to the original color.

$$\max |h(p) - h'(p)| \leq H_{max} \tag{10}$$

$$\max |s(p) - s'(p)| \leq S_{max} \tag{11}$$

$$\max |v(p) - v'(p)| \leq V_{max} \tag{12}$$

Under three abovementioned constraints, recolored images are generated. Several typical recolor images are shown in Fig. 3(c). As shown in Fig. 3, we can observe that there exists the slight perception difference for people with normal vision when seeing the original images and the recolored images.

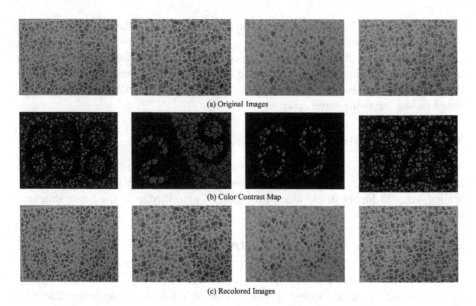

(a) Original Images

(b) Color Contrast Map

(c) Recolored Images

Fig. 3. Comparison of three types of images: (a) original images; (b) color contrast map; and (c) recolored images.

4 Experimental Results and Discussion

4.1 User Visual Perception Characteristics

To make our algorithm more robust to the different characteristics of color blind people's visual perception, we need to conduct an experiment before starting the algorithm. The results of this test will be used in the algorithm in Sect. 2.

The unrecognizable colors of subjects can be found through the subjective experiment. We classify the type of color vision deficiency of users by the Ishihara's plate [21]. Subjects' performances and their previous yearly physical examination results in hospital provided reference for us to determine their visual perception.

Subsequently, we use images with clear structural area as our test images. The test and resulted images are available in the website[1]. The images are processed by shifting the hue channel in HSV color space. Subjects are asked to point out the most unrecognizable image in the series of images with same saturation and value channel (the hue channel is different). The results derived from subjects with CVD and subjects with normal vision are compared. By analyzing results, we can find the most unrecognizable hue for each subject. The most unrecognizable hue is the parameter h_0 described in Sect. 2.

---

[1] https://figshare.com/articles/Dataset_and_code_Matlab_for_recoloring_images/9742337.

4.2 Subjective Experiment

The parameters of the algorithm is chosen as follows: $\alpha = 0.01$, $\sigma = 0.001$, $h_0 = 0.12$. Notice that h_0 can be set different for different subjects, but in this study we used 0.12 for all the subjects.

To measure the performance of the algorithm, the subjective experiment is conducted by requiring subjects to view a series of original images and their recolored images. Both subjects with normal vision and subjects with CVD participate in this experiment. Subjects with CVD are asked to describe their feelings about the conversion. Considering the relatively high prevalence of deuteranomaly (Table 1), a total of 13 subjects who are already verified to have deuteranomaly participate in the experiment. We set four indicators for the conversion towards the original image: structural information, naturalness, detail preserving, and authenticity. There are 28 Ishihara's plates in the subjective experiment. The recognition efficiency of these subjects on original images and recolored images is compared. We observe that the rate of accuracy increases after processing the images, which indicates the effectiveness of our algorithm.

Table 2. Experiment results of subjects with CVD

Subject ID	Accuracy before	Accuracy after	Structural information	Naturalness	Detail	Authenticity
1	14.29%	60.71%	6.14	5.04	5.54	5.07
2	25.00%	46.43%	6.04	5.29	5.54	5.25
3	21.43%	64.29%	5.54	4.64	4.96	4.96
4	10.71%	50.00%	5.54	4.75	4.96	4.25
5	17.86%	53.57%	6.07	5.04	5.96	4.82
6	50.00%	75.00%	6.39	5.25	4.82	5.00
7	25.00%	57.14%	5.93	5.04	4.43	4.46
8	39.29%	67.86%	6.04	5.68	6.18	4.96
9	39.29%	75.00%	6.00	4.71	3.54	4.11
10	60.71%	78.57%	6.11	4.93	4.46	4.71
11	28.57%	71.43%	5.68	4.79	5.14	4.57
12	57.14%	78.57%	7.00	5.36	4.50	4.57
13	17.86%	53.57%	5.64	5.79	5.68	5.75
Average	**31.32%**	**64.01%**	**6.01**	**5.10**	**5.05**	**4.81**

As can be seen from Table 2, there are four indicators namely structural information, naturalness, detail, and authenticity. Each subject reports feelings about the difference between the original image and recolored image. Structural information depends on whether they can see more shapes in recolored images; Naturalness depends on whether they think the color of recolored images are natural and comfortable for them; Detail depends on whether they can see more details in recolored images and Authenticity depends on their evaluation of the traces of artificial adjustment in recolored images. The baseline is 5 points. If they think the recolored image is better, they score higher than 5 points. Conversely, they score lower than 5 points if they think the recolored image becomes worse.

The experimental results show that for the group of normal vision subjects, the recognizing accuracy is above 90% for both original and recolored images (results not shown). While for the group of CVD subjects, the accuracy of recognizing shapes rises from 31.32% to 64.01% and three indicators are not lower than 5 points while score of Authenticity is 4.81. The abovementioned results indicate that our recoloring algorithm contribute to the visual perception quality for the viewers with CVD with little sacrifice in image quality.

4.3   Discussion

From Table 2, it could be observed that the effectiveness of our proposed algorithm varies from person to person, although it is generally helpful for people with CVD. Each subject's accuracy was improved, and they were able to get more structural information from the recolored images. This proves the effectiveness of the algorithm. However, some subjects improved their accuracy by only about 20% while some subjects improved by more than 40%. Besides, the score of the quality of recolored images varies from person to person. In addition, the performance of our algorithm on natural images remains to be verified by future experiments.

The reason for subjective evaluation difference may be the value of h_0, which we used 0.12 in our algorithm. One of the advantages of our algorithm is that we can set different values for h_0 in formula (4) and (7) according to the situation of different subjects, which represents their most unrecognizable hue. Nevertheless, we used 0.12 for all subjects because they are all verified to have deuteranomaly. We believe that adjusting the h_0 value for each type of color blindness will make our algorithm works very well in the actual application. In the real world, if we want to develop an assistive device based on our algorithm for people with CVD, and the value of h_0 can be adjusted based on each subject's prior experiment to guarantee the excellent performance of the assistive device.

Our recoloring method can be adopted in personal multimedia equipment. After the short prediction test such as selecting series images on smartphones, the most unrecognized hue can be found, and the customized h_0 value is therefore determined for CVD users. More suitable h_0 can ensure more structural information and higher image quality, so the short time of preparation for each user is worthy.

Another possible application scenario is to assist people with CVD driving. Visual perception is important for safe driving, because the drivers obtain more than 80% information from traffic activities using their visual function (the remaining 20% information are obtained by sensory function, memory function, etc.). However, people with CVD may have difficulty in understanding some traffic information such as traffic lights and taillights. Special lenses [22] may help but they are specific to a particular disability (Deuteranopia and Protanopia) and require the adaptation of the eye to be effective [23]. Our algorithm may be more adaptable to different types of CVD drivers, and it can be applied to wearable vision devices [23] or to display screens [24] that recolor traffic information. Our algorithm can be embedded in various types of assistive devices for people with CVD [25] containing cane, glasses, hat, belt, bracelet, jacket, flashlight, hand-held cube, glove. Based on the characteristics of our algorithm, the one-time prior test is essential for each driver with CVD to ensure the acquisition of accurate recolored traffic information.

5 Conclusion

In this paper, we proposed an algorithm to recolor images for people with color vision deficiency (CVD). The algorithm is based on non-local recoloring method and prefers a user specified parameter according to the visual characteristics. Subjective experiments are conducted among normal vision subjects (results not shown) and color blind subjects. We found the CVD people's accuracy of Ishihara's plate testing arises by using our algorithm. Experimental results also show that our algorithm is robust, detail preserving and maintains naturalness. In future work, more researches in visual characteristics of people with CVD is needed for improving algorithm performance, such as the method of automatically setting the specified parameter for the same type of CVD people.

Acknowledgement. This work is sponsored by the Shanghai Sailing Program (No. 19YF1414100), the National Natural Science Foundation of China (No. 61831015, No. 61901172), the STCSM (No. 18DZ2270700), and the China Postdoctoral Science Foundation funded project (No. 2016M600315).

References

1. Young, T.: II. The Bakerian lecture. On the theory of light and colours. Philos. Trans. R. Soc. Lond. **92**, 12–48 (1802)
2. Svaetichin, G.: Spectral response curves from single cones. Acta Physiol. Scand. Suppl. **39** (134), 17–46 (1956)
3. Cisco Systems, Inc.: Cisco visual networking index: forecast and trends (2017–2022). https://www.cisco.com/c/en/us/solutions/collateral/service-provider/visual-networking-index-vni/white-paper-c11-741490.html. Accessed 27 Feb 2019
4. Huang, et al.: Enhancing color representation for the color vision impaired. In: Workshop on Computer Vision Applications for the Visually Impaired (2008)
5. Brettel, H., et al.: Computerized simulation of color appearance for dichromats. JOSA A **14** (10), 2647–2655 (1997)
6. Yaguchi, H., et al.: Computerized simulation of color appearance for anomalous trichromats using the multispectral image. JOSA A **35**(4), B278–B286 (2018)
7. Pendhari, N., et al.: Color modification system for barrier free vision. In: 2017 International Conference on Innovations in Information, Embedded and Communication Systems (ICIIECS), pp. 1–4 (2017)
8. Jenny, B., et al.: Color design for the color vision impaired. Cartogr. Perspect. **58**, 61–67 (2007)
9. Bischof, et al.: BLINDSCHEMES: Stata module to provide graph schemes sensitive to color vision deficiency (2019). https://econpapers.repec.org/software/bocbocode/s458251.htm

10. Huang, J.B., et al.: Image recolorization for the colorblind. In: 2009 IEEE International Conference on Acoustics, Speech and Signal Processing, pp. 1161–1164 (2009)
11. Rasche, K., et al.: Detail preserving reproduction of color images for monochromats and dichromats. IEEE Comput. Graph. Appl. **25**(3), 22–30 (2005)
12. Doliotis, P., et al.: Intelligent modification of colors in digitized paintings for enhancing the visual perception of color-blind viewers. In: IFIP International Conference on Artificial Intelligence Applications and Innovations, pp. 293–301 (2009)
13. Ruminski, J., et al.: Color transformation methods for dichromats. In: 3rd International Conference on Human System Interaction, pp. 634–641 (2010)
14. Huang, J.B., et al.: Information preserving color transformation for protanopia and deuteranopia. IEEE Signal Process. Lett. **14**(10), 711–714 (2007)
15. Hassan, M.F., et al.: Naturalness preserving image recoloring method for people with red–green deficiency. Sig. Process. Image Commun. **57**, 126–133 (2017)
16. Xu, Q., Zhang, X., Zhang, L., Zhu, G., Song, J., Shen, P.: An efficient recoloring method for color vision deficiency based on color confidence and difference. In: Yang, J., et al. (eds.) CCCV 2017. CCIS, vol. 771, pp. 270–281. Springer, Singapore (2017). https://doi.org/10.1007/978-981-10-7299-4_22
17. Zhu, Z., et al.: Naturalness-and information-preserving image recoloring for red–green dichromats. Sig. Process. Image Commun. **76**, 68–80 (2019)
18. Doron, R., et al.: Spatial visual function in anomalous trichromats: Is less more? PLoS ONE **14**(1), e0209662 (2019)
19. Jeong, J.Y., et al.: An efficient re-coloring method with information preserving for the color-blind. IEEE Trans. Consum. Electron. **57**(4), 1953–1960 (2011)
20. Buades, A., et al.: A non-local algorithm for image denoising. In: 2005 IEEE Computer Society Conference on Computer Vision and Pattern Recognition (CVPR 2005), vol. 2, pp. 60–65 (2005)
21. Ishihara, S.: Ishihara's Test for Colour-Blindness. Kanehara Shuppan Company, Tokyo (1985)
22. Chen, X., et al.: Method and eyeglasses for rectifying color blindness. U.S. Patent 5,369,453 (1994)
23. Melillo, P., et al.: Wearable improved vision system for color vision deficiency correction. IEEE J. Transl. Eng. Health Med. **5**, 1–7 (2017)
24. Wing, T.: Colorblind vehicle driving aid. U.S. Patent Application 10/799,112 (2005)
25. Hu, M., et al.: An overview of assistive devices for blind and visually impaired people. Int. J. Robot. Autom. **34**(5), 580–598 (2019)

Missing Elements Recovery Using Low-Rank Tensor Completion and Total Variation Minimization

Jinglin Zhang[1], Mengjie Qin[2], Cong Bai[2], and Jianwei Zheng[2(✉)]

[1] School of Atmospheric Science,
Nanjing University of Information Science, Nanjing, China
[2] College of Computer Science,
Zhejiang University of Technology, Hangzhou, China
zjw@zjut.edu.cn

Abstract. The Low-rank (LR) and total variation (TV) are two most popular regularizations for image processing problems and have sparked a tremendous number of researches, particularly for moving from scalar to vector, matrix or even high-order based functions. However, discretization schemes commonly used for TV regularization often ignore the difference of the intrinsic properties, which is not effective enough to exploit the local smoothness, let alone the problem of edge blurring. To address this issue, in this paper, we consider the color image as three-dimensional tensors, then measure the smoothness of these tensors by TV norm along the different dimensions. The three-order tensor is then recovered by Tucker decomposition factorization. Specifically, we propose integrating Shannon total variation (STV) into low-rank tensor completion (LRTC). Moreover, due to the suboptimality of nuclear norm, we propose a new nonconvex low-rank constraint for closer rank approximation, namely truncated γ-norm. We solve the cost function using the alternating direction method of multipliers (ADMM) method. Experiments on color image inpainting tasks demonstrate that the proposed method enhances the details of the recovered images.

Keywords: Tensor completion · Low-rank · Shannon total variation

1 Introduction

In the fields of computer vision and image processing [1], image inpainting is a vital research topic which can be regarded as a missing value estimation problem. The main problem of missing value estimation is how to build up bridge between the known elements and the unknown ones. Some methods often used in imaging inpainting [2], i.e., PDEs [2] and belief Propagation [3], mainly pay great attention to the local relationship. The fundamental assumption is that the missing entries lie on adjacent elements. In other words, the further apart the two elements are, the less dependent they are on each other. However, occasions are quite common in the natural images that is the missing item depends

G. Zhai et al. (Eds.): IFTC 2019, CCIS 1181, pp. 35–48, 2020.
https://doi.org/10.1007/978-981-15-3341-9_4

on those which are far away from itself. Thus, it is necessary to develop a novel algorithm that can acquire global information in the image.

Matrix completion [4], namely the second-order tensor completion problem [5], has been reported to be able to effectively estimate the missing values in a matrix from a small number of know items. To solve this ill-posed problem, the matrix completion method assumes that the restored matrix is low-rank, and then minimizes the difference between the given incomplete matrix and the estimation matrix by using this constraint condition. Matrix completion has been widely used in image/video inpainting, denoising [6] and decoding problems. There are indeed numerous solutions because of ill-posed problem with less constraints [7]. And it is a general assumption that when a matrix needs to be completed, its rank should be as low as possible. Unfortunately, the disadvantage of "$rank(\cdot)$" is non-convex, NP-hard and discrete. Therefore, it can be substituted by nuclear norm which is continuous, convex and easy to optimize [8]. In fact, it has been proved that it is the tightest convex approximation of the rank of matrix among the existing method [9].

In implementation, the nuclear norm is set as the ℓ_1-norm on the vector of singular values. It gets low-rankness through encouraging the sparsity of singular values. As Fan and Li pointed out [10], the ℓ_1-norm is a loose approximation to the ℓ_0-norm and it over penalizes large entries of vectors. So, the mere fact that we can draw an analogy between the ℓ_0-norm of vectors and the rank function of matrices, we see that the large entries of vectors are also over penalized. The Schatten-p quasi-norm ($0 < p < 1$) [11,12] is suggested to replace the nuclear norm to better approximate the rank function. Recently, some novel low rank approximation methods have been proposed, e.g., the weighted Schatten-p norm. Besides, Ref. [13] proposes a γ-norm based low-rank regularize that is totally different from Schatten-p norm and holds more preferable low-rank property.

Tensor is a generalization of matrix and vector, which is convictive to show multidimensional data or interactions relevant to multiple factors. Firstly, a third-order tensor is well suited to describe the height, the width and the color channel of a color image. In recent years, low-rank constraint is also used to recover higher-order tensors from local observations. The difference is that the rank of tensors is not explicitly defined. Therefore, many methods of low-rank tensor completion (LRTC) [14] are accomplished by extending the definition of matrix rank into the tensor form. In Ref. [14], the nuclear norm of a tensor is defined as an average of all the nuclear norms of its unfolded matrices. The LRTC is accomplished by minimizing the nuclear norm of the restored tensor. However, since it shares the same entries for all the unfolded matrices in each mode, their nuclear norms are interdependent consequently. Thus, the defined tensor nuclear norm is difficult to minimize. To remedy this issue, Ref. [14] introduces several auxiliary matrices for different modes to separate the interdependent terms in optimization. As a result, they present two enhanced methods, FaLRTC and HaLRTC. With the help of unfolding formulation, the two LRTC solvers are also improved by using ADMM [15]. In addition, Tomioka et al. [16] developed three relaxations to estimate low-rank tensors, which respectively denote tensor as a matrix, constraint and mixture models. Moreover, in recent studies [17,18], the authors proposed their approach involve using of tensor decomposition technique. There are two decomposition models: Tucker decomposition and polyadic decomposition. Tensor decomposition is a method used to decompose a tensor of Nth-order into another tensor of smaller size, termed as N factor matrices and the core tensor. As the same as the singular value decomposition (SVD) of matrix factorization, the minimum size of a core tensor is similar to the rank of the original tensor.

In our paper, we consider that low-rank constraints, albeit useful, are not sufficient to effectively use some potential local structures of tensors for completion. This point is in particular obviously for image inpainting. Due to the existence of objects or edges, visual data tend to show smooth and segmented structure in spatial dimension. Without special considerations on the local structures, the recovered results may be barely satisfactory. Total variation (TV) regularization was proposed for image recovery [18], it has proven extremely useful for many applications like image inpainting or interpolation [7]. More recently, Abergel and Moisan proposed the Shannon Total Variation (STV) [19], which performs better in the fields of artifact removal, isotropy and sub-pixel accuracy. Specifically, according to the Riemann sum of corresponding integrals, the continuous TV of the Shannon interpolate can be approximated by STV. So, as to utilize most of low-rank complementary information, in this paper, STV is introduced into tensor recovery for employing local piecewise smooth structure.

The main contributions of the paper can be summarized as follows: (1) we propose an image inpainting approach using direct tensor modeling techniques, which can infer the multichannel factors and the predictive distribution over missing entries given an incomplete tensor. (2) Taking both local smoothness and global structure into consideration, we propose our Tucker-based low rank tensor completion method with truncated γ-norm and Shannon total variation regularization. (3) To solve the nonconvex optimization problem in our algorithm, we introduce the alternating direction method of multipliers (ADMM) method to deliver the completion results. (4) Extensive experiments on color image inpainting tasks demonstrate that the proposed method enhances the details of the recovered images.

2 Related Work

2.1 A Simple Formulation of Tensor Completion

Let $X, M \in \mathbb{R}^{p \times q}$ be the observed matrix, the elements of M in the set Ω are given while the remaining elements are missing. A general matrix completion problem can be written as follows [20]:

$$\min_{X} \frac{1}{2} \|X_{\Omega} - M_{\Omega}\|_F^2 + \tau \|X\|_* \tag{1}$$

Where X is the target matrix to be recovered and τ is a constant. Ref. [14] extended the matrix nuclear norm to the tensor case and proposed to recover the missing entries in a low rank tensor by solving a nuclear norm minimization problem.

$$\min_{\mathcal{X}} \frac{1}{2} \|\mathcal{X}_\Omega - \mathcal{Y}_\Omega\|_F^2 + \tau \|\mathcal{X}\|_* \qquad (2)$$

Where $\mathcal{X}, \mathcal{Y} \in \mathbb{R}^{I_1 \times \cdots I_n}$ are n-mode tensors with identical size in each mode, the nuclear norm of tensor $\mathcal{X}$ is defined in $\|\mathcal{X}_{(i)}\|_* := \sum_{i=1}^n \alpha_i \|\mathcal{X}_{(i)}\|_*$, and α_is are constants satisfying $\alpha_i \geq 0, \sum_{i=1}^n \alpha_i = 1$. Under this definition, the optimization in (2) can be written as

$$\min_{\mathcal{X}} \sum_{i=1}^n \alpha_i \|\mathcal{X}_{(i)}\|_* \quad s.t. \mathcal{X}_\Omega = \mathcal{Y}_\Omega \qquad (3)$$

The problem in (3) is difficult to solve due to the interdependent matrix nuclear norm terms, i.e., while we optimize the sum of multiple matrix nuclear norms, the matrices share the same entries and cannot be optimized independently. The key motivation of simplifying this original problem is how to split these interdependent terms so that they can be solved independently. In related studies, Ji Liu el.at. proposed High Accuracy Low Rank Tensor Completion (HaLRTC) [14], they introduced additional matrices $M_1, \ldots, M_n$ and obtained the following equivalent formulation:

$$\min_{\mathcal{X}, \{M_{(i)}\}_{i=1}^n} \sum_{i=1}^n \alpha_i \|M_{(i)}\|_* s.t. \mathcal{X}_\Omega = \mathcal{Y}_\Omega, \{\mathcal{X}_{(i)} = M_{(i)}\}_{i=1}^n \qquad (4)$$

Where $M_{(i)} \in \mathbb{R}^{I_i \times (\prod_{k \neq i} I_k)}$, and $\mathcal{X}, \mathcal{Y} \in \mathbb{R}^{I_1 \times \cdots \times I_n}$.

2.2 Low-Rank Tensor Completion with TV

Theoretically, within lower bound of the rank function of matrices, the nuclear norm is the tightest convex norm. Thus, the nuclear norm is adopted as a measure of rank, and it can be used to represent the low rank prior. According to the case discussion in the introduction, this work also considers the sparse gradient regularization. Altogether we have the following formulation:

$$\min_{\mathcal{X}, \{M_{(i)}\}_{i=1}^n} \sum_{i=1}^n \alpha_i \|M_{(i)}\|_* + \sum_{i=1}^n \lambda_i \|\nabla M_{(i)}\|_0 \quad s.t. \mathcal{X}_\Omega = \mathcal{Y}_\Omega, \{\mathcal{X}_{(i)} = M_{(i)}\}_{i=1}^n$$
$$(5)$$

Noticing the second term in Eq. (5) is the ℓ_0-norm of gradient. Minimization corresponding to the ℓ_0-norm is usually approximated to ℓ_1-norm, thus the ℓ_0 gradient can become convex by relaxing to total variation [22]. The existed LRTV scheme is used in image inpainting can be formulated that

$$\min_{\mathcal{X}, \{M_{(i)}\}_{i=1}^n} \sum_{i=1}^n \alpha_i \|M_{(i)}\|_* + \sum_{i=1}^n \lambda_i \mathrm{TV}(M_{(i)}) \quad s.t. \mathcal{X}_\Omega = \mathcal{Y}_\Omega, \{\mathcal{X}_{(i)} = M_{(i)}\}_{i=1}^n$$
$$(6)$$

Compared with the cost function only has the low rank regularization term, the TV regularization can improve the restoration effect. However, there are some draw-backs of TV regularization, which smooths out the true depth of the edges and it can be observed that TV norm always becomes close or even lower than truth in the depth of painting effects. And even with a lower than ground-truth TV, the image is still visually noisy.

3   LRTC with Shannon Total Variation for Image Inpainting

3.1   The Shannon Total Variation

For the total variation property, most algorithms choose to approximate the continuous TV by a sum (over all pixels) of the ℓ_2-norm of a discrete finite difference estimate in the image gradient, that is,

$$\text{TV}^d(u) = \sum_{(i,j)\in\Omega} \sqrt{(\partial_1 u(i,j))^2 + (\partial_2 u(i,j))^2} \tag{7}$$

Where

$$\begin{cases} \partial_1 u(i,j) = u(i+1.j) - u(i,j) \\ \partial_2 u(i,j) = u(i,j+1) - u(i,j) \end{cases} \tag{8}$$

In some situations, an anisotropic scheme (ℓ_2-norm) may be used [24], leading to the anisotropic discrete TV that can be written as

$$\text{TV}^d_{ani}(u) = \sum_{(i,j)\in\Omega} |\partial_1 u(i,j)| + |\partial_2 u(i,j)| \tag{9}$$

In general, the performance of the anisotropic TV constrain is unsatisfactory both in pixel-level and subpixel-level. Because interpolating on the image is difficult, whether by minimizing TV^d-based energies or sampling based on Shannon theory. Recently, Remy Abergel and Lionel Moisan proposed a new total variation which called the Shannon Total Variation (STV) [19].

Let $|\cdot|$ denote the ℓ^2 norm, and $\Omega = I_p \times I_Q$ denote a 2-D discrete domain of size $P \times Q$. $u \in \mathbb{R}^\Omega$ is a discrete gray-level image. The definition of the Shannon total variation is

$$\text{STV}_\infty(u) = \int_{[0,P]\times[0,Q]} |\nabla U(x,y)| dx dy \tag{10}$$

In which U is the Shannon interpolation of u, and the gradient of the trigonometric polynomial U is denoted by $\nabla U : \mathbb{R}^2 \to \mathbb{R}^2$. Therefore, no closed-form formula exists for (10), then Remy Abergel and Lionel Moisan approximated this continuous integral with the Riemann sum:

$$\text{STV}_n(u) = \frac{1}{n^2} \sum_{(i,j)\in\Omega_n} |\nabla u(i,j)| \tag{11}$$

Where $n \in \mathbb{N}^*, \Omega_n = I_{nP} \times I_{nQ}$, and $\forall (i,j) \in \Omega_n, \nabla u(i,j) = \nabla U(\frac{i}{n}, \frac{j}{n})$.

The studies of Remy Abergel and Lionel Moisan showed that it is difficult to interpolation on the processed images based on the variational TV when the TV is discretized by the classical finite difference scheme. Among them, STV successfully addresses this issue. Figure 1 shows the recovery results of (a) by TV and STV. When magnifying the images, compared with (c), (d) is more blurred. So, we can absolutely believe that STV will also perform better than traditional TV regularization in our algorithm model.

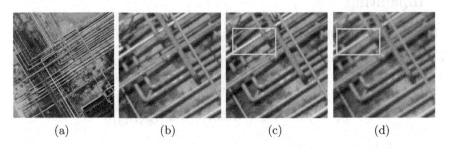

| (a) | (b) | (c) | (d) |

Fig. 1. A simple image restoration instance on 60% random missing. (a–b) the original. (c) recovery results by STV. (d) recovery results by TV.

3.2 LRTC with STV and Truncated γ-norm

In the previous section we have augmented the low rank tensor completion and the STV. We can summarize as follows, using low-rank regularization to help retrieve useful information from remote regions and using total variation regularization can keep better local consistency. As discussed in the end of the previous section, we consider replace TV constraint with STV constraint in Eq. (6). Moreover, instead of apply a low-rank regularization on the unfolded matrices of tensor $\mathcal{X}$, we except its Tucker-based factors to be low-rank, which can be represented by the following

$$\min_{\mathcal{X}, \mathcal{G}, \{M_{(i)}, U^{(i)}\}_{i=1}^n} \beta_i \sum_{i=1}^n \lambda_1 \text{STV}(M_{(i)}) + \frac{1}{n} \sum_{i=1}^n \left\| U^{(i)} \right\|_* + \lambda_2 \|\mathcal{G}\|_F^2 \quad (12)$$

$$s.t.\{\mathcal{X}_{(i)} = M_{(i)}\}_{i=1}^n, \mathcal{X}_\Omega = \mathcal{Y}_\Omega, \mathcal{X} = \mathcal{G} \times_1 U^{(1)} \times \ldots \times_n U^{(n)}$$

About the second constraint (nuclear norm) in Eq. (12) cannot obtain well-pleasing performance. Most of the existed researches use the convex kernel norm instead of the rank function, and replace the ℓ_0-norm with the ℓ_1-norm. However, the simple replacement does not involve the essence of the problem. The nuclear norm is defined as the sum of singular values in mathematical, and when the rank function is treated as several nonzero singular values, each singular value contributes equally to the function. There are similar differences between the ℓ_0-norm and ℓ_1-norm when performing a theoretical analysis. Therefore, the effect achieved by the solution to the relaxation problem may be far from the expected effect of the original solution. Instead, some researchers consider trying to solve such problems with non-convex substitution functions.

By minimizing the low-rank norm, all the singular values are simultaneously minimized, and thus the rank cannot be well approximated in practice. So, in this paper, we propose novel truncated γ-norm by only minimizing the smallest $\min(p,q) - r$ singular values, where $\min(p,q)$ is the number of singular values and r is the rank of the matrix, which is defined as

$$\text{rank}(\boldsymbol{X}) \approx \|\boldsymbol{X}\|_{\gamma,r} = \sum_{i=r+1}^{\min(p,q)} \left(1 - e^{-\sigma_i(\boldsymbol{X})/\gamma}\right) \tag{13}$$

Where $\gamma > 0$. Here, we refer to (13) as truncated γ-norm.

In this paper, we propose a new formulation with the truncated γ-norm and Shannon total variation regularized for recovering the Tucker-based low-rank tensor. The proposed Tucker-based low rank tensor completion via truncated γ-norm joint Shannon total variation regularized (LRRTV) method is formulated as follow:

$$\min_{\boldsymbol{X},\mathcal{G},\{M_{(i)},U^{(i)}\}_{i=1}^n} \beta_i \sum_{i=1}^n \lambda_1 \text{STV}(M_{(i)}) + \frac{1}{n} \sum_{i=1}^n \left\|U^{(i)}\right\|_{\gamma,r} + \lambda_2 \|\mathcal{G}\|_F^2 \tag{14}$$
$$s.t.\{\boldsymbol{X}_{(i)} = M_{(i)}\}_{i=1}^n, \boldsymbol{X}_\Omega = \mathcal{Y}_\Omega, \boldsymbol{X} = \mathcal{G} \times_1 U^{(1)} \times \ldots \times_n U^{(n)}$$

Where tensor $\boldsymbol{X}$ represents the recovery result; λ_1, λ_2 are tunable parameters; $\beta_1, \beta_2, \ldots, \beta_n$ are 0 or 1, which indicate there is a smooth piecewise priori on the n-th mode of tensors recovered by our method. When $\mathcal{Y}$ is a tensor of color image, we set $\beta_1, \beta_2 = 1$ and $\beta_3 = 0$, because only spatial dimensions are expected to have smooth and piecewise priors.

4   LRRTC Optimization

In our model, because of the three terms in Eq. (14) are interdependent, we apply the alternating direction method of multipliers (ADMM) for optimization process. Hence, we introduce other matrices $\{R_{(i)}\}_{i=1}^n$ and $\{V^{(i)}\}_{i=1}^n$ as auxiliary variables, which can make our optimization process more simply. We split the interdependencies and rewrite the optimization problem as

$$\min_{\boldsymbol{X},\mathcal{G},\{N_{(i)},U^{(i)}\}_{i=1}^n} \beta_i \sum_{i=1}^n \lambda_1 \text{STV}(M_{(i)}) + \frac{1}{n} \sum_{i=1}^n \left\|U^{(i)}\right\|_{\gamma,r} + \lambda_2 \|\mathcal{G}\|_F^2 \tag{15}$$
$$s.t.\{V^{(i)} = U^{(i)}, M_{(i)} = R_{(i)}, R_{(i)} = X_{(i)}\}^n$$
$$\boldsymbol{X}_\Omega = \mathcal{Y}_\Omega, \boldsymbol{X} = \mathcal{G} \times_1 U^{(1)} \times_2 U^{(2)} \times \ldots \times_n U^{(n)}$$

We follow ADMM algorithm to solve the cost function in Eq. (15) which can prove to be efficient for solving optimization problems with multiple non-smooth terms in the cost function. By using the augmented Lagrange formulation, the optimization problem is changed into:

$$
\mathcal{L} = \sum_{i=1}^{n} \lambda_1 \left(\beta_i \mathrm{STV}(\boldsymbol{M}_{(i)}) + \frac{\rho_1}{2} \left\| \boldsymbol{M}_{(i)} - \boldsymbol{R}_{(i)} + \frac{\Lambda_{1(i)}}{\rho_1} \right\|_F^2 \right)
$$

$$
+ \sum_{i=1}^{n} \lambda_1 \left(\frac{\rho_2}{2} \left\| \boldsymbol{R}_{(i)} - \boldsymbol{\mathcal{X}}_{(i)} + \frac{\Lambda_{2(i)}}{\rho_2} \right\|_F^2 \right)
$$

$$
+ \sum_{i=1}^{n} \left(\frac{1}{n} \| \boldsymbol{U}^{(i)} \|_{\gamma,r} + \frac{\rho_3}{2} \left\| \boldsymbol{V}^{(i)} - \boldsymbol{U}^{(i)} + \frac{\Lambda_{3(i)}}{\rho_3} \right\|_F^2 \right) \tag{16}
$$

$$
+ \lambda_2 \| \mathcal{G} \|_F^2 + \frac{\rho_4}{2} \left\| \boldsymbol{\mathcal{X}} - \mathcal{G} \times_1 \boldsymbol{V}^{(1)} \times_2 \boldsymbol{V}^{(2)} \times \dots \times_n \boldsymbol{V}^{(n)} + \frac{\boldsymbol{Z}}{\rho_4} \right\|_F^2
$$

$$
s.t.\{ \boldsymbol{V}^{(i)} = \boldsymbol{U}^{(i)}, \boldsymbol{M}_{(i)} = \boldsymbol{R}_{(i)}, \boldsymbol{R}_{(i)} = \boldsymbol{\mathcal{X}}_{(i)} \}_{i=1}^{n}
$$

$$
\boldsymbol{\mathcal{X}}_\Omega = \boldsymbol{\mathcal{Y}}_\Omega, \boldsymbol{\mathcal{X}} = \mathcal{G} \times_1 \boldsymbol{U}^{(1)} \times_2 \boldsymbol{U}^{(2)} \times \dots \times_n \boldsymbol{U}^{(n)}
$$

Where matrices $\{\Lambda_{1(i)}\}_{i=1}^{n}, \{\Lambda_{2(i)}\}_{i=1}^{n}, \{\Lambda_{3(i)}\}_{i=1}^{n}$ are Lagrange multipliers; $\| \cdot \|_F^2$ is Frobenius norm of a matrix or a tensor. Next, we derive the update formulae of $\{\boldsymbol{M}_{(i)}\}_{i=1}^{n}, \{\boldsymbol{U}^{(i)}\}_{i=1}^{n}, \{\boldsymbol{V}^{(i)}\}_{i=1}^{n}, \{\boldsymbol{R}_{(i)}\}_{i=1}^{n}, \{\boldsymbol{\mathcal{X}}_{(i)}\}_{i=1}^{n}$ and $\mathcal{G}$ to be solved in particularly Eq. (16), the cost function is convex if the remaining other matrices are kept fixed. Equation (16) can be solved iteratively via the following subproblems.

Fixing $\{\boldsymbol{U}^{(i)}\}_{i=1}^{n}, \{\boldsymbol{V}^{(i)}\}_{i=1}^{n}, \{\boldsymbol{R}_{(i)}\}_{i=1}^{n}, \{\boldsymbol{\mathcal{X}}_{(i)}\}_{i=1}^{n}$ and $\mathcal{G}$, for $\{\boldsymbol{M}_{(i)}\}_{i=1}^{n}$ by the following problem,

$$
\min_{\{\boldsymbol{M}_{(i)}\}_{i=1}^{n}} \lambda_1 \left(\beta_i \mathrm{STV}(\boldsymbol{M}_{(i)}) + \frac{\rho_1}{2} \left\| \boldsymbol{M}_{(i)} - \boldsymbol{R}_{(i)} + \frac{\Lambda_{1(i)}}{\rho_1} \right\|_F^2 \right) \tag{17}
$$

according to the dual formulation of the Shannon total variation, and Eq. (17) can be written as follows:

$$
\min_{\{M_{(i)}\}_{i=1}^{n}} \lambda_1 \left(\left\langle \frac{\beta_i}{n^2} \nabla_n \boldsymbol{M}_{(i)}, p_{(i)} \right\rangle - \delta_{\| \cdot \|_{\infty,2} \leq 1}(p_{(i)}) + \frac{\rho_1}{2} \left\| \boldsymbol{M}_{(i)} - \boldsymbol{R}_{(i)} + \frac{\Lambda_{1(i)}}{\rho_1} \right\|_F^2 \right)
$$
$$\tag{18}$$

Due to the complexity of Eq. (18) the Chambolle-Pock (CP) algorithm [23] is proposed and which can be used to address various TV-based image processing tasks and it also comes with nice convergence theorems. The CP algorithm can be shown below

$$
\min_x \max_y \{ \langle Kx, y \rangle + G(x) - F^*(y) \} \tag{19}
$$

Our Eq. (18) has exactly the form of (19) with $(x, y) = (\boldsymbol{M}_{(i)}, p_{(i)})$, $k = \frac{\beta}{n^2} \nabla_n$, $G(\boldsymbol{M}_{(i)}) = \frac{\rho_1}{2} \left\| \boldsymbol{M}_{(i)} - \boldsymbol{R}_{(i)} + \frac{\Lambda_{1(i)}}{\rho_1} \right\|_F^2$, and $F^*(p_{(i)}) = \delta_{\| \cdot \|_{\infty,2} \leq 1}(p_{(i)})$. Then, we can solve the problem (17) as Algorithm 1.

Algorithm 1. Chambolle-Pock(CP) resolvent Algorithm for Problem (17).

1: $\tau, \sigma > 0; \theta \in [0,1]; k = 0$
2: Initialize $M_{(i)}, p_{(i)}$ to zero values
3: $\bar{M}_{(i)}^0 = N_{(i)}^0$
4: Repeat
5: $p_{(i)}^{k+1} = \text{prox}_\sigma[F^*]\left(p_{(i)}^k + \sigma K \bar{M}_{(i)}^k\right)$
6: $M_{(i)}^{k+1} = \text{prox}_\sigma[G]\left(M_{(i)}^k - \tau K^T p_{(i)}^{k+1}\right)$
7: $\bar{M}_{(i)}^{k+1} = M_{(i)}^{k+1} + \theta\left(M_{(i)}^{k+1} - M_{(i)}^k\right)$
8: $k = k + 1$

The proximal mapping of prox_σ and prox_τ

$$\text{prox}_\sigma[F^*](z) = \arg\min_{z'}\left\{F^*(z') + \frac{\|z - z'\|_F^2}{2\sigma}\right\} \tag{20}$$

Fixing $\{M_{(i)}\}_{i=1}^n, \{V^{(i)}\}_{i=1}^n, \{R_{(i)}\}_{i=1}^n, \{\mathcal{X}_{(i)}\}_{i=1}^n$ and $\mathcal{G}$, for $\{U^{(i)}\}_{i=1}^n$ by the following problem,

$$\min_{\{U^{(i)}\}_{i=1}^n} \frac{1}{n}\sum_{i=1}^n\left(\frac{1}{n}\|U^{(i)}\|_{\gamma,r} + \frac{\rho_3}{2}\left\|U^{(i)} - V^{(i)} + \frac{\Lambda_{3(i)}}{\rho_3}\right\|_F^2\right) \tag{21}$$

Which is a non-convex problem, let $f(U) = \frac{1}{2}\|U^{(i)} - G_k\|_F^2$ with $G_k = V^{(i)} - \frac{\Lambda_{3(i)}}{\rho_3}$. It is foolproof to prove that the gradient of $f(U)$ is Lipschitz continuous by setting the Lipschitz constant being 1. U can be represented as $U^{(i)} = \sum_{i=r+1}^{\min(p,q)} \sigma_i(U^{(i)})u_i v_i^T$, denote the gradient of ϕ at σ_i. Because of the non-ascending order of singular values, and the gradient of a non-convex function is the opposite of it, we can obtain that

$$U^{(i)} = \frac{1}{\rho_3}\sum_{i=r+1}^{\min(p,q)}\nabla\phi(\sigma_i)\sigma_i(U^{(i)}) + \frac{1}{2}\left\|U^{(i)} - V^{(i)} + \frac{\Lambda_{3(i)}}{\rho_3}\right\|_F^2 \tag{22}$$

Fixing $\{M_{(i)}\}_{i=1}^n, \{U^{(i)}\}_{i=1}^n, \{V^{(i)}\}_{i=1}^n, \{\mathcal{X}_{(i)}\}_{i=1}^n$ and $\mathcal{G}$, for $\{R_{(i)}\}_{i=1}^n$ by the following problem,

$$\min_{\{R_{(i)}\}_{i=1}^n} \lambda_1\sum_{i=1}^n\frac{\rho_1}{2}\left\|M_{(i)} - R_{(i)} + \frac{\Lambda_{1(i)}}{\rho_1}\right\|_F^2 + \lambda_1\sum_{i=1}^n\frac{\rho_2}{2}\left\|R_{(i)} - \mathcal{X}_{(i)} + \frac{\Lambda_{2(i)}}{\rho_2}\right\|_F^2 \tag{23}$$

Hence, the following update formula is derived by solving the minimization problem:

$$R_{(i)} = \lambda_1(\rho_1 I + \rho_2 I)^{-1}(\Lambda_{1(i)} + \rho_1 M_{(i)} + \rho_2\mathcal{X}_{(i)} - \Lambda_{2(i)}) \tag{24}$$

Where I stand for the identify matrix.

Similarly, the optimization problem $\{\boldsymbol{V}^{(i)}\}_{i=1}^n$ can be represented as:

$$
\begin{aligned}
\boldsymbol{V}^{(i)} = &\left(-\Lambda_{3(i)} + \rho_3 \boldsymbol{U}^{(i)} + \left(\boldsymbol{Z}_{(i)} + \rho_4 \boldsymbol{X}_{(i)}\right) \boldsymbol{V}^{(-i)} \mathcal{G}_{(i)}^T\right) \\
&\times \left(\rho_3 I + \rho_4 \boldsymbol{G}_{(i)} \boldsymbol{V}^{(-i)^T} \boldsymbol{V}^{(-i)} \mathcal{G}_{(i)}^T\right)^{-1}
\end{aligned}
\tag{25}
$$

The update formulae of $\boldsymbol{X}$ is computed as:

$$
[\boldsymbol{X}]_{\bar{\Omega}} = \left[\frac{\sum_{i=1}^n \beta_n \left(-\mathrm{fold}_{(i)}(\Lambda_{2(i)} + \rho_2 \boldsymbol{R}_{(i)})\right) - \boldsymbol{Z} + \rho_4 \hat{\boldsymbol{X}}}{\sum_{i=1}^n \beta_n \rho_2 + \rho_4}\right]
\tag{26}
$$

Where $\otimes$ is the Kronecker product, $\boldsymbol{V}^{(-i)} = \boldsymbol{V}^{(1)} \otimes \boldsymbol{V}^{(2)} \otimes \ldots \otimes \boldsymbol{V}^{(i-1)} \otimes \boldsymbol{V}^{(i+1)} \otimes \ldots \boldsymbol{V}^{(i)}$, and in equation (26) $\hat{\boldsymbol{X}} = \mathcal{G} \times_1 \boldsymbol{V}^{(1)} \times_2 \boldsymbol{V}^{(2)} \times \ldots \times \times_n \boldsymbol{V}^{(n)}$.

Algorithm 2. The LRRTC for tensor completion.

Input: an incomplete tensor $\boldsymbol{\mathcal{Y}}$, iteration number $K, \lambda, \rho_1, \rho_2, \rho_3, \rho_4$ and $\mu \in [1, 1.5]$.
Output: a recovery tensor $\boldsymbol{X}$.
1: $[\boldsymbol{X}]_\Omega = [\boldsymbol{\mathcal{Y}}]_\Omega, [\boldsymbol{X}]_{\bar{\Omega}} = 0$,randomly initialize $\{\boldsymbol{N}_{(i)}\}_{i=1}^n, \{\boldsymbol{R}_{(i)}\}_{i=1}^n, \{\boldsymbol{M}_{(i)}\}_{i=1}^n$.
2: **For** $k = 1$ to K **do**
3:   Update $\{\boldsymbol{M}_{(i)}\}_{i=1}^n, \{\boldsymbol{U}^{(i)}\}_{i=1}^n, \{\boldsymbol{R}_{(i)}\}_{i=1}^n, \{\boldsymbol{V}^{(i)}\}_{i=1}^n, \boldsymbol{X}$ and $\mathcal{G}$ by equation (17),
    (22), (24), (25),(26),(27) respectively.
4:   $\{\Lambda_{1(i)}\}_{i=1}^n = \{\Lambda_{1(i)} + \rho_1 (\boldsymbol{M}_{(i)} - \boldsymbol{R}_{(i)})\}_{i=1}^n$
5:   $\{\Lambda_{2(i)}\}_{i=1}^n = \{\Lambda_{2(i)} + \rho_2 (\boldsymbol{R}_{(i)} - \boldsymbol{X}_{(i)})\}_{i=1}^n$
6:   $\{\Lambda_{3(i)}\}_{i=1}^n = \{\Lambda_{3(i)} + \rho_3 (\boldsymbol{V}^{(i)} - \boldsymbol{U}^{(i)})\}_{i=1}^n$
7:   $\boldsymbol{Z} = \boldsymbol{Z} + \mathcal{G} \times_1 \boldsymbol{V}^{(1)} \times_2 \boldsymbol{V}^{(2)} \times \ldots \times \times_n \boldsymbol{V}^{(n)}$
8:   $\rho_1 = \mu \rho_1, \rho_2 = \mu \rho_2, \rho_3 = \mu \rho_3, \rho_4 = \mu \rho_4$.
9:   $k = k + 1$
10: **Return** $\boldsymbol{X}$.

Inspired by these update formulae, the solving procedure of our method is given in Algorithm 2. Following the framework of ADMM, our solving procedure through iteration. According to our derivations, our solving procedure updates the auxiliary matrices $\{\boldsymbol{M}_{(i)}\}_{i=1}^n, \{\boldsymbol{U}^{(i)}\}_{i=1}^n, \{\boldsymbol{R}_{(i)}\}_{i=1}^n, \{\boldsymbol{V}^{(i)}\}_{i=1}^n$, and core tensor $\mathcal{G}$ and target output variable $\boldsymbol{X}$, which are shown in the third line. In the next 4 lines, the Lagrange multipliers, $\Lambda_{1(i)}, \Lambda_{2(i)}, \Lambda_{3(i)}$ and $\boldsymbol{Z}$ are updated as the standard ADMM. In line 8, to accelerate convergence, ρ_1, ρ_2, ρ_3 and ρ_4 are adaptively increased.

5   Experimental Results

We apply our method to a variety of natural images with different inpainting tasks, i.e. text removal and randomly missing pixels filling. All the samples are shown in Fig. 1. We compare our approach with some recently presented algorithms, i.e., HaLRTC [14], BCPF [24], BCPF-MF, LRTC-TV-II [25]. The eight ground-truth images for the experiment are shown in Fig. 2 Each of them is three color channels image and the resolution of each one is 256-by-256 pixels. And, they can be represented as 256-by-256-by-3 tensors. The well-known evaluation metric, Peak Signal to Noise Ratio (PSNR), is adopted to demonstrate the performance of all the competing methods. The PSNR function is expressed as follows:

$$\text{PSNR} = 10 \log_{10} \frac{\hat{\boldsymbol{\mathcal{X}}}_{true}^2}{\frac{1}{\prod_{i=1}^{n} I_n} \| \boldsymbol{\mathcal{X}} - \boldsymbol{\mathcal{X}}_{true} \|_F^2} \tag{27}$$

Where $\boldsymbol{\mathcal{X}}$, $\boldsymbol{\mathcal{X}}_{true}$, and $\hat{\boldsymbol{\mathcal{X}}}_{true}$ represent the recovered tensor, ground-truth tensor, and the maximum value in the ground-truth tensor, respectively. The larger PSNR indicates better recovery performance. To make a fair comparison between algorithms, the tunable parameters for all the competing methods are manually selected to report the best performance in terms of quantitative criteria, visual assessment, and computational cost. Furthermore, the maximum iteration number and the stop tolerance are set as 100 and $1e-5$, respectively. It has been proved by our experiments that 200 iterations or more will perform better. We set $\beta_1 = \beta_2 = 1, \beta_3 = 0$ and the parameter $\lambda_1 = 0.5, \lambda_2 = 100$.

Experimental environment: CPU for Intel (R) Core i7-7500U 2.70GHz, 8GB of memory capacity, the system is 64-bit Microsoft Windows 10, the software version is MATLAB R2016a.

Fig. 2. Ground truth of eight color images. (Color figure online)

5.1   Experiments on Random Pixels Missing

In this subsection, to testing the inpainting performance, a color image with 60% missing pixels under two noise conditions, i.e., noise free and SNR = 40 dB, which were considered as observations. The performances of all algorithms are shown in Fig. 3. The worst perceptual result is the recovered image by HaLRTC, it is non-smooth and contains vertical or horizontal noisy lines obviously. The reason of the worst performance is that the local smooth and piecewise property of visual data are neglected in HaLRTC. LRTC-TV-II is an enhanced model obtained by HaLRTC with TV regularization. This is attributed to the additionally introduced variables for achieving better recovery effectiveness. The proposed method LRRTC performs the best result among all the approaches.

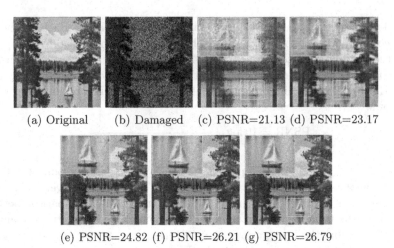

(a) Original   (b) Damaged   (c) PSNR=21.13 (d) PSNR=23.17

(e) PSNR=24.82 (f) PSNR=26.21 (g) PSNR=26.79

Fig. 3. Recovery results of (c) HaLRTC, (d) FBCP, (e) FBCP-MF, (f) LRTC-TV-II, and (g) LRRTC methods on the 60% random missing.

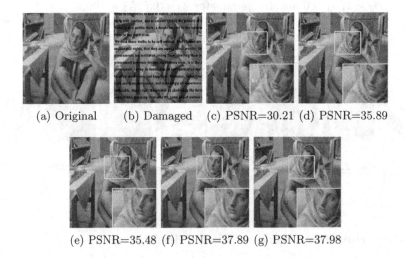

(a) Original   (b) Damaged   (c) PSNR=30.21 (d) PSNR=35.89

(e) PSNR=35.48 (f) PSNR=37.89 (g) PSNR=37.98

Fig. 4. Recovery results of (c) HaLRTC, (d) FBCP, (e) FBCP-MF, (f) LRTC-TV-II, and (g) LRRTC methods on the text removal.

5.2 Experiments on Text Removal

The effects comparison of the text removal for barbara color image by different algorithms are shown in Fig. 4. The image with damaged structure and texture information can be recovered by the above algorithm effectively according to the qualitative evaluation of the experience. It can be observed in Fig. 4 that although the damaged image can be repaired by HaLRTC. In general, there are still some serious error or ambiguities in the restoration of the texture information, especially the damaged region. According to the evaluation index such as PSNR, the performance of our method is better than others when the color images damaged by text. And from a subjective visual perspective, the region of removed text is much closer to the original image by our method. It can be seen in the experiment results that the algorithm LRRTC is best for the recovery of the edge of the chair.

6   Conclusion

In this study, we propose a novel method which integrates Shannon total variation and truncated γ-norm into Tucker-based low-rank tensor completion for modeling the local smoothness and piecewise priors of visual data. Experiments mentioned above exhibit that the algorithm mentioned in this paper has advantages in both qualitative and quantitative indexes when repairing damaged images. The proposed methods are very basic and have great potential to improve performance. The current disadvantage of the algorithms are that they run slowly and time-consuming because STV requests much more time. This is also a question we will consider in the future that whether there is a closed form solution to make our algorithm more efficient.

Acknowledgment. This research is funded by Natural Science Foundation of China under Grant Nos. 61702275, 61976192, 61602413, 41775008, and by Zhejiang Provincial Natural Science Foundation of China under Grant Nos. LY18F020032 and LY19F030016.

References

1. Guillemot, C., Le Meur, O.: Image inpainting: overview and recent advances. IEEE Sig. Process. Mag. **31**(1), 127–144 (2014)
2. Bertalmio, M., Sapiro, G., Caselles, V., Ballester, C.: Image inpainting. In: Proceedings of the 27th Annual Conference on Computer Graphics and Interactive Techniques, pp. 417–424. ACM Press/Addison-Wesley Publishing Co. (2000)
3. Komodakis, N.: Image completion using global optimization. In: 2006 IEEE Computer Society Conference on Computer Vision and Pattern Recognition, CVPR 2006, vol. 1, pp. 442–452. IEEE (2006)
4. Candès, E.J., Tao, T.: The power of convex relaxation: near-optimal matrix completion. arXiv preprint arXiv:0903.1476
5. Gandy, S., Recht, B., Yamada, I.: Tensor completion and low-n-rank tensor recovery via convex optimization. Inverse Probl. **27**(2), 025010 (2011)
6. Jidesh, P., Febin, I.: Estimation of noise using non-local regularization frameworks for image denoising and analysis. Arab. J. Sci. Eng. **44**(4), 3425–3437 (2019)
7. Bini, A.: Image restoration via dost and total variation regularisation. IET Image Process. **13**(3), 458–468 (2018)
8. Fan, J., Li, R.: Variable selection via nonconcave penalized likelihood and its oracle properties. J. Am. Stat. Assoc. **96**(456), 1348–1360 (2001)
9. Markovsky, I.: Applications of structured low-rank approximation. IFAC Proc. Vol. **42**(10), 1121–1126 (2009)

10. Candés, E.J., Recht, B.: Exact matrix completion via convex optimization. Found. Comput. Math. **9**(6), 7–17 (2009)
11. Chartrand, R.: Exact reconstruction of sparse signals via nonconvex minimization. IEEE Sig. Process. Lett. **14**(10), 707–710 (2007)
12. Shang, F., Liu, Y., Cheng, J.: Scalable algorithms for tractable Schatten quasi-norm minimization. In: Thirtieth AAAI Conference on Artificial Intelligence (2016)
13. Chen, Y., Guo, Y., Wang, Y., et al.: Denoising of hyperspectral images using nonconvex low rank matrix approximation. IEEE Trans. Geosci. Remote Sens. **55**(9), 5366–5380 (2017)
14. Liu, J., Musialski, P., Wonka, P., Ye, J.: Tensor completion for estimating missing values in visual data. IEEE Trans. Pattern Anal. Mach. Intell. **35**(1), 208–220 (2013)
15. Boyd, S., Parikh, N., Chu, E., Peleato, B., Eckstein, J., et al.: Distributed optimization and statistical learning via the alternating direction method of multipliers. Found. Trends R Mach. Learn. **3**(1), 1–122 (2011)
16. Tomioka, R., Hayashi, K., Kashima, H.: Estimation of low-rank tensors via convex optimization. arXiv preprint arXiv:1010.0789
17. Li, X., Ye, Y., Xu, X.: Low-rank tensor completion with total variation for visual data inpainting. In: Proceedings of the Thirty-First AAAI Conference on Artificial Intelligence, vol. 419, pp. 2210–2216 (2017)
18. Chen, Y.-L., Hsu, C.-T., Liao, H.-Y.: Simultaneous tensor decomposition and completion using factor priors. IEEE Trans. Pattern Anal. Mach. Intell. **36**(3), 577–591 (2014)
19. Recht, B., Fazel, M., Parrilo, P.A.: Guaranteed minimum-rank solutions of linear matrix equations via nuclear norm minimization. SIAM Rev. **52**(3), 471–501 (2010)
20. Abergel, R., Moisan, L.: The shannon total variation. J. Math. Imaging Vis. **59**(2), 341–370 (2017)
21. Xu, J., Zhang, L., Zhang, D., Feng, X.: Multi-channel weighted nuclear norm minimization for real color image denoising, pp. 1096–1104. CoRR (2017)
22. Ji, T.Y., Huang, T.Z., Zhao, X.L., Ma, T.H., Liu, G.: Tensor completion using total variation and low-rank matrix factorization. Inf. Sci. **326**, 243–257 (2016)
23. Zhou, L., Tang, J.: Fraction-order total variation blind image restoration based on l_1-norm. Appl. Math. Model. **51**, 469–476 (2017)
24. Chambolle, A., Pock, T.: A first-order primal-dual algorithm for convex problems with applications to imaging. J. Math. Imaging Vis. **40**(1), 120–145 (2011)
25. Zhao, Q., Zhang, L., Cichocki, A.: Bayesian CP factorization of incomplete tensors with automatic rank determination. IEEE Trans. Pattern Anal. Mach. Intel. **37**(9), 1751–1763 (2015)
26. Li, X., Ye, Y., Xu, X.: Low-rank tensor completion with total variation for visual data inpainting. In: 2017 Proceedings of the Thirty-First AAAI Conference on Artificial Intelligence, pp. 2210–2216 (2017)

Hyperspectral Image Super-Resolution Using Multi-scale Feature Pyramid Network

He Sun, Zhiwei Zhong, Deming Zhai, Xianming Liu, and Junjun Jiang[(✉)]

School of Computer Science and Technology, Harbin Institute of Technology,
Harbin, China
`jiangjunjun@hit.edu.cn`

Abstract. Hyperspectral (HS) images are captured with rich spectral information, which have been proved to be useful in many real-world applications, such as earth observation. Due to the limitations of HS cameras, it is difficult to obtain HS images with high-resolution (HR). Recent advances in deep learning (DL) for single image super-resolution (SISR) task provide a powerful tool for restoring high-frequency details from low-resolution (LR) input image. Inspired by this progress, in this paper, we present a novel DL-based model for single HS image super-resolution in which a feature pyramid block is designed to extract multi-scale features of the input HS image. Our method does not need auxiliary inputs which further extends the application scenes. Experiment results show that our method outperforms state-of-the-arts on both objective quality indices and subjective visual results.

Keywords: Hyperspectral image processing · Image super-resolution · Deep learning

1 Introduction

Hyperspectral (HS) images are of vital importance for various remote sensing and computer vision tasks. Compared to RGB images, HS images carry a more complete spectrum ranging from visible bands to shortwave infrared bands [1]. As a result of the extended spectrum, HS images contain more information about material characteristics of objects, so they are widely applied to tasks like classification [2,3], tracking [4], land degradation detection [5] and natural disaster accessing [6].

However, HS cameras are not able to capture HS images at high spatial resolution (HR) due to the limitations of the HS imaging system. Hence the need arises to perform super-resolution (SR) on the input low spatial resolution (LR) HS image. The main difficulty of SR task for HS images lies in the fact that HS images usually have over 100 bands and the correlation of each band needs to be considered. Most of the existing methods [7–12] require one additional HR image as an auxiliary input. Moreover, they are based on the assumption

© Springer Nature Singapore Pte Ltd. 2020
G. Zhai et al. (Eds.): IFTC 2019, CCIS 1181, pp. 49–61, 2020.
https://doi.org/10.1007/978-981-15-3341-9_5

that the HS image and the auxiliary HR image are well-registered, which is not satisfied during real world observations. Therefore, this paper mainly focuses on the problem of single HS image SR (SHSISR).

To accomplish this task, we use convolutional neural network (CNN) to perform end-to-end SHSISR task directly on the LR HS input. We find that the core of this task is to effectively extract multi-scale features from the successive channels of the input. To this end, we propose a feature pyramid block that can effectively extract the multi-scale features of HS images. Equipped with this block, our proposed model is able to accomplish the SHSISR task for HS images. Experiment results show that the whole network achieves better performance compared to present state-of-the-arts.

2 Related Works

Hyperspectral Image SR. There are two main streams in HS image superresolution task: fusion-based methods and single-HS-image-based methods. The former one, which is also referred as HS pan-sharpening [13], fuses a LR HS image with a HR panchromatic (PAN), RGB or multispectral (MS) image taken from the same scene. Fusion-based approaches involve Bayesian-based methods [7,8], matrix/tensor factorization-based methods [9,10] and DL-based methods [11,12]. However, in the real world, it is hard to obtain a HS image and its auxiliary MS image of high quality at the same time. The captured MS/HS image pairs are noisy, unregistered and it may even occur that they are taken at different time. The latter SHSISR methods naturally avoid these problems. Early attempts involves [14], but they fail to achieve the same image quality for the sake of no auxiliary HR information. It is not until recent that, with the help of the representation power of CNNs and the collect of large HS datasets, several CNN-based methods [15–17] gradually advance this kind of method for HS images. Our work follows this trend.

Deep Learning for SISR. SISR task is an ill-posed inverse problem which aims to recover high frequency details from a single LR input image. Dong et al. [18] first proposed a 3-layer SRCNN to conduct SISR. Following this pioneering work, Kim et al. [19] and Lim et al. [20] designed much deeper networks with skip connections to perform SISR task. Most recent advances on SISR task focus on techniques in network design. For example, Lai et al. [21] built a Laplacian pyramid to progressively upsample a LR image. Zhang et al. [22,23] introduced a channel attention layer [24] and densely-connected structure to further improve the network performance. Haris et al. [25] used feedback mechanism to help feature extraction in network. However, we can not directly apply the above SISR methods because existing SISR networks are not designed to handle so many input channels. Therefore, we use a different multi-scale extractor at the beginning of the network to help feature extraction.

Multi-scale Feature Extraction in CNN. The concept of multi-scale pyramid is important in computer vision. Methods using image-level pyramids like SIFT [26] have long been developed and widely used in vision tasks. However, the concept of feature-level pyramid has not received enough attention until recently Lin *et al.* [27] introduced a Feature Pyramid Network (FPN) for objection detection. Later the block was used in instance segmentation model Mask R-CNN [28] and achieved the state-of-the-art result on that task. Inspired by their work, we design a feature pyramid block to extract multi-scale features from the input HS image.

In this work, we propose a new Feature Pyramid Network for Super-Resolution (FPNSR). Specifically, we use the feature pyramid concept in high-level vision tasks and redesign the structure to meet the needs of low-level SHSISR task. The proposed feature pyramid block in the network exactly satisfies the nature of HS images and achieves the best result among comparison methods in all experiments.

3 The Proposed Method

3.1 Network Architecture

The overall framework of our FPNSR model is illustrated in Fig. 1. The input LR HS image is first upsampled via bicubic upsampling and then fed into a Feature Pyramid Block (FPBlock) for multi-scale feature extraction. The FPBlock forms a feature-level pyramid through pooling and upsampling operations. Detailed structure about FPBlock can be found in Sect. 3.2. Then the feature maps are fed into the next K successive Residual Blocks (ResBlock) to perform feature extraction. Lastly, a 2D convolution layer is added so as to adjust the channels to the same number of input channels.

Fig. 1. An illustration of the overall network architecture for FPNSR. Detailed structure of FPBlock and ResBlock can be found in Sects. 3.2 and 3.3.

3.2 FPBlock Design

FPBlock is the core of this network as it extracts multi-scale features from the input HS image. Inspired by the work [27], we use convolution and pooling operations to build a feature-level pyramid block for feature extraction. The detailed structure of the block is illustrated in Fig. 2. The FPBlock consists of a convolution layer at the beginning and N following stacked stages, each of which is 2 times spatially smaller than the previous one.

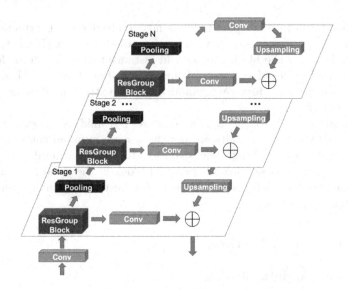

Fig. 2. An illustration of the proposed FPBlock. Design of the FPBlock can be found in Sect. 3.2.

The block can be viewed as two paths going upward and downward respectively in the feature pyramid. FPBlock first uses a 1×1 convolution layer to adjust the input channels. We denote the input LR HS image as X and the function of a 1×1 convolution layer as f_{conv}. This can be viewed as the stage 0 of the upward path of the pyramid:

$$Y_{up}^0 = f_{conv}(X). \tag{1}$$

In the upward path, we aim to gradually extract features at different scale through Residual Group (ResGroup) Blocks (Detailed structure illustrated in Fig. 3) and pooling layers to build the pyramid from bottom to top. The operations of upward path of stage i is as follows:

$$F_{up}^i = f_{res}(Y_{up}^{i-1}), \ Y_{up}^i = g_{pool}(F_{up}^i). \tag{2}$$

Here we use f_{res} to denote the ResGroup Block and g_{pool} to denote the pooling layer. F_{up}^i is the extracted features of stage i and Y_{up}^i is the output of stage i during upward path. The relation of upward path and downward path at the final stage N (top of the pyramid) is as follows:

$$Y_{down}^N = f_{conv}(Y_{up}^N). \tag{3}$$

In the downward path, we use upsampling blocks to obtain feature maps of each stage. Note that at each stage we not only use information from all below stages $(Y_{down}^{i+1} \sim Y_{down}^N)$, but also information at the same stage stored when going upward (F_{up}^i), which goes through a 1×1 convolution layer and is then merged to the features from the output of the upsampling block (h_{up}) by an element-wise add operation:

$$Y_{down}^i = f_{conv}(F_{up}^i) + h_{up}(Y_{down}^{i+1}). \tag{4}$$

3.3 ResBlock and ResGroup Block Design

As the multi-scale features of HS image have been extracted by the preceding FPBlock, we can now view the problem as a common SISR task and introduce best practices in feature extraction as in [20]. Here we build K ResBlocks, and each of them consists of two convolutional layers after a ReLU activation function. A skip connection path joins the beginning to the end of the block. We set a scale factor 0.1 to each ResBlock at the end to stabilize the training in deep networks. For the feature extraction part inside FPBlock, we use a variant of ResBlock to replace 2D convolutions with group convolutions (ResGroup Block). This is because group convolution allows the layer to have more channels with a relative low cost of extra parameters compared to 2D convolution. An illustration of these two blocks is in Fig. 3.

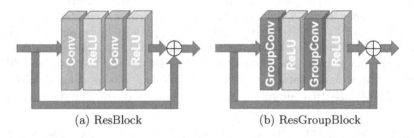

(a) ResBlock   (b) ResGroupBlock

Fig. 3. An illustration of two kinds of ResBlocks. The only difference between ResGroup Block and ResBlock is the type of convolution layer.

For other settings, we adopt L1 loss as our loss function. We choose average pooling as the pooling layer and the ESPCN [29] as the upsampling block in our FPNSR model. Note that the FPBlock is a general framework that pooling and upsampling operation can be replaced by any other structure with the same function.

4   Experiments

In this section, the performance of the proposed FPNSR is evaluated on HS image dataset for different scale factors. Both objective quality indices and subjective visual results are provided in comparison with recent DL-based SHSISR and SISR methods.

Table 1. Quantitative comparison results of FPNSR with other methods on Chikusei dataset for scale factor 4.

	CC	SAM	RMSE	ERGAS	MPSNR	MSSIM
Bicubic	0.9410	3.5699	0.0154	7.3865	37.4147	0.9005
VDSR [19]	0.9227	3.6642	0.0148	6.8708	37.7755	0.9065
EDSR [20]	0.9510	2.5580	0.0121	5.3708	39.8289	0.9354
CNN-CNMF [16]	0.9202	3.8580	0.0150	6.7234	37.7438	0.9003
3D-FCNN [15]	0.9355	3.1174	0.0140	6.0026	38.6091	0.9127
GDRRN [17]	0.9369	2.5000	0.0137	5.9540	38.7198	0.9193
DeepPrior [32]	0.9293	3.5590	0.0147	6.2096	38.1923	0.9010
FPNSR (Proposed)	**0.9549**	**2.3348**	**0.0116**	**5.1113**	**40.1784**	**0.9400**

Dataset. We evaluate our method on Chikusei dataset [30], which contains a HS image of 2517×2335 pixels for spatial resolution and 128 spectral bands ranging from 363 nm to 1018 nm taken in Chikusei, Ibaraki, Japan. In Chikusei dataset, we cropped 4 non-overlap 512×512 images from the top of the original image as test set. For scale factor 4, the rest of the image is cropped into 64×64 patches with a stride of 32. The 64×64 patch is then spatially downsampled 4 times using MATLAB *imresize* function to generate a LR-HR image pair. Around 13% of the image pairs are selected as evaluation set and the rest are left for training. The training data is augmented by flipping and rotating, resulting in a total 24,000 LR-HR training pairs. The generation of training data for other scale factors follows the same procedure.

Network Training Details. The FPNSR model used for experiment has $K = 12$ ResBlocks. The stages of FPBlock is set to $N = 3$. The whole network is trained using Adam [31] optimizer with a start learning rate of 0.0001 decayed by 0.1 every 10 epochs. For Chikusei dataset, we use a batch size of 16 and run 15 epochs. It took roughly 3.5 h to train the network on two NVIDIA 1080Ti GPUs.

Comparison Methods. To demonstrate the performance of our method, we compared FPNSR with several recent single HS image SR methods, namely CNN-CNMF [16], 3D-FCNN [15], GDRRN [17] and DeepPrior [32]. CNN-CNMF [16] is a matrix factorization method using transfer learning method as a pre-processing step. 3D-FCNN [15] builds a CNN by using 3D convolutions instead of common 2D ones. GDRRN [17] designs a network using recursive blocks and group convolutions. DeepPrior [32] is a method basing on the principles of deep image prior [33] to restore a HS image. We also include 2 well-known SISR methods for RGB images: VDSR [19] and EDSR [20] for comparison.

Table 2. Quantitative comparison results of FPNSR with other methods on Chikusei dataset for scale factor 8.

	CC	SAM	RMSE	ERGAS	MPSNR	MSSIM
Bicubic	0.8314	5.0436	0.0224	4.8488	34.5049	0.8228
VDSR [19]	0.8344	5.1778	0.0216	4.9052	34.5661	0.8305
EDSR [20]	0.8636	4.4205	0.0201	4.5091	35.4217	0.8501
CNN-CNMF [16]	0.8249	5.3041	0.0224	4.8843	34.3488	0.8215
3D-FCNN [15]	0.8428	4.8432	0.0215	4.5964	34.8375	0.8313
GDRRN [17]	0.8421	4.3160	0.0214	4.5879	34.8153	0.8357
DeepPrior [32]	0.8366	5.3386	0.0219	4.6789	34.6692	0.8126
FPNSR (Proposed)	**0.8690**	**4.1236**	**0.0197**	**4.4819**	**35.5724**	**0.8566**

4.1 Objective Results

In this section, we demonstrate quality indices for our FPNSR method. We use 4 widely adopted quality indices [13,34] in HS fusion task: *cross correlation* (CC), *spectral angle mapper* (SAM), *root mean squared error* (RMSE) and *erreur relative globale adimensionnelle de Synthese* (ERGAS). We also include 2 standard SISR indices MPSNR and MSSIM for evaluation. Table 1 shows the comparison results on Chikusei dataset for scale factor 4. We use the code released by the authors to train GDRRN [17], 3D-FCNN [15], DeepPrior [32] and EDSR [20]. As the code of CNN-CNMF [16] is not provided, we implemented the model according to the settings of the original paper. The results of VDSR [19] is obtained in a per channel manner by using pre-trained model for RGB images.

Table 2 demonstrates the comparison results for scale factor 8. Implementation of the methods is the same as scale factor 4 except we use ×2 × 4 to obtain ×8 results for VDSR [19] and the transfer learning part of CNN-CNMF [16] as the original models do not consider a ×8 situation.

According to the quality indices results, for scale factor 4, CNN-CNMF [16] and VDSR [19] yield similar results that are only slightly better than traditional bicubic upsampling as their training part is done using RGB images in a per channel manner. The DeepPrior [32] method is better than these two methods but still not satisfying because it does not need training data and requires manual intervention to decide when to stop running. 3D-FCNN [15] and GDRRN [17] are two specially designed networks for SHSISR, and they achieve comparable better results on all 6 indices except SAM. Note that GDRRN [17] achieves the second best SAM result, this is partially because it uses SAM as a regularization term in loss function. EDSR [20] improves the results remarkably as it uses much deeper network and requires longer training time. Our FPNSR model, achieves all the highest on 6 indices than the second best EDSR [20], which should be owed to the FPBlock.

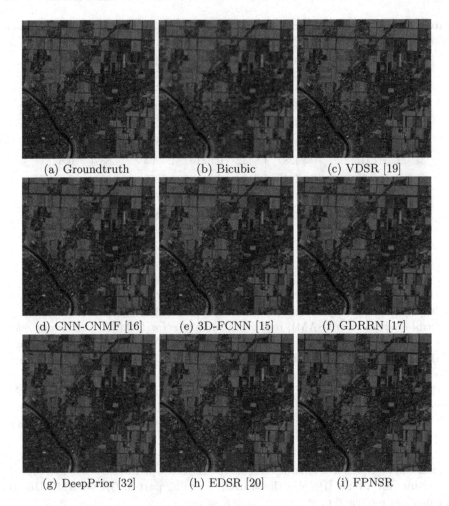

(a) Groundtruth   (b) Bicubic   (c) VDSR [19]

(d) CNN-CNMF [16]   (e) 3D-FCNN [15]   (f) GDRRN [17]

(g) DeepPrior [32]   (h) EDSR [20]   (i) FPNSR

Fig. 4. Pseudo-RGB visualization of super-resolution results for scale factor 4. We use 70, 100, 36 bands as RGB bands. (a) is the ground truth sample image from our test set. (b)–(h) are 8 SR results from comparison methods. (i) is the result of our proposed model.

It can also be observed that same relations of the comparison methods hold when scale factor is 8. In Table 2, the MPSNR value of all methods are below 35 dB except EDSR [20] and FPNSR, and FPNSR performs slightly better than EDSR [20].

Therefore, we conclude that our FPNSR model produces the best results among all methods in different scale factors, leaving the second highest method EDSR [20] 0.35 dB for scale factor 4 and 0.15 dB for scale factor 8 on quality index MPSNR.

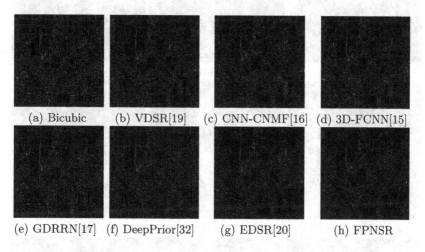

(a) Bicubic (b) VDSR[19] (c) CNN-CNMF[16] (d) 3D-FCNN[15]

(e) GDRRN[17] (f) DeepPrior[32] (g) EDSR[20] (h) FPNSR

Fig. 5. Pseudo-RGB visualization of error maps for scale factor 4. Error maps are scaled 10 times for better visualization.

4.2 Visual Comparison Results

In this section we give visual comparison results for different methods. As the total spectral information can not be demonstrated in RGB color, we show a pseudo-RGB image generated by selecting according bands to approximate the RGB bands of a HS image. Figures 4 and 5 show SR images and its corresponding error maps from our test set for scale factor 4, respectively.

According to the results, we find that our proposed FPNSR model produces more details and is much closer to the ground truth image. For example, there is a small rectangular roof in the right bottom corner of the image in Fig. 4 whose shape is poorly demonstrated with blurry edges except our model. This situation is more obvious when the scale factor is switched to 8 in Figs. 6 and 7, one can notice that there is a road along the right side of the river in the middle of the scene, which can only be observed in EDSR [20] and our model.

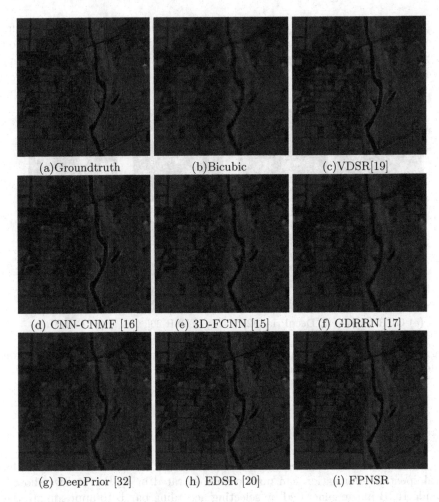

Fig. 6. Pseudo-RGB visualization of super-resolution results for scale factor 8. We use 70, 100, 36 bands as RGB bands. (a) is the ground truth sample image from our test set. (b)–(h) are 8 SR results from comparison methods. (i) is the result of our proposed model.

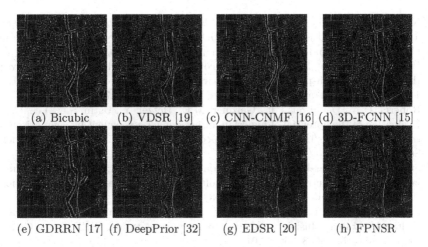

(a) Bicubic   (b) VDSR [19]   (c) CNN-CNMF [16] (d) 3D-FCNN [15]

(e) GDRRN [17] (f) DeepPrior [32]   (g) EDSR [20]   (h) FPNSR

Fig. 7. Pseudo-RGB visualization of error maps for scale factor 8. Error maps are scaled 10 times for better visualization.

5   Conclusion

In this work, we proposed a new model for SHSISR. Substantial experiments have shown that our new FPNSR model can effectively extract multi-scale features from HS images and achieve better SR results than present methods. The proposed FPBlock can also be easily generalized and plugged into other DL-based methods as a pre-processing block for HS image restoration. However, our method does not explicitly consider the grouping of bands in HS images, which is important for HS image restoration tasks and further analysis. In the future, we will focus on automatic grouping of the bands in HS images by using DL-based methods. We may also use grouped bands to design lightweight network models to perform HS image tasks for mobile devices.

Acknowledgements. This work is supported by the National Science Foundation under Grant Nos. 61971165, 61672193, and 61922027, and is also supported by the Fundamental Research Funds for the Central Universities.

References

1. Bioucas-Dias, J.M., et al.: Hyperspectral unmixing overview: geometrical, statistical, and sparse regression-based approaches. IEEE J. Sel. Top. Appl. Earth Obs. Remote Sens. **5**, 354–379 (2012)
2. Haut, J.M., Paoletti, M.E., Plaza, J., Li, J.Y., Plaza, A.J.: Active learning with convolutional neural networks for hyperspectral image classification using a new bayesian approach. IEEE Trans. Geosci. Remote Sens. **56**, 6440–6461 (2018)
3. Camps-Valls, G., Bruzzone, L.: Kernel-based methods for hyperspectral image classification. IEEE Trans. Geosci. Remote Sens. **43**, 1351–1362 (2005)

4. Nguyen, H.V., Banerjee, A., Chellappa, R.: Tracking via object reflectance using a hyperspectral video camera. In: 2010 IEEE Computer Society Conference on Computer Vision and Pattern Recognition - Workshops, pp. 44–51 (2010)

5. Huete, A.R., Miura, T., Gao, X.: Land cover conversion and degradation analyses through coupled soil-plant biophysical parameters derived from hyperspectral EO-1 hyperion. IEEE Trans. Geosci. Remote Sens. **41**, 1268–1276 (2003)

6. Roberts, D.A., Dennison, P.E., Gardner, M.E., Hetzel, Y., Ustin, S.L., Lee, C.T.: Evaluation of the potential of hyperion for fire danger assessment by comparison to the airborne visible/infrared imaging spectrometer. IEEE Trans. Geosci. Remote Sens. **41**, 1297–1310 (2003)

7. Simões, M., Bioucas-Dias, J.M., Almeida, L.B., Chanussot, J.: A convex formulation for hyperspectral image superresolution via subspace-based regularization. IEEE Trans. Geosci. Remote Sens. **53**, 3373–3388 (2015)

8. Wei, Q., Bioucas-Dias, J.M., Dobigeon, N., Tourneret, J.Y.: Hyperspectral and multispectral image fusion based on a sparse representation. IEEE Trans. Geosci. Remote Sens. **53**, 3658–3668 (2014)

9. Yokoya, N., Yairi, T., Iwasaki, A.: Coupled nonnegative matrix factorization unmixing for hyperspectral and multispectral data fusion. IEEE Trans. Geosci. Remote Sens. **50**, 528–537 (2012)

10. Zhang, K., Wang, M., Yang, S.Y., Jiao, L.: Spatialspectral-graph-regularized low-rank tensor decomposition for multispectral and hyperspectral image fusion. IEEE J. Sel. Top. Appl. Earth Obs. Remote Sens. **11**, 1030–1040 (2018)

11. Qu, Y., Qi, H., Kwan, C.: Unsupervised sparse Dirichlet-Net for hyperspectral image super-resolution. In: 2018 IEEE/CVF Conference on Computer Vision and Pattern Recognition, pp. 2511–2520 (2018)

12. Xie, Q., Zhou, M., Zhao, Q., Meng, D., Zuo, W., Xu, Z.: Multispectral and hyperspectral image fusion by MS/HS fusion net. ArXiv abs/1901.03281 (2019)

13. Loncan, L., et al.: Hyperspectral pansharpening: a review. IEEE Geosci. Remote Sens. Mag. **3**, 27–46 (2015)

14. Akgun, T., Altunbasak, Y., Mersereau, R.M.: Super-resolution reconstruction of hyperspectral images. IEEE Trans. Image Process. **14**, 1860–1875 (2005)

15. Mei, S., Yuan, X., Ji, J., Zhang, Y., Wan, S., Du, Q.: Hyperspectral image spatial super-resolution via 3D full convolutional neural network. Remote Sens. **9**, 1139 (2017)

16. Yuan, Y., Zheng, X., Lu, X.: Hyperspectral image superresolution by transfer learning. IEEE J. Sel. Top. Appl. Earth Obs. Remote Sens. **10**(5), 1963–1974 (2017)

17. Li, Y., Zhang, L., Ding, C., Wei, W., Zhang, Y.: Single hyperspectral image super-resolution with grouped deep recursive residual network. In: 2018 IEEE Fourth International Conference on Multimedia Big Data (BigMM), pp. 1–4 (2018)

18. Dong, C., Loy, C.C., He, K., Tang, X.: Image super-resolution using deep convolutional networks. IEEE Trans. Pattern Anal. Mach. Intell. **38**, 295–307 (2014)

19. Kim, J., Lee, J.K., Lee, K.M.: Accurate image super-resolution using very deep convolutional networks. In: 2016 IEEE Conference on Computer Vision and Pattern Recognition (CVPR), pp. 1646–1654 (2015)

20. Lim, B., Son, S., Kim, H., Nah, S., Lee, K.M.: Enhanced deep residual networks for single image super-resolution. In: 2017 IEEE Conference on Computer Vision and Pattern Recognition Workshops (CVPRW), pp. 1132–1140 (2017)

21. Lai, W.S., Huang, J.B., Ahuja, N., Yang, M.H.: Deep Laplacian pyramid networks for fast and accurate super-resolution. In: 2017 IEEE Conference on Computer Vision and Pattern Recognition (CVPR), pp. 5835–5843 (2017)

22. Zhang, Y., Tian, Y., Kong, Y., Zhong, B., Fu, Y.: Residual dense network for image super-resolution. In: 2018 IEEE/CVF Conference on Computer Vision and Pattern Recognition, pp. 2472–2481 (2018)
23. Zhang, Y., Li, K., Li, K., Wang, L., Zhong, B., Fu, Y.: Image super-resolution using very deep residual channel attention networks. ArXiv abs/1807.02758 (2018)
24. Hu, J., Shen, L., Sun, G.: Squeeze-and-excitation networks. In: 2018 IEEE/CVF Conference on Computer Vision and Pattern Recognition, pp. 7132–7141 (2017)
25. Haris, M., Shakhnarovich, G., Ukita, N.: Deep back-projection networks for super-resolution. In: IEEE/CVF Conference on Computer Vision and Pattern Recognition, pp. 1664–1673 (2018)
26. Lowe, D.G.: Distinctive image features from scale-invariant keypoints. Int. J. Comput. Vis. **60**, 91–110 (2004)
27. Lin, T.Y., Dollár, P., Girshick, R.B., He, K., Hariharan, B., Belongie, S.J.: Feature pyramid networks for object detection. In: 2017 IEEE Conference on Computer Vision and Pattern Recognition (CVPR), pp. 936–944 (2016)
28. He, K., Gkioxari, G., Dollar, P., Girshick, R.B.: Mask R-CNN. IEEE Trans. Pattern Anal. Mach. Intell. (2018)
29. Shi, W., et al.: Real-time single image and video super-resolution using an efficient sub-pixel convolutional neural network. In: 2016 IEEE Conference on Computer Vision and Pattern Recognition (CVPR), pp. 1874–1883 (2016)
30. Yokoya, N., Iwasaki, A.: Airborne hyperspectral data over Chikusei. Technical report SAL-2016-05-27, Space Application Laboratory, University of Tokyo, Japan (May 2016)
31. Kingma, D.P., Ba, J.: Adam: a method for stochastic optimization. CoRR abs/1412.6980 (2014)
32. Sidorov, O., Hardeberg, J.Y.: Deep hyperspectral prior: denoising, inpainting, super-resolution. ArXiv abs/1902.00301 (2019)
33. Ulyanov, D., Vedaldi, A., Lempitsky, V.S.: Deep image prior. In: 2018 IEEE/CVF Conference on Computer Vision and Pattern Recognition, pp. 9446–9454 (2017)
34. Wald, L.: Data Fusion, Definitions and Architectures - Fusion of Images of Different Spatial Resolutions. Les Presses de l'Ecole des Mines, Paris (2002)

Machine Learning

Unsupervised Representation Learning Based on Generative Adversarial Networks

Shi Xu and Jia Wang[✉]

Department of Electronic Engineering, Shanghai Jiao Tong University,
Shanghai, China
{xushi2017,jiawang}@sjtu.edu.cn

Abstract. This paper introduces a novel model for learning disentangled representations based on Generative Adversarial Networks. The training model is unsupervised without identity information. Unlike Info-GAN in which the disentangled representation is learnt by getting the variational lower bound of the mutual information indirectly, our method introduces a direct way by adding predicting networks and encoder into GANs and measuring the correlation among the encoder outputs. Experiment results on MNIST demonstrate that the proposed model is more generalizable and robust than InfoGAN. With experiments on Celeba-HQ, we show that our model can extract factorial features with complicate datasets and produce results comparable to supervised models.

Keywords: Unsupervised representation learning · Disentangled features · Generative Adversarial Networks

1 Introduction

Over the past few years, we have witnessed the great success of generative adversarial networks (GANs) [10] for a variety of applications. GANs are powerful framework for learning generative models, which are expressed as a zero-sum game between two neural networks. The generator network produces samples from the arbitrary given distribution, while the adversarial discriminator tries to distinguish between real data and the generator network's output. Meanwhile, the generator network tries to fool the discriminator network by producing plausible samples which are close to real samples. Unlike other deep generative models, GANs can be trained end to end and do not need approximation. GANs There are many works inspired by the original GANs, such as WGAN [2], EBGAN [27], BEGAN [5]. It is important for GANs to learn good representations for real data, which can help understand the progress of generating samples. Some generative models [20,24] try to learn sparse representations since a single sample should not have so many attributes at the same time, so it is reasonable that most of the dimensions in the latent codes are zero. Some

© Springer Nature Singapore Pte Ltd. 2020
G. Zhai et al. (Eds.): IFTC 2019, CCIS 1181, pp. 65–78, 2020.
https://doi.org/10.1007/978-981-15-3341-9_6

generative models [26] learn robust representations by exposing samples to noise and restituting samples. Some generative models [7] aim to make the dimensions of the latent codes independent of each other, which means the representations are disentangled. Disentangled representations are expected to be semantically well aligned with observation (*e.g.* the width and rotation of figures).

InfoGAN [7] proposes an unsupervised approach based on GANs to learn disentangled representations for real data, by enforcing the generator network to increase the mutual information between the generated samples and real data. Although InfoGAN can learn disentangled representations of data, the indirect way of learning disentangled representations tend to be inefficient and sometimes useless, as shown in Sect. 4.1.

In this paper, we propose to learn disentangled representations directly based on GANs without supervision. We do so by adding k predicting networks into the structure of GANs and encourage the predicting networks to learn disentangled features. We find our method is able to discover meaningful hidden representations on the image datasets. The method is even robust when the dataset is irregular, containing images that are rotated and inconsistent in size.

The rest of the paper is organized as follows. In Sect. 2, we briefly review some of the related concepts and GAN technologies. We present our approach in Sect. 3, and experiment results are given in Sect. 4. Section 5 concludes the paper.

2   Related Work

2.1   Representation Learning

Representation learning is used for extracting value from unlabelled data which lies in huge quantities and use learning representations that exposes significant features as decodable factors [3,4]. There have been a large amount of related work on learning representations of data. Many representation learning models are supervised or semi-supervised, which learn features from labeled data. The data label allows the model to compute an error term, the degree to which the system fails to produce the label, which can then be used as feedback to correct the learning process [22,25]. Various supervised or semi-supervised models based on the autoencoder (AE) framework [6,11,16,18,21] proposes the method with an encoder first extracting features from the data and a decoder maps from feature space back into input space. There are also a number of unsupervised feature learning methods. Unsupervised feature learning is learning features from unlabeled data. The unsupervised feature learning approaches include K-means clustering [19], principal component analysis (PCA) [12], information maximizing GANs (InfoGAN) [7].

2.2   Generative Adversarial Networks

Generative Adversarial Networks (GANs) [10] establish a min-max adversarial game between two neural networks, a generator network G and a discriminator D. The goal of the generator work is to produce a realistic sample that is indistinguishable from the true sample. The discriminator network aims to distinguish well between the real samples and the samples created by the generator network. The adversarial game is formulated as below:

$$\min_{G} \max_{D} L(D,G) = E_{p(x)}[log(D(x))] + E_{p(z)}[log(1 - D(G(z)))] \tag{1}$$

Under an optimal discriminator D^*, Eq. 1 theoretically involves with the Jensen-Shannon divergence [9] between the synthetic and the true sample distribution: $JS(G(z)||P(x))$.

2.3   Information Maximizing GANs

Information maximizing Generative Adversarial Network (InfoGAN) [7] proposes an unsupervised approach based on GANs to learn disentangled representations for real data. InfoGAN introduces an additional latent code c and encourages the latent code to learn the semantic features of the data. InfoGAN adds a mutual information (MI) maximization term between the latent code c and the generated samples $G(z,c)$, the training objective of InfoGAN is:

$$\min_{G,Q} \max_{D} L_{InfoGAN}(D,G,Q) = L(D,G) - \lambda I(c; G(z,c)) \tag{2}$$

where $L(D,G)$ is the training objective of GANs in Eq. 1, $I(c; G(z,c))$ denotes the mutual information and λ is a weight coefficient. InfoGAN uses variational information maximization [1] to optimize the training objective and gets the variational lower bound of the mutual information, $L_I(G,Q)$:

$$
\begin{aligned}
L_I(G,Q) &= E_{c\sim P(c), x\sim G(z,c)}[\log Q(c|x)] + H(c) \\
&= E_{x\sim G(z,c)}[E_{c'\sim P(c|x)}[\log Q(c'|x)]] + H(c) \\
&\leq I(c; G(z,c))
\end{aligned}
\tag{3}
$$

2.4   Bidirectional GANs

Bidirectional Generative Adversarial Network (BiGANs) [8] is a structure of GANs with encoder and decoder added. However, instead of combining the encoder and decoder, BiGAN separates the encoder and decoder, saves the input and output of the encoder and decoder separately, and send them together to the discriminator. More specifically, the input image x is passed into the encoder and the code $E(x)$ is mapped. Meanwhile, the code z from an arbitrary distribution is passed into the decoder, which is the generator in the initial GANs, and the output image $G(z)$ is generated. Then the image and the code will be concatenated separately, resulting in the data $(x, E(x))$ and $(G(z), z)$. The paired data will be passed into the discriminator and the discriminator will try to distinguish the paired data comes from the encoder or the generator. The structure of BiGAN is shown in Fig. 1. The BiGAN training objective is also defined as a minimax objective:

$$\min_{G,E} \max_{D} L(D,E,G) = E_{p(x)}[log(D(x,z))] + E_{p(z)}[log(1 - D(G(z),z))] \tag{4}$$

The encoder induces a distribution $p_E(z|x) = \delta(z - E(x))$ mapping data points x into the latent feature space of the generator. The discriminator is modified to take input from the latent space, predicting $P_D(Y|x, z)$, where $Y = 1$ if X is sampled from the real data distribution p_x and $Y = 0$ if x is generated from the output of $G(z)$, $z \sim p_z$.

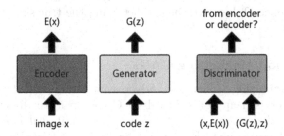

Fig. 1. The structure of BiGAN.

3   Our Approach

We now present a direct way of learning factorial representation. Our method is shown in Fig. 2. Similar to BiGAN in Fig. 1, we use x to denote the input image, and the code z is sampled from an arbitrary distribution. Unlike BiGAN, part of z is separated into the latent code c. From Fig. 2, the output of the encoder is the initial code $E(x)$ and the latent code $C(x)$; the input of the generator is the initial code z and the latent code c; the output of the generator is $G(z,c)$. The paired data $(x, E(x), C(x))$ and $(G(z,c), z, c)$ will still be passed into the discriminator and the discriminator will try to distinguish the paired data comes from the encoder or the generator. Besides, there are n predicting networks. For the i-th latent code c_i, the input of the predicting network is z and $c_{j\neq i}$ and the

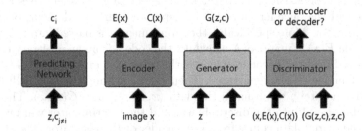

Fig. 2. The structure of our method based on BiGAN and predicting networks.

output of the predicting network is c'_i. The loss of the i-th predicting network is $L(f_i) = E[c_i - c'_i]^2$ and the training objective of our method is shown below:

$$\min_{G,E,F} \max_D L(D, E, G, F) = E_{p(x)}[log(D(x, z))]$$

$$+ E_{p(z)}[log(1 - D(G(z), z))]$$

$$- \lambda E_{p(x)}[C(x) - C'(x)]^2 \quad (5)$$

where λ is a hyper parameter which can be adjusted. In practice, each module G, D, E and F is a parametric function with parameters θ_G, θ_D, θ_E and θ_F respectively. The minimax objective in Eq. 5 will be optimized using the same alternating gradient based optimization as [10]. In one iteration, the discriminator parameters θ_D are updated by taking some steps with the positive gradient direction $\nabla_{\theta_D} L(D, E, G, F)$, then the encoder parameters θ_E, the generator parameters θ_G and predicting network parameters θ_F are updated by taking a step in the negative gradient direction $-\nabla_{\theta_E, \theta_G, \theta_F} L(D, E, G, F)$.

Suppose there are n dimensions of features to be learned, the disentanglement of features can be formulated as:

$$P(c_1, c_2, \cdots, c_n | X) = \prod_{i=1}^{n} P(c_i | X) \quad (6)$$

where X means the overall samples and c_i means the i-th dimension of the features. In the sequel we omit the conditioning on X for the purpose of brevity. Equivalently, we can change Eq. 6 into n conditional independent expressions,

$$P(c_i | c_{\backslash i}^n) = P(c_i), \quad i = 1, \cdots, n \quad (7)$$

where we use $c_{\backslash i}^n$ to denote $\{c_1, c_2, \cdots, c_{i-1}, c_{i+1}, \cdots, c_n\}$. Suppose the conditional expectation of c_i given $c_{\backslash i}^n$ is c'_i, i.e. $c'_i = E(c_i | c_{\backslash i}^n)$, $i = 1, \cdots, n$. Then if c_i is independent of all other latent codes, $E[c_i - c'_i]^2$ equals to the variance of c_i, which achieves the maximum value of the prediction error. Denote

$$L(f) = \sum_{i=1}^{n} E[c_i - c'_i]^2. \quad (8)$$

We add $L(f)$ into the structure of BiGAN and force the network to maximize $L(f)$. Then the loss function of our approach is

$$\min_{G,E} \max_D L(D, E, G) = E_{p(x)}[log(D(x, z))]$$

$$+ E_{p(z)}[log(1 - D(G(z), z))]$$

$$- \lambda L(f) \quad (9)$$

To simulate $E[c_i - c'_i]^2$, we add n predicting networks for all latent codes, as shown in Fig. 2. The input of the i-th predicting network is $c_{\backslash i}^n$ and we denote the output c_i''. Suppose the output of the i-th predicting network is the conditional expectation of c_i, the loss of the entire structure can be further changed into

$$\min_{G,E,F} \max_{D} L(D,E,G,F) = E_{p(x)}[log(D(x,z))]$$
$$+ E_{p(z)}[log(1 - D(G(z),z))]$$
$$- \lambda \sum_{i=1}^{n} E[c_i - c_i'']^2 \qquad (10)$$

where F means the n predicting networks. The mechanism behind the procedure can be explained as follows. Suppose c_i is not disentangled from other latent codes, then the expectation of the i-th output of predicting networks is not equal to 0, while the latent codes are normalized to subject to the Gaussian distribution, then $E[c_i - c_i'']^2$ is smaller than $E[c_i]^2$, which is the variance of c_i. Thus, the entire network will carry on to force c_i to be disentangled from other latent codes. Theoretically, the entire network is equivalent to that by cascading a de-correlating linear transform after the encodes, as shown in Fig. 3. However, the progress of de-correlating linear transform is carried out in an implicit way, absorbed into the networks. In fact, explicitly carrying out de-correlating linear transformation on the encodes like PCA directly cannot extract disentangled features with human observation, as shown Sect. 4.3. It is worth mentioning that similar to [16], the adversarial training process is necessary. Without the adversarial discriminator, the encoder network can produce non-informative outputs, which means c_i'' can be arbitrary distributed, without any

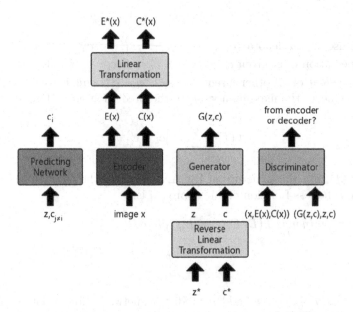

Fig. 3. Carrying out linear transformation on the encodes.

connection with c_i, then $L(f)$ may also become very large. However, in this case, there is no guarantee that c_i will extract useful features. The difference between out method and [16] is, [16] is supervised and the adversarial loss is provided by the identity classifier while our method is unsupervised and the adversarial loss is provided by the discriminator.

4   Experiments

To evaluate the effectiveness of our method, we first carry on our experiments on the MNIST dataset [15] and the CelebA-HQ dataset [13], which is base on the CelebA dataset [17].

4.1   Robustness of Our Method

For the MNIST dataset, instead of using full connected networks in the initial GANs [10], in which the 28×28 digit image must be reshaped into a 784-D vector, we take convolutional neural networks as the base network structure. We choose to model 28 random codes, $z \sim \mathcal{N}(0, 1)$, and 2 continuous latent codes, $c \sim \mathcal{N}(0, 1)$.

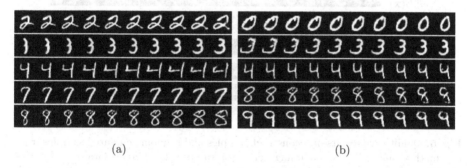

(a)   (b)

Fig. 4. The effect of varying latent codes on our method with MNIST (a) are the results of varying c_1 on our method with MNIST; (b) are the results of varying c_2 on our method with MNIST.

In Fig. 4, we show that how our continuous latent codes have effects on the generated images. The two continuous latent codes capture continuous variations in the style of the MNIST digits like the latent codes in InfoGAN [7]: c_1 controls the width of the generated digits while c_2 controls the rotation of the generated digits. The quality of our results is similar to that of InfoGAN, which proves the effectiveness of our method. Besides, our method is more generalizable than InfoGAN since c_1 and c_2 are sampled from the normal distribution, which means c_1 and c_2 can vary from very small values to very large values like -10 to 10. However, the continuous latent codes in InfoGAN just vary from -2 to 2 [7].

Because our method is based on BiGAN [8], we also carry on similar experiments like that in BiGAN. We encode random real samples and reconstruct them with the generator and get qualitative results, as shown in Fig. 5. From Fig. 5, we find that our method can reconstruct real samples from the output of the encoder, as well as generate random samples, which means the generator can learn the features of the dataset well. The generated samples and the reconstructed samples are even better than that in BiGAN. Also, the reconstructed image is not exactly the same as the original image, which means that the model is not simply remembering all the pictures, which is overfitting. Since our method is more generalizable, our method can tolerate a certain degree of irregularity in the dataset, such as the rotation and scaling of the images in the dataset. Hence, we modify the original MNIST dataset by rotating and zooming the images in it and get an irregular dataset. Some of the images from the irregular dataset are shown in Fig. 6. All the images in the original MNIST dataset are randomly rotated from $-90°$ to $90°$. The images are also scaled with a random scaling factor between 0.6 and 1.

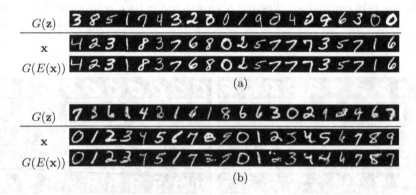

Fig. 5. Qualitative results of reconstruct samples and random generated samples from our method and BiGAN, including generator samples $G(z)$, real data x and corresponding reconstructed samples $G(E(x))$. (a) Are qualitative results from our method; (b) are qualitative results from BiGAN.

Fig. 6. Random samples from the irregular MNIST dataset.

In Figs. 7 and 8, we show the effects of varying c_1 and c_2 on the irregular dataset using our method and InfoGAN, without changing the network structure. The two continuous latent codes are still useful for the irregular dataset with our method, with c_1 controlling the size of the generated digits and c_2 controlling the rotation of the generated digits. However, the results using InfoGAN is not as good as ours. The influence of c_1 and c_2 in InfoGAN is not explicit, some images change sharply while some images are almost the same. This shows that our method is more generalizable and robust, even if the dataset is irregular, within a certain degree of rotation and scaling. It is worth noting that the more irregular the data set, the worse the quality of the generated images. In the third row of Fig. 7, when the width of the digit "9" is too small, it will be converted to the digit "7". Also, in the first row of Fig. 8, when the angle of the generated image is too large, the content of the image will be difficult to identify.

| (a) | (b) |

Fig. 7. The effect of varying c_1 on the irregular MNIST dataset with our method with on InfoGAN, respectively. (a) Are the results from our method; (b) are the results from InfoGAN.

| (a) | (b) |

Fig. 8. The effect of varying c_2 on the irregular MNIST dataset with our method with on InfoGAN, respectively. (a) Are the results from our method; (b) are the results from InfoGAN.

4.2   Experiments on Complicate Dataset

For the experiments on the CelebA-HQ [13] dataset, we take convolutional neural networks as the base network structure. We choose to use the images with the size of 64×64 and model 128 random codes, $z \sim \mathcal{N}(0,1)$, and 2 continuous latent codes, $c \sim \mathcal{N}(0,1)$.

In Fig. 9, we show the results of varying c_1 and c_2. From the results, we can find that c_1 controls the smiling of the face while c_2 controls the eyes of the face, which can be easily distinguished by humans. It is worth mentioning that when we carry out numerous experiments, the effects of varying c_1 and c_2 can be different. Usually, the latent codes can control the smiling, hair color, face angle, size of eyes. This means that these features can be easily learned by the predicting networks. Hence, we increase the number of latent code to 4 while keep

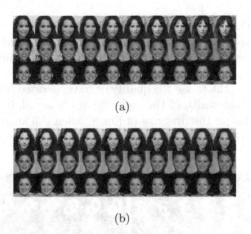

(a)

(b)

Fig. 9. The effect of varying c_1 and c_2 on our method with CelebA-HQ. (a) Are the results varying c_1, which controls the smiling of the face; (b) are the results varying c_2, which controls the eyes of the face.

the dimensions of z with 128, and the results are shown in Fig. 10. The four latent codes control width of face, smiling, side angle of face and eyes, respectively. This shows that for our method, the learning difficulty of the disentangled feature is different from each other. Hence, when the amount of latent codes increase, the features may become more difficult to learn, which results in non-decoupled features.

Fig. 10. The effect of increasing the number of latent codes with CelebA-HQ. c_1, c_2, c_3, c_4 control the smiling, hair color, face angle, size of eyes respectively.

Also, we present results from training infoGAN, BiGAN and our method on CelebA-HQ to compare different methods on complicate datasets, as shown in Fig. 11. All the generators and discriminators of three methods have the same network structure. From Fig. 11, both BiGAN and our method can generate recognizable face images, while the results of infoGAN are unreal and infoGAN is faced with mode collapse. But we find that the faces of BiGAN are more likely to be abnormal and the structure of faces from our method is more complete, which also proves that our method is more generalizable than infoGAN and BiGAN.

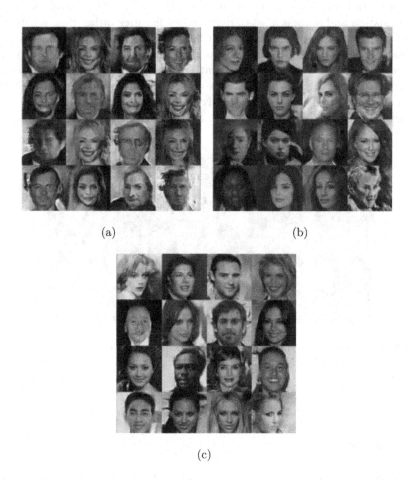

(a)   (b)

(c)

Fig. 11. The results from training BiGAN and our method on CelebA-HQ. (a) Are the results of infoGAN on CelebA-HQ; (b) are the results of BiGAN on CelebA-HQ; (c) are the results of our method on CelebA-HQ.

4.3   Comparsion with PCA

Since the function of predicting networks is produce factorial representations, we consider using principal component analysis (PCA) [12] to extract features from the outputs of the encoder network. Then we vary the results of principal components and reversely change the codes and change the reconstructed images. The results are shown in Fig. 12. From Fig. 12, we can find that PCA can also extracts features such as gender, background color, hair color and face color. However, the extracted features are not decoupled with human observation, which is quite different from our method. When the principal components change, it may change the gender and hair color at the same time. From [23], PCA involves feature transformation and obtains a set of transformed features rather than a subset of the original features. In other words, PCA extracts features that contain most information, however, the features are not disentangled with human observation. While our method extracts features that are disentangled with discriminator, which extracts spatial information with convolutional neural networks, similar to human observation. Also, PCA mainly produces color-related features, while our method produces size-related and angle-related features.

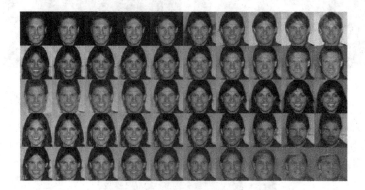

Fig. 12. The effect of varying the first five principal components on the outputs of encoder with CelebA-HQ. The extracted features are not decoupled with human observation

5   Conclusions

This paper introduces a representation learning algorithm base on BiGAN. The method is completely unsupervised and learns interpretable and disentangled representations on the MNIST and CelebA-HQ dataset. Unlike other representation learning algorithms based on GANs like InfoGAN, the training procedure is direct and easy. Also, the method is generalizable and robust when the dataset is irregular and complicate. The idea of using predicting networks to learn disentangled representations can also applied to other methods like other work based on GANs [10] or VAE [14]. Also, this work can be used for improving semi-supervised learning.

Acknowledgements. We thank the reviewers. This work is supported in part by the Chinese Science Foundation under grant 61771305.

References

1. Agakov, D.B.F.: The IM algorithm: a variational approach to information maximization. In: Advances in Neural Information Processing Systems, vol. 16, p. 201 (2004)
2. Arjovsky, M., Chintala, S., Bottou, L.: Wasserstein generative adversarial networks. In: International Conference on Machine Learning, pp. 214–223 (2017)
3. Bengio, Y., Courville, A., Vincent, P.: Representation learning: a review and new perspectives. IEEE Trans. Pattern Anal. Mach. Intell. **35**(8), 1798–1828 (2013)
4. Bengio, Y., et al.: Learning deep architectures for AI. Found. Trends Mach. Learn. **2**(1), 1–127 (2009)
5. Berthelot, D., Schumm, T., Metz, L.: Began: boundary equilibrium generative adversarial networks. arXiv preprint arXiv:1703.10717 (2017)
6. Bourlard, H., Kamp, Y.: Auto-association by multilayer perceptrons and singular value decomposition. Biol. Cybern. **59**(4–5), 291–294 (1988)
7. Chen, X., Duan, Y., Houthooft, R., Schulman, J., Sutskever, I., Abbeel, P.: Info-GAN: interpretable representation learning by information maximizing generative adversarial nets. In: Advances in Neural Information Processing Systems, pp. 2172–2180 (2016)
8. Donahue, J., Krähenbühl, P., Darrell, T.: Adversarial feature learning. arXiv preprint arXiv:1605.09782 (2016)
9. Endres, D.M., Schindelin, J.E.: A new metric for probability distributions. IEEE Trans. Inf. Theory **49**, 1858–1860 (2003)
10. Goodfellow, I., et al.: Generative adversarial nets. In: Advances in Neural Information Processing Systems, pp. 2672–2680 (2014)
11. Hinton, G.E., Zemel, R.S.: Autoencoders, minimum description length and Helmholtz free energy. In: Advances in Neural Information Processing Systems, pp. 3–10 (1994)
12. Jolliffe, I.: Principal Component Analysis. Springer, New York (2011)
13. Karras, T., Aila, T., Laine, S., Lehtinen, J.: Progressive growing of GANs for improved quality, stability, and variation. arXiv preprint arXiv:1710.10196 (2017)
14. Kingma, D.P., Welling, M.: Auto-encoding variational Bayes. arXiv preprint arXiv:1312.6114 (2013)
15. LeCun, Y., Bottou, L., Bengio, Y., Haffner, P., et al.: Gradient-based learning applied to document recognition. Proc. IEEE **86**(11), 2278–2324 (1998)
16. Liu, Y., Wei, F., Shao, J., Sheng, L., Yan, J., Wang, X.: Exploring disentangled feature representation beyond face identification. In: Proceedings of the IEEE Conference on Computer Vision and Pattern Recognition, pp. 2080–2089 (2018)
17. Liu, Z., Luo, P., Wang, X., Tang, X.: Deep learning face attributes in the wild. In: Proceedings of the IEEE International Conference on Computer Vision, pp. 3730–3738 (2015)
18. Maaløe, L., Sønderby, C.K., Sønderby, S.K., Winther, O.: Improving semi-supervised learning with auxiliary deep generative models. In: NIPS Workshop on Advances in Approximate Bayesian Inference (2015)
19. MacQueen, J., et al.: Some methods for classification and analysis of multivariate observations. In: Proceedings of the Fifth Berkeley Symposium on Mathematical Statistics and Probability, Oakland, CA, USA, vol. 1, pp. 281–297 (1967)
20. Makhzani, A., Frey, B.: k-Sparse autoencoders. Comput. Sci. (2014)
21. Makhzani, A., Shlens, J., Jaitly, N., Goodfellow, I., Frey, B.: Adversarial autoencoders. arXiv preprint arXiv:1511.05644 (2015)

22. Mikolov, T., Chen, K., Corrado, G., Dean, J.: Efficient estimation of word representations in vector space. arXiv preprint arXiv:1301.3781 (2013)
23. Mitra, P., Murthy, C., Pal, S.K.: Unsupervised feature selection using feature similarity. IEEE Trans. Pattern Anal. Mach. Intell. **24**(3), 301–312 (2002)
24. Ng, A., et al.: Sparse autoencoder. In: CS294A Lecture Notes, vol. 72, pp. 1–19 (2011)
25. Rasmus, A., Berglund, M., Honkala, M., Valpola, H., Raiko, T.: Semi-supervised learning with ladder networks. In: Advances in Neural Information Processing Systems, pp. 3546–3554 (2015)
26. Vincent, P., Larochelle, H., Bengio, Y., Manzagol, P.A.: Extracting and composing robust features with denoising autoencoders. In: Proceedings of the 25th International Conference on Machine Learning, pp. 1096–1103. ACM (2008)
27. Zhao, J., Mathieu, M., LeCun, Y.: Energy-based generative adversarial network. arXiv preprint arXiv:1609.03126 (2016)

Interactive Face Liveness Detection Based on OpenVINO and Near Infrared Camera

Nana Zhang[1](✉), Jun Huang[2], and Hui Zhang[2]

[1] Shanghai Jian Qiao University, Shanghai, China
nanazhang2004@163.com
[2] Shanghai Ocean University, Shanghai, China

Abstract. For the security of face recognition, this paper proposes an interactive face liveness detection method based on OpenVINO and near infrared camera. Firstly, the face feature points are normalized and the faces are aligned in the environment of OpenVINO and near infrared camera. Secondly, the Euclidean Distance between the mouth feature vectors is calculated. When the distance is greater than a threshold, the system will judge it as a smile. Finally, the system will send random smile commands to the authenticated users to realize liveness detection. According to the results, the proposed method can effectively distinguish between real people and printed photos, and the running time of the liveness detection system based on OpenVINO can reach 14–30 ms, the recognition accuracy can reach 0.977, which has outstanding generalization ability in practical project applications.

Keywords: Near infrared ray · Smile detection · OpenVINO · Liveness detection · Face fraud · Feature point location

1 Introduction

With the development of image processing, deep learning, and computer vision, Face recognition technology has also been widely used, for example, attendance checking, access control unlocking, mobile payment, financial business processing, online remote examination and other aspects. Compared with fingerprint recognition, iris recognition and voice recognition, face recognition is convenient, fast and stable, so it can stand out in biometrics. However, it still faces some defects, compared with fingerprints and irises, the face is relatively simple and convenient to obtain, So it is easy to be attacked by video, photos and other ways. In order to solve the security problem of face recognition system, it is necessary to increase the step of face liveness detection before face recognition, and perform a "fake face" judgment on the obtained face image. At present, the research has received the attention of scholars at home and abroad, there are mainly research methods based on image texture features [1], deep learning [2] and interactive facial movements [3, 4]. In the above method, image texture feature extraction is difficult and computational complexity is high, while deep learning requires a large amount of data. Moreover, it has high requirements on computer hardware and slow training speed, which is not conducive to the experiment. On the contrary, interactive facial movements have the advantages of less computation, less time consumption and high

© Springer Nature Singapore Pte Ltd. 2020
G. Zhai et al. (Eds.): IFTC 2019, CCIS 1181, pp. 79–90, 2020.
https://doi.org/10.1007/978-981-15-3341-9_7

system robustness, and it can be applied to a variety of scenarios, so most of the products in the current market are based on interactive facial movements.

K.Kollreider [5] asked the user to be authenticated to read any number between 0 and 9, and then judged whether the target was a liveness face according to the recognized lip change information. Jee [6] used Gaussian filter to preprocess the face image in a published article, and obtained the eye position by gradient descent method, collected five consecutive frames of face images and normalized them, then calculated the hamming distance of left and right eyes, and set the threshold value of liveness face, finally, measure the threshold range to achieve accurate detection of the liveness face. Smiatacz [7] calculates the optical flow generated when the face is rotated and the optical flow generated when the photo is rotated, and saves them separately. Then, SVM classifier is used to train these optical flow data to obtain a liveness detector. In Singh [8] proposed a liveness detection algorithm based on eye and mouth movements, in that paper, the Haar feature classifier was used to detect the face, locate the eyes and mouth area, and determine the mouth movement based on the HSV value of the teeth.

For interactive facial movements, in this paper, a near infrared camera is used to collect face images, and extracting the face ROI region through the face detection model, The next step is to use feature points to locate the ROI image of face, and complete face alignment and size normalization. Finally, the aligned face images are marked with feature points again, and the Euclidean Distance between the two mouth points is calculated. Once the Euclidean distance is greater than the set threshold, it can be defined as a smile. On this basis, the system will send a smile command in different time periods. the authenticated user makes a smile according to the instructions issued by the system, once the actions of the user to be authenticated conform to the system requirements, the object is a liveness, otherwise it is a fraud attack. Figure 1 is the basic flow chart of the liveness detection system.

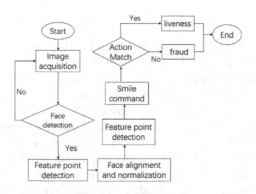

Fig. 1. System flow chart

2   Face Alignment

2.1   Camera Reading

OpenCV 4.1.0 encapsulates a class that encapsulates methods for reading video streams, including reading file streams and camera streams [9], the default camera of the computer can be called by passing in the parameter "0" in the class function. For experimental purposes, the computer is equipped with a 940 nm near infrared camera, which is used to obtain image data.

2.2 Face Detection and ROI Image

The face detection model used in the paper uses the SSD_MobileNet_V2 pre-training model, which is obtained through "Transfer Learning" in the deep learning environment of TensorFlow. SSD-MobileNet is a combination of MobileNet [10] and SSD [11] networks, Mobilenet is called a lightweight convolutional neural network, the biggest feature of which is the depthwise separable convolution. It reduces the amount of parameters by splitting the traditional convolution operation into a deep convolution and a 1×1 convolution. According to the document [10], MobileNet is very small, very fast, and has a very high accuracy, which is suitable for mobile devices. Assuming that the input feature map size is "$C_{in} \times H_{in} \times W_{in}$", the convolution kernel size is "$K \times K$", and the output feature map size is "$C_{out} \times H_{out} \times W_{out}$", then the calculation amount of the standard convolution kernel is:

$$ K \times K \times C_{in} \times C_{out} \times H_{out} \times W_{out} \tag{1} $$

The calculation amount after using the depth decomposable convolution is:

$$ K \times K \times C_{in} \times H_{out} \times W_{out} + C_{in} \times C_{out} \times H_{out} \times W_{out} \tag{2} $$

The ratio of Eqs. (1) to (2) is:

$$ \frac{K \times K \times C_{in} \times H_{out} \times W_{out} + C_{in} \times C_{out} \times H_{out} \times W_{out}}{K \times K \times C_{in} \times C_{out} \times H_{out} \times W_{out}} = \left(\frac{1}{C_{out}} + \frac{1}{K^2} \right) < 1 \tag{3} $$

It can be obtained from Eq. (3), the amount of parameters in the network is indeed reduced.

The target detection network proposed in document [11] is named SSD. The network adds five convolution layers to VGG16 [12] to obtain more feature maps for detection. For each feature map, SSD introduces the concept of the default box, which sets a series of default boxes with different scales and sizes in the center of each feature map. These default boxes will be reverse mapped to a certain position in the original image. If the position of a default box is very close to the position of the truth box, the category of the default box is predicted by the loss function. The loss function equation is as follows:

$$ L_{(x,c,l,g)} = \frac{1}{N} \left(L_{conf}(x,c) + \alpha L_{loc}(x,l,g) \right) \tag{4} $$

In Eq. (4), "N" is the number of matching to the label box, and the "α" parameter is used to adjust the ratio between "loc(location loss)" and "conf(confidence loss)", the default value is "1".

Because SSD-MobileNet uses MobileNet network instead of VGG network as the basic network structure, it has a small amount of parameters and high speed, which is very suitable for real-time environment. The training process of SSD-MobileNet is shown below (Fig. 2):

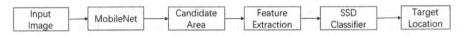

Fig. 2. SSD-MoboileNet training process

In the experiment, the trained model is used to implement face detection in the system environment of OpenCV and OpenVINO, and the face area of each frame is extracted as the input data of feature point location. Figure 3 shows the detection effect of real people and photos under the near infrared camera and the ROI image captured by the face detection model.

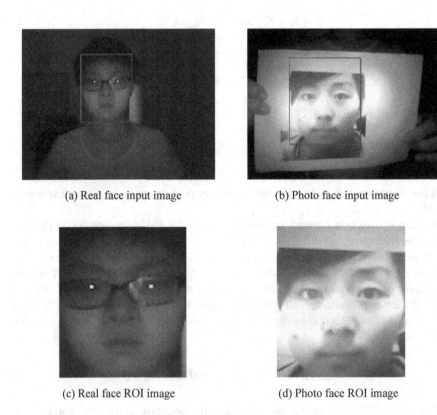

(a) Real face input image   (b) Photo face input image

(c) Real face ROI image   (d) Photo face ROI image

Fig. 3. Face detection and ROI extraction

2.3   Feature Point Detection and Face Alignment

Feature Point Location. Before performing face alignment, feature point positioning needs to be performed on the extracted ROI image. Feature point localization is the key to the whole liveness detection algorithm, and its accuracy directly affects the subsequent detection effect. The feature point model used in the paper is trained in the PyTorch deep learning framework, the model includes 3×3 convolution kernels, batch normalization, PRelu activation functions, pooling layers, and fully connected layers. In the network structure, each parameter of the convolutional layer is forward-propagated, and network parameters can be optimized to extract image features. The computational equation [13] of convolution layer is as follows:

$$y_i = \max\left(0, b_j + \sum_i w_{ij} * x_i\right) \qquad (5)$$

In Eq. (5), "x_i" means the ith input feature map, "y_i" means the jth output feature map, "b_j" means the bias term, "w_{ij}" means the convolution kernel, "$*$" means the convolution operator. As the number of network layers increases, the training speed may be slower or the gradient may explode. Therefore, using the batch normalization operation before the activation function can alleviate the symptoms of overfitting, the equation is as follows:

$$x_k = \frac{x_k - E[x_k]}{\sqrt{Var[x_k]}} \qquad (6)$$

$$y_k = \gamma x_k + \beta_k \qquad (7)$$

In Eq. (6), "x_k" means the kth dimension of the input data, and "$E[x_k]$" means the average of the dimension. "$\sqrt{Var[x_k]}$" means the standard deviation, If the normalization operation in this way may cause the feature to disappear. Therefore, two variables "γ" and "β_k" are added to the Eq. (7) to restore the feature. Because the expression ability of linear model is insufficient, after the convolution operation, a nonlinear operation needs to be added to the feature map. So the activation function used in the network model is PRelu function. The model also uses the max pooling operation, that is, takes the max value in the neighborhood as the output, the purpose is to reduce the dimension of features. The max pooling operation is shown in Fig. 4.

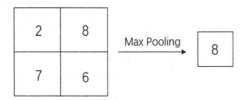

Fig. 4. Max pooling operation

The last layer in the network is the full connection layer, in order to integrate the extracted local features to achieve feature point location. Five feature points of the face can be predicted by using the model in this paper. The rendering is shown in Fig. 5.

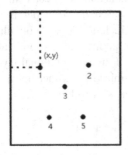

(a) Number of five feature points   (b) Feature point effect map

Fig. 5. Feature point positioning details

Affine Transformation. Assuming that there is only one face in the extracted ROI region, and the affine transformation [14] operation is performed by calculating the rotation angle of the face. Thereby a face correction image is obtained. The specific steps includes the following process:

1. Take the upper left corner of the face ROI area as the origin, and set the coordinate to O (0, 0).
2. The coordinates of the five feature points in the ROI image are defined as:
   There are five coordinate points, such as $P1(x_1, y_1), P2(x_2, y_2), P3(x_3, y_3),$ $P4(x_4, y_4), P5(x_5, y_5)$.
3. Based on the two coordinate points $(P1, P2)$ of both eye, a linear function between the two points is obtained, and the angle between the linear function and the X-axis of the image coordinate axis is calculated:

$$\theta = \tan^{-1}\left(\frac{y_2 - y_1}{x_2 - x_1}\right) \tag{8}$$

Because the deflected face images need to be aligned, the rotation matrix needs to be calculated, and the equation of the rotation matrix is as follows:

$$A = \begin{bmatrix} a_{11} & a_{12} \\ a_{21} & a_{22} \end{bmatrix} \quad B = \begin{bmatrix} b_{11} \\ b_{21} \end{bmatrix} \tag{9}$$

$$R = \begin{bmatrix} a_{11} & a_{12} & b_{11} \\ a_{21} & a_{22} & b_{21} \end{bmatrix} \tag{10}$$

In Eq. (9), "A" means a matrix of coordinate axis rotation and scaling, and "B" means a matrix of coordinate axis translation. In Eq. (10), "R" means the rotation matrix and is used for coordinate transformation of each point in the image. The above equation has been realized in OpenCV, so this paper uses OpenCV to calculate rotation matrix.

4. Once the rotation matrix "R" is obtained, the coordinates of each point in the image can be rotated. The relevant calculation equation is as follows:

$$\begin{bmatrix} x' \\ y' \\ 1 \end{bmatrix} = \begin{bmatrix} a_{11} & a_{12} & b_{11} \\ a_{21} & a_{22} & b_{21} \\ 0 & 0 & 1 \end{bmatrix} \begin{bmatrix} x \\ y \\ 1 \end{bmatrix} \tag{11}$$

In Eq. (11), (x, y) is the coordinates of each pixel of the original image, and (x', y') is the coordinates of the pixel points after the rotation of the image. If the Eq. (11) is converted into a corresponding function, it can be expressed as:

$$\text{src}(a_{11}x + a_{12}y + b_{11}, a_{21}x + a_{22}y + b_{21}) = \text{dst}(x', y') \tag{12}$$

In Eq. (12), "src" is the original image, and "dst" is the image after rotation. Similarly, the affine transformation function in OpenCV is used for face alignment.

3 Smile Detection

3.1 Image and Feature Point Normalization

Before calculating the mouth eigenvectors, the aligned images need to be normalized. Because the position of the face in the camera is unstable. If the face is too far or too close to the camera, it will interfere with the distance between the feature points of the mouth, which seriously affects the effect of smile detection. The normalization method is as follows:

$$k = \frac{200}{y_{face.rows}} \tag{13}$$

In the above equation, "200" is the height of the image normalized, and "$y_{face.rows}$" is the height of the aligned face image. The normalized coordinates of the five feature points are:

$$rx = k * rotate\_x, ry = k * rotate\_y \tag{14}$$

Then:

$$[M1(rx_1, ry_1), M2(rx_2, ry_2), \ldots\ldots, M5(rx_5, ry_5)] \tag{15}$$

The above operations are all aimed at face aligned images. In Eq. (14), "$rotate\_x$" and "$rotate\_y$" are unnormalized original coordinates. In Eq. (15), "rx" and "ry" are the normalized feature point coordinates.

3.2 Mouth Feature Vector Algorithm

The image after face alignment and normalization can reduce the noise between mouth feature vectors, which can greatly improve the robustness of the model. This paper makes a smile judgment by using the Euclidean Distance between the mouth feature vectors, that is, calculating the Euclidean Distance between points M4 and M5. The specific equation is as follows:

$$distantce_{4-5} = \left| \sqrt{(rx_4 - rx_5)^2 + (ry_4 - ry_5)^2} \right| \tag{16}$$

In Eq. (16), "$distance_{4-5}$" is the distance between the two feature vectors of the mouth.

3.3 Threshold Setting

Because the distance between two corners of the mouth is different in each authenticated user's natural state, it is impossible to set a standard value for smiling judgment. In the experiment, calculate the average distance between the feature vectors of the mouth of the user in the top 1000 frames, the distance calculation equation is as follows:

$$avg\_distant_{4-5} = \frac{1}{n} \sum_{i=1}^{n} (distant_{4-5})_i, (n = 1000) \tag{17}$$

In the above equation, "$(distant_{4-5})_i$" is the mouth vector distance of each frame image, "$avg\_distant_{4-5}$" is the average distance, which is taken as the standard distance. During the experiment, we found 15 volunteers and tested each of them 4 times, totaling 60 times. We obtained the mouth distance of the volunteers under normal and smiling state respectively, and took the average of 60 times. From the results of the records, when the authenticated user smiles, the Euclidean Distance between points M4 and M5 is about 1.3–1.4 times greater than when it is natural. Therefore, 1.35 is used as the threshold coefficient in the paper. Assuming that "$smile\_distance_{4-5}$" is the Euclidean Distance after the smile, there are:

$$smile\_distance_{4-5} = 1.35 \times distance_{4-5} \tag{18}$$

4 Experimental Results and Analysis

In this paper, the experimental platform is Windows10, the CPU is i5-8300H, using C++ programming language, based on OpenCV and OpenVINO deep learning acceleration toolkit, using USB near infrared camera 640 * 480 to capture image data, the test photos in the paper are all from the NUAA liveness database (http://parnec.nuaa.edu.cn/xtan/data/nuaaimposterdb.html).

Because this paper uses 940 nm near infrared camera to capture images, it prevents attacks from handheld devices, under the instruction of a smile, a real person can make a smile while a paper photograph can't make a corresponding action, for the 3D face print model, because the cost is too high, this paper will not discuss anything. According to this theory, liveness detection can be achieved. The experimental results show: under the laboratory conditions, if the authenticated user is within the effective detection distance, the operating efficiency of the system can reach 14–25 ms per frame, which has very high real-time performance. The experimental results are shown in Fig. 6.

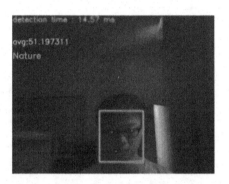

(a) The effect of the normal person

(b) Liveness smile result

(c) The normal effect of photos

(d) Effect under Photo Smile Command

Fig. 6. Experimental results under system command

Figure 6 shows the results of the judgment of the real person and the photo under the system smile command. When the user's smiling action conforms to the system command, the system will pass the authentication, otherwise, the authentication will fail.

The curve in Fig. 7 shows the change of the mouth distance of the user under the smile command and the photo under the smile command in the past 50 frames, the figure shows, when the user smiles (25–39 frames), the distance between the mouth is obviously larger, which basically conforms to the definition of smile detection in the paper (the Euclidean Distance in the smiling state is about 1.35 times the normal state). Because the photo can not make a smile action, so the change of mouth distance is basically smooth.

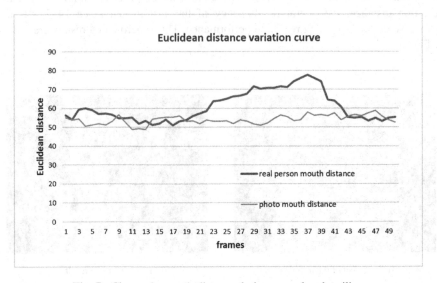

Fig. 7. Change in mouth distance during normal and smiling

In order to test the generalization capabilities of the final system, 300 liveness face authentication and paper photo attacks were performed in the laboratory environment. The experimental data are shown in Table 1:

Table 1. Liveness test data

Testing object	Number of tests	Number of passes	Number of rejections	Accuracy
Total	Number of system recognition success			Total accuracy
Real person	300	292	8	0.973
Paper photo	300	6	294	0.980
600	586			0.977

According to the above table, the recognition accuracy of the system for real people and photos can reach 0.977. In order to understand the performance of the system, It is necessary to compare with other methods of face liveness detection. The document [9] uses texture feature algorithm, the document [15] uses facial optical flow method, the document [16] uses face area algorithms in near infrared and binocular cameras, and the document [17] uses the method of blink detection, background detection and random instructions for interactive liveness detection. By comparing the accuracy of the algorithm, the following figure can be obtained (Fig. 8):

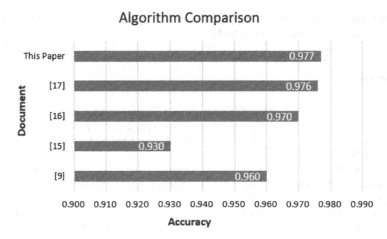

Fig. 8. Algorithm comparison results

From the above figure, it can be seen that the algorithm proposed in this paper is slightly higher than the document algorithm in the table and has good system performance.

5   Conclusion

This paper proposes an interactive liveness detection method for OpenVINO and near infrared cameras. When the system gives a smile command, once the authenticated user makes a smile, the system will determine that it is a real person. Experiments show that this method can effectively prevent handheld device attacks and paper photo attacks. At the same time, the system runs in real time and can be effectively applied in many occasions. Not only that, we also encountered a small problem during the experiment.

Of course, there are some shortcomings in this paper. When we set the smile threshold, because the number of samples collected is small, we can't cover the mouth threshold of all people. Therefore, we will collect more sample data in the following experimental links to enhance the persuasive power of the experimental results. When the criminals rotated the photos (the angle was greater than 30), they could still simulate the smiling effect. Therefore, the angle judgment is made in the experiment. Once the deflection angle of the photos or real faces is greater than 30, the system will stop authentication. Taking the above measures, the system can well prevent the occurrence of the above situations.

Because the number of face feature points used in this paper is small, the application range of interactive face liveness detection is relatively shallow, so the next step of our research will focus on 106 facial feature points, using convolution neural network to train 106 facial feature points, based on 106 feature points. Interactive face liveness detection can be realized by combining open and close eye detection, open and close eye detection, head pose estimation and other methods.

Acknowledgment. This work was supported by the Shanghai Municipal Education Commission's "Morning Plan" project (NO. AASH1702).

References

1. Chingovska, I., Anjos, A., Marcel, S.: On the effectiveness of local binary patterns in face anti-spoofing. In: Biometrics Special Interest Group, pp. 1–7. IEEE (2012)
2. Yang, J., Lei, Z., Li, S.Z.: Learn convolutional neural network for face anti-spoofing. Comput. Sci. **9218**, 373–384 (2014)
3. Yeh, C.H., Chang, H.H.: Face liveness detection with feature discrimination between sharpness and blurriness. In: Fifteenth IAPR International Conference on Machine Vision Applications, pp. 398–401. IEEE (2017)
4. Singh, M., Arora, A.S.: A robust anti-spoofing technique for face liveness detection with morphological operations. Opt. Int. J. Light Electron Opt. **139**(4), 347–354 (2017)
5. Kollreider, K., Fronthaler, H., Bigun, J.: Evaluating liveness by face images and the structure tensor. In: IEEE Workshop on Automatic Identification Advanced Technologies, pp. 75–80 (2005)
6. Jee, H.K., Jung, S.U., Yoo, J.H.: Liveness detection for embedded face recognition system. Enformatika **1**, 235–238 (2006)
7. Smiatacz, M.: Liveness measurements using optical flow for biometric person authentication. Metrol. Meas. Syst. **19**(2), 257–268 (2012)
8. Singh, A.K., Joshi, P., Nandi, G.C.: Face recognition with liveness detection using eye and mouth movement. In: International Conference on Signal Propagation and Computer Technology, pp. 592–597. IEEE (2014)
9. Zhang, G., Feng, R.: Liveness detection system based on human face. Comput. Syst. Appl. **26**(12), 37–42 (2017). (in Chinese)
10. Howard, A.G., Zhu, M., Chen, B., et al.: MobileNets: efficient convolutional neural networks for mobile vision applications (2017)
11. Liu, W., et al.: SSD: single shot multibox detector. In: Leibe, B., Matas, J., Sebe, N., Welling, M. (eds.) ECCV 2016. LNCS, vol. 9905, pp. 21–37. Springer, Cham (2016). https://doi.org/10.1007/978-3-319-46448-0_2
12. Simonyan, K., Zisserman, A.: Very deep convolutional networks for large-scale image recognition [EB/OL], 10 April 2015. https://arxiv.org/pdf/1409.1556.pdf
13. He, F., Zhao, Q.: Head pose estimation based on deep learning. Comput. Technol. Dev. **26**(11), 1–4 (2016). (in Chinese)
14. Song, Z., Zhou, S., Guan, J.: A novel image registration algorithm for remote sensing under affine transformation. IEEE Trans. Geosci. Remote Sens. **52**(8), 4895–4912 (2014)
15. Kim, G., Eum, S., Suhr, J.K., et al.: Face liveness detection based on texture and frequency analyses. In: IAPR International Conference on Biometrics, pp. 67–72. IEEE (2012)
16. Qiu, C.: Research on human face detection based on binocular camera. Mod. Comput. **2018**(35), 41–44 + 66 (2018). (in Chinese)
17. Ma, Y., Tan, L., Dong, X., Yu, C.-C.: Interactive liveness detection algorithm for VTM. Comput. Eng. **45**(03), 256–261 (2019). (in Chinese)

Multi-scale Generative Adversarial Learning for Facial Attribute Transfer

Yicheng Zhang[1], Li Song[1,2(✉)], Rong Xie[1], and Wenjuan Zhang[1]

[1] Institute of Image Communication and Network Engineering,
Shanghai Jiao Tong University, Shanghai, China
{ironic,song_li,xierong,zhangwenjun}@sjtu.edu.cn
[2] Shanghai Institute for Advanced Communication and Data Science,
Shanghai 200240, China

Abstract. Generative Adversarial Network (GAN) has shown its impressive ability on facial attribute transfer. One crucial part in facial attribute transfer is to retain the identity. To achieve this, most of existing approaches employ the L1 norm to maintain the cycle consistency, which tends to cause blurry results due to the weakness of the L1 loss function. To address this problem, we introduce the Structural Similarity Index (SSIM) in our GAN training objective as the measurement between input images and reconstructed images. Furthermore, we also incorporate a multi-scale feature fusion structure into the generator to facilitate feature learning and encourage long-term correlation. Qualitative and quantitative experiments show that our method has achieved better visual quality and fidelity than the baseline on facial attribute transfer.

Keywords: Generative Adversarial Network · Facial attribute transfer · Multi-scale feature fusion

1 Introduction

Facial attribute transfer is a significant and complicated task where the goal is to transfer the facial attributes (e.g. hair color, age, gender) of a person in an input image while keeping irrelevant information such as the background and identity unchanged (see Fig. 1). This task can be extended to a lot of new applications in different areas including the portrait beautifying, film-making industry, fashion business etc. The past few years have witnessed tremendous advances in this task with Generative Adversarial Networks [1–3] being introduced and improved. StarGAN [4] is proposed as a scalable approach capable of transferring several attributes simultaneously given discrete binary labels. Pumarola et al. [5] synthesizes anatomically-aware facial expressions according to the magnitude of different action units. However, these existing approaches are insufficient in two aspects: (1) They both maintain the unconcerned information by enforcing the cycle consistency, which in practice is a L1 loss function between the original

© Springer Nature Singapore Pte Ltd. 2020
G. Zhai et al. (Eds.): IFTC 2019, CCIS 1181, pp. 91–102, 2020.
https://doi.org/10.1007/978-981-15-3341-9_8

image and the reconstructed image. As L1 norm measures the element-wise content discrepancy, it is prone to causing the overall smoothness in images, even for high-frequency texture parts. In addition, L1 norm may impose too strong a constraint on the transferring process, which makes some attribute transfer involving significant changes in shape (e.g. adding eyeglasses) even harder. (2) Most of the generators are designed to be standard single-scale CNN structures. On one hand, due to the locality of conventional convolution, it fails to obtain long-term correlation in images. On the other hand, single-scale features are unable to acquire necessary context information for fine-grained synthesis.

Fig. 1. Examples of our facial attribute transfer results on CelebA dataset. The goal is to transfer the facial attributes of a person in an input image while keeping irrelevant information unchanged.

To address these problems, we propose our novel multi-scale GAN model for facial attribute transfer. Instead of the L1 norm, we introduce the Multi-Scale Structural Similarity Index (MS-SSIM) [33] as the cycle consistency loss, which is more aligned with the characteristics of human visual system, thus making the overall structure seemingly unchanged. Moreover, MS-SSIM alleviates the cycle consistency constraint so that more plausible hallucination is encouraged and it does not cause global smoothing as well. Inspired by the success of atrous spatial pyramid pooling proposed in [6], we incorporate a multi-scale feature fusion module in the up-sampling stage of the generator, which comprises of several atrous convolutions with different atrous rates. This module serves as a complement to the conventional convolution to capture long-range correlation on feature maps and increase the receptive field dispensing with additional computation. Furthermore, the multi-scale feature fusion module empirically proves to be an effective global contextual prior to blend multi-scale features, which contributes to fine-grained texture synthesis.

Overall, our contributions in this work are three-fold:

- We propose our novel multi-scale GAN model for facial attribute transfer, which fuses multi-scale features to obtain long-term correlation on feature maps and acquires global contextual information for fine-grained synthesis.
- We introduce the Multi-Scale Structural Similarity Index (MS-SSIM) as the cycle consistency loss to avoid the global smoothing and encourage more plausible hallucination.
- We provide both qualitative and quantitative results to demonstrate the superiority of our method over the state-of-the-art method on the CelebA dataset.

2   Related Work

Generative Adversarial Networks. Generative Adversarial Networks (GAN) has shown remarkable success in various tasks including image translation [7–13], image inpainting [14–17], super-resolution [18] etc. The typical GAN is composed of a generator and a discriminator. The discriminator acts as a critic to differentiate fake samples from real samples while the generator learns to generate fake samples to fool the discriminator. As GANs are well-known for its training instability and suffering from the mode collapse, many researches have endeavored to alleviate the problems by modifying the loss function [19–21], employing training tricks [22] and adding the regularization constrains [23,24].

Conditional GANs. An effective way to constrain the generated samples is feeding condition information into the generator. Mirza et al. [25] first advocate that auxiliary condition information enables the network to generate the samples matching with the condition input, which has been extended to synthesize high-resolution images [26]. Recent studies have incorporated several forms of conditions to the generator such as language descriptions [27,28], label information [4,5] and images [8,30]. Other works focus on reinforcing the matching between generated samples and input conditions by employing different architectures of the discriminator [27,31,32].

Image-to-Image Translation. Image-to-image translation is an active research area on GANs, which is aimed at learning a mapping across different domains. Pix2pix [8] tackles the problem using paired data with conditional GANs. In consideration of the difficulties of acquiring paired training data, several researches seek to train the networks in an unsupervised manner. CycleGAN [12] introduces a cycle consistency loss and trains two pairs of generators and discriminators to learn a bijection between two domains. UNIT [10] and MUNIT [7] assume that two domains share the same latent space and develop a framework where two generators partially share weights to learn the joint distribution.

Facial Attribute Transfer. One of the most successful applications with GANs is the facial attribute transfer. ELEGANT [29] is able to transfer the facial attributes of two input images by exchanging their latent encodings in

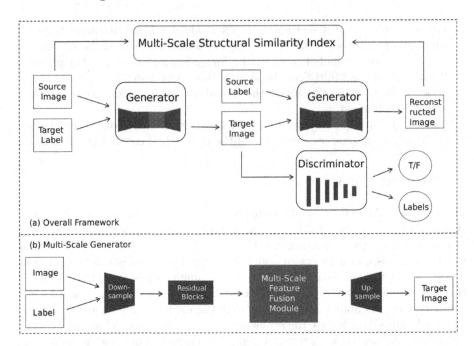

Fig. 2. (a) Overall framework. We train the generator to learn the multi-domain image translation among different attributes, and the discriminator learns to recognize two sources of error: unrealistic generated samples and mismatch between images and attribute annotations. (b) Multi-scale generator. The generator is basically an encoder-decoder structure which is composed of the down-sampling stage, several residual blocks, the multi-scale feature fusion module and the up-sampling stage.

a disentangled manner. StarGAN [4] proposes an unified GAN framework for learning mappings among multiple domains conditioned on a discrete binary label. Pumarola et al. [5] introduces a novel conditioning scheme based on the magnitude of action units to generate anatomically-aware expressions. However, these methods only focus on the single-scale feature which is not sufficient for acquiring necessary context information and they not able to find the long-range dependencies in images. To address the issue, we incorporate a multi-scale feature fusion module into the generator to enlarge the receptive fields and obtain global contextual information for fine-grained synthesis. In addition, the L1 cycle consistency loss may be too strong a constraint for some transfer involving significant changes in shape. Consequently, we introduce the Multi-Scale Structural Similarity Index (MS-SSIM) which is more aligned with the human visual system, thus significantly improving the visual quality and fidelity of generated images.

3   Multi-scale Feature Fusion GAN

In this section, we present our multi-scale feature fusion GAN framework for facial attribute transfer. The overall framework of our model is shown in Fig. 2(a). We train the generator to learn the multi-domain image translation among different attributes, and the discriminator learns to recognize two sources of error: unrealistic generated samples and mismatch between images and attribute annotations.

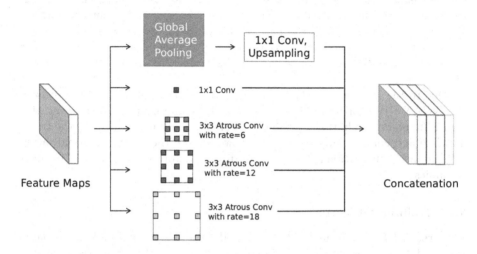

Fig. 3. Multi-scale feature fusion module. The module is composed of five parallel branches. Atrous convolutions with different rates enable the generator to enlarge the receptive fields and obtain long-range dependencies, dispensing with additional parameters.

3.1   Network Architecture

Multi-scale Generator. The structure of multi-scale generator is demonstrated in Fig. 2(b). For simplicity, the input image and the target attribute label can be denoted as $I_r \in \mathbb{R}^{H \times W \times 3}$ and $c_g \in \mathbb{R}^{K \times 1}$ respectively. Our goal is to train a generator G approximating a mapping function to translate I_r to an desired image I_g conditioned on c_g, namely $G(I_r|c_g) \rightarrow I_g$. Moreover, we leverage G once again to obtain the reconstructed image I_{rec}, $G(I_g|c_r) \rightarrow I_{rec}$.

Inspired by the success of the multi-scale training scheme in semantic segmentation [6], we incorporate a multi-scale feature fusion module in our generator to enlarge the receptive fields and obtain long-range dependencies. As shown in Fig. 3, this module is composed of five parallel branches. The first branch is a global average pooling, followed by a 1×1 convolution, which serves as a extractor for image-level features. The second branch is a normal 1×1 convolution to learn local features for each sub-region. The remaining branches are multi-scale 3×3 atrous convolutions with rates $= \{6, 12, 18\}$ respectively. Subsequently, these feature maps of five branches are concatenated together and fed to the next stage as the multi-scale fused features. Furthermore, the atrous convolution is an efficient way for multi-scale learning because the kernel size is the same as normal convolution, namely dispensing with additional parameters.

Multitasking Discriminator. The goal of facial attribute transfer is twofold: the fidelity as a photo and close match with target class labels. To this end, we employ a multitasking discriminator with an auxiliary classifier [32] to recognize the facial attributes on both real images and generated images. We adopt PatchGAN [8] as the basic architecture for the discriminator, which empirically proves to accelerate convergence of models. The auxiliary classifier is on top of the discriminator to predict the probability distribution $\hat{c} \in \mathbb{R}^{K \times 1}$ for attributes.

Due to the instability of GAN training, many researches [4,5] have utilized the WGAN-GP algorithm [23] to satisfy the Lipschitz continuity. However, this algorithm requires additional gradients calculation and hence increase the training time and computation cost. Instead, we impose the spectral normalization [24] on each convolutional layer in the discriminator to stabilize the training dynamics.

3.2   Training Objective

Adversarial Loss. To make the generated data distribution indistinguishable from real data, we adopt the least squares loss function [20] for the discriminator which allows to generate higher quality images and performs more stable during training. In detail, the adversarial loss is defined as:

$$\mathcal{L}_D^{adv} = \mathbb{E}_{I_r}[(D_{adv}(I_r) - 1)^2] + \mathbb{E}_{I_r, c_g}[D_{adv}(G(I_r|c_g))^2] \tag{1}$$

$$\mathcal{L}_G^{adv} = \mathbb{E}_{I_r, c_g}[(D_{adv}(G(I_r|c_g)) - 1)^2] \tag{2}$$

where $G(I_r, c_g)$ denotes the transferred images from the input image I_r conditioned on the target label c_g. In practice, we leverage the label smoothing trick in the loss function. More specifically, the expectation for true samples is set to a uniform distribution from 0.9 to 1.0, meanwhile, from 0.0 to 0.1 for false samples.

Classification Loss. In our work, the discriminator is able to not only differentiate real samples from false samples, but also classify images to their corresponding labels. Therefore, we add an auxiliary classifier on top of the discriminator D and introduce the classification loss:

$$\mathcal{L}_D^{cls} = \mathbb{E}_{I_r}[-log(D_{cls}(c_r|I_r))] \tag{3}$$

$$\mathcal{L}_G^{cls} = \mathbb{E}_{I_r, c_g}[-log(D_{cls}(c_g|G(I_r|c_g)))] \tag{4}$$

Reconstruction Loss. To preserve the content of the original image I_r in the facial attribute transfer process, we utilize the generator again on the generated image I_g to obtain the reconstruction image I_{rec} and apply a cycle consistency loss to the generator. Many works [4,5] have adopted the L1 norm as the reconstruction loss, but we argue that the L1 norm may cause the overall smoothing for high-frequency parts. Moreover, the L1 constraint may be too strong for some attribute transfer involving significant changes in shape. Consequently, we introduce the Multi-Scale Structural Similarity Index (MS-SSIM) as the reconstruction loss:

$$\mathcal{L}_{rec} = MSSSIM(I_r, I_rec) \tag{5}$$

Assuming that X, Y are two images, the SSIM between them is defined as:

$$SSIM(X,Y) = \frac{(2\mu_X\mu_Y + C_1)(2\sigma_{XY} + C_2)}{(\mu_X^2 + \mu_Y^2 + C_1)(\sigma_X^2 + \sigma_Y^2 + C_2)} \tag{6}$$

where μ and σ represent the mean value and variance of an image and σ_{XY} denotes the covariance. C_1 and C_2 are both hyper-parameters. Subsequently, MS-SSIM can be calculated by combining SSIM of different scales of images. MS-SSIM defines the structure information as the combination of luminance (mean value), contrast (variance) and structural similarity (covariance). In this way, generated images and reconstructed images are more aligned with the human visual system and more plausible hallucination is encouraged.

Full Objective. Finally, the full objective for training the generator G and the discriminator D is:

$$\mathcal{L}_D = \mathcal{L}_D^{adv} + \lambda_{cls}\mathcal{L}_D^{cls} \tag{7}$$

$$\mathcal{L}_G = \mathcal{L}_G^{adv} + \lambda_{cls}\mathcal{L}_G^{cls} + \lambda_{rec}\mathcal{L}_{rec} \tag{8}$$

where $\lambda_{cls} = 1$ and $\lambda_{rec} = 10$ are hyper-parameters to decide how much the classification loss and the reconstruction loss weigh in the overall loss function.

4   Implementation Details

Network Architecture. In the generator, we utilize two convolutional layers with stride $= 2$ for down-sampling, six residual blocks for feature extraction and two transposed convolutional layers for up-sampling. We use instance normalization as the norm layer and ReLU as the activation unit in the generator. There are six convolutional layers with stride $= 2$ in the discriminator and spectral normalization is applied to each of them. The activation unit used in the discriminator is Leak ReLU with negative slope $= 0.01$.

Training Details. We train our model on CelebA dataset. The optimizer is Adam with learning rate $= 0.0001$, beta1 $= 0.5$, beta2 $= 0.999$ for both generator and discriminator. We set the batch size to be 16 and alternatively train the discriminator and the generator for 20,000 iterations. The learning rate linearly decays to zero over the last 10,000 iterations.

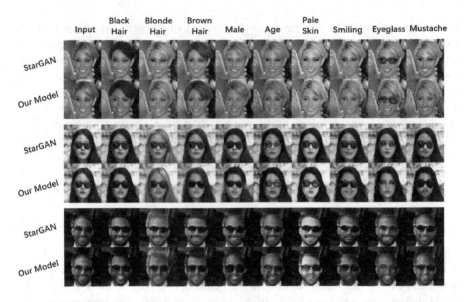

Fig. 4. Qualitative results on the CelebA dataset. Each row shows results from the same input images and each column shows results conditioned on the same labels.

5   Experiments

In this section, we conduct both qualitative and quantitative results on the CelebA dataset to validate the effectiveness of our model. We adopt StarGAN [4] as our baseline model, which uses a unified framework to learn multiple attribute transfer. First, we choose nine attributes from CelebA label annotations and demonstrate qualitative results compared with StarGAN. Next, we introduce the Fréchet Inception Distance (FID) [34] as the evaluation metrics to show quantitatively the superiority of our model over the baseline.

CelebA Dataset. The CelebA dataset [35] contains 202599 face images, and we randomly choose 2000 of them as the test set. The nine attributes we choose are *Black Hair, Blond Hair, Brown Hair, Male, Age, Pale Skin, Smiling, Eyeglasses* and *Mustache*. All the experiments in our paper are conducted on the test set.

Evaluation Metrirc. We introduce the Fréchet Inception Distance (FID) [34] as the quantitative evaluation metric. FID measures the Fréchet Distance of the feature maps between two distributions, which signifies the similarity from the feature level and lower FID represents larger similarity.

5.1   Qualitative Results

To demonstrate the effectiveness of our model, we compare our model with StarGAN through qualitative evaluation. In Fig. 4, each row shows results from the same input images and each column shows results conditioned on the same labels. As shown in Fig. 4, our model is able to deal with multiple attribute transfer while keeping the content unchanged by utilizing the MS-SSIM as the reconstruction loss instead of the L1 norm. The MS-SSIM alleviates the cycle consistency constraint and thus encourages more plausible changes. For examples, in the sixth column of the third group, our result seems older than that of StarGAN because more aging features are facilitated to emerge. Furthermore, by virtue of the multi-scale feature fusion module, our model succeeds in making reasonable hallucination as well. For instance, in the ninth column of the second and third group, our model is capable of generating the basic sketch of real eyes after removing glasses while StarGAN generates great artifacts or simply preserves partial eyeglasses regions of the original image. In addition, due to the large receptive fields of multi-scale atrous convolutions, our generator is allowed to capture long-term dependencies in images and hence generates images with higher fidelity and less artifacts. For example, in the sixth column and eighth column of the second group, there are obvious artifacts on the forehand and mouth with StarGAN, which were not seen in our results.

5.2 Quantitative Results

We randomly choose 2000 target labels and generate 2000 samples with different methods. After that, we calculate the FID between test data and generated images. Note that MFF means the multi-scale feature fusion module. As shown in Table 1, the FID of our model without MFF is a bit lower than that of StarGAN, which proves that using the MS-SSIM alone may not contribute too much to generation. But we believe that alleviating the cycle consistency constraint is non-trivial for high quality generation. Moreover, the lowest FID of our full model validates the effectiveness of the multi-scale feature fusion module and the superiority over the baseline.

Table 1. FID of our model and StarGAN.

Model	FID
StarGAN	35.94
Our model W/O MFF	35.11
Our full model	**33.18**

5.3 User Study

To compare our model with the baseline from human judgment, we conduct a user study with 50 volunteers. We randomly choose 5 real images and 10 target labels to obtain 50 generated images using StarGAN, our model without MFF and our full model respectively. Volunteers are asked to rank the images generated by three different models according to their visual quality and matching with the target labels. As seen in Table 2, our full model acquires the highest first rank rate while StarGAN obtains the most third rank, which further demonstrates the effectiveness of our method compared with the baseline.

Table 2. User study

Model	Rank 1	Rank 2	Rank 3
StarGAN	24.37%	27.54%	48.09%
Our model W/O MFF	25.70%	43.18%	31.12%
Our model	49.93%	29.28%	20.79%

6   Conclusion

In this paper, we propose a multi-scale GAN framework for facial attribute transfer. To alleviate the cycle consistency constraint, we introduce the Multi-Scale Structural Similarity Index (MS-SSIM) as the reconstruction loss function. Moreover, we incorporate a multi-scale feature fusion module in our generator to obtain long-range correlation on feature maps and thus generate images with high quality and less artifacts. Qualitative and Quantitative results demonstrates the superiority of our model over the baseline.

References

1. Denton, E.L., Chintala, S., Fergus, R., et al.: Deep generative image models using a Laplacian pyramid of adversarial networks. In: Advances in Neural Information Processing Systems, pp. 1486–1494 (2015)
2. Goodfellow, I., et al.: Generative adversarial nets. In: Advances in Neural Information Processing Systems, pp. 2672–2680 (2014)
3. Radford, A., Metz, L., Chintala, S.: Unsupervised representation learning with deep convolutional generative adversarial networks. arXiv preprint arXiv:1511.06434 (2015)
4. Choi, Y., Choi, M., Kim, M., Ha, J.-W., Kim, S., Choo, J.: StarGAN: unified generative adversarial networks for multi-domain image-to-image translation. In: Proceedings of the IEEE Conference on Computer Vision and Pattern Recognition, pp. 8789–8797 (2018)
5. Pumarola, A., Agudo, A., Martinez, A.M., Sanfeliu, A., Moreno-Noguer, F.: GANimation: anatomically-aware facial animation from a single image. In: Proceedings of the European Conference on Computer Vision (EC-CV), pp. 818–833 (2018)
6. Chen, L.-C., et al.: Rethinking atrous convolution for semantic image segmentation. arXiv preprint arXiv:1706.05587 (2017)

7. Huang, X., Liu, M.-Y., Belongie, S., Kautz, J.: Multimodal unsupervised image-to-image translation. In: Proceedings of the European Conference on Computer Vision (ECCV), pp. 172–189 (2018)
8. Isola, P., Zhu, J.-Y., Zhou, T., Efros, A.A.: Image-to-image translation with conditional adversarial networks. In: Proceedings of the IEEE Conference on Computer Vision and Pattern Recognition, pp. 1125–1134 (2017)
9. Kim, T., Cha, M., Kim, H., Lee, J.K., Kim, J.: Learning to discover cross-domain relations with generative adversarial networks. In: Proceedings of the 34th International Conference on Machine Learning, vol. 70, pp. 1857–1865. JMLR.org (2017)
10. Liu, M.-Y., Breuel, T., Kautz, J.: Unsupervised image-to-image translation networks. In: Advances in Neural Information Processing Systems, pp. 700–708 (2017)
11. Taigman, Y., Polyak, A., Wolf, L.: Unsupervised cross-domain image generation. arXiv preprint arXiv:1611.02200 (2016)
12. Zhu, J.-Y., Park, T., Isola, P., Efros, A.A.: Unpaired image-to-image translation using cycle-consistent adversarial networks. In: Proceedings of the IEEE International Conference on Computer Vision, pp. 2223–2232 (2017)
13. Zhu, J.-Y., et al.: Toward multimodal image-to-image translation. In: Advances in Neural Information Processing Systems, pp. 465–476 (2017)
14. Pathak, D., Krahenbuhl, P., Donahue, J., Darrell, T., Efros, A.A.: Context encoders: feature learning by inpainting. In: Proceedings of the IEEE Conference on Computer Vision and Pattern Recognition, pp. 2536–2544 (2016)
15. Yeh, R., Chen, C., Lim, T.Y., Hasegawa-Johnson, M., Do, M.N.: Semantic image inpainting with perceptual and contextual losses, vol. 2. arXiv preprint arXiv:1607.07539 (2016)
16. Yu, J., Lin, Z., Yang, J., Shen, X., Lu, X., Huang, T.S.: Free-form image inpainting with gated convolution. arXiv preprint arXiv:1806.03589 (2018)
17. Yu, J., Lin, Z., Yang, J., Shen, X., Lu, X., Huang, T.S.: Generative image inpainting with contextual attention. In: Proceedings of the IEEE Conference on Computer Vision and Pattern Recognition, pp. 5505–5514 (2018)
18. Ledig, C., et al.: Photo-realistic single image super-resolution using a generative adversarial network. In: Proceedings of the IEEE Conference on Computer Vision and Pattern Recognition, pp. 4681–4690 (2017)
19. Arjovsky, M., Chintala, S., Bottou, L.: Wasserstein GAN. arXiv preprint arXiv:1701.07875 (2017)
20. Mao, X., Li, Q., Xie, H., Lau, R.Y.K., Wang, Z., Paul Smolley, S.: Least squares generative adversarial networks. In: Proceedings of the IEEE International Conference on Computer Vision, pp. 2794—2802 (2017)
21. Zhao, J., Mathieu, M., LeCun, Y.: Energy-based generative adversarial network. arXiv preprint arXiv:1609.03126 (2016)
22. Salimans, T., Goodfellow, I., Zaremba, W., Cheung, V., Radford, A., Chen, X.: Improved techniques for training GANs. In: Advances in Neural Information Processing Systems, pp. 2234–2242 (2016)
23. Gulrajani, I., Ahmed, F., Arjovsky, M., Dumoulin, V., Courville, A.C.: Improved training of wasserstein GANs. In: Advances in Neural Information Processing Systems, pp. 5767–5777 (2017)
24. Miyato, T., Kataoka, T., Koyama, M., Yoshida, Y.: Spectral normalization for generative adversarial networks. arXiv preprint arXiv:1802.05957 (2018)
25. Mirza, M., Osindero, S.: Conditional generative adversarial nets. arXiv preprint arXiv:1411.1784 (2014)

26. Wang, T.-C., Liu, M.-Y., Zhu, J.-Y., Tao, A., Kautz, J., Catanzaro, B.: High-resolution image synthesis and semantic manipulation with conditional GANs. In: Proceedings of the IEEE Conference on Computer Vision and Pattern Recognition, pp. 8798–8807 (2018)

27. Reed, S., Akata, Z., Yan, X., Logeswaran, L., Schiele, B., Lee, H.: Generative adversarial text to image synthesis. arXiv preprint arXiv:1605.05396 (2016)

28. Zhang, H., et al.: StackGAN: text to photo-realistic image synthesis with stacked generative adversarial networks. In: Proceedings of the IEEE International Conference on Computer Vision, pp. 5907–5915 (2017)

29. Xiao, T., Hong, J., Ma, J.: ELEGANT: exchanging latent encodings with GAN for transferring multiple face attributes. In: Proceedings of the European Conference on Computer Vision (ECCV), pp. 168–184 (2018)

30. Lin, J., Xia, Y., Qin, T., Chen, Z., Liu, T.-Y.: Conditional image-to-image translation. arXiv e-prints, arXiv:1805.00251, May 2018

31. Miyato, T., Koyama, M.: cGANs with projection discriminator. arXiv preprint arXiv:1802.05637 (2018)

32. Odena, A., Olah, C., Shlens, J.: Conditional image synthesis with auxiliary classifier GANs. In: Proceedings of the 34th International Conference on Machine Learning, vol. 70, pp. 2642–2651. JMLR.org (2017)

33. Wang, Z., Simoncelli, E.P., Bovik, A.C.: Multiscale structural similarity for image quality assessment. In: The Thrity-Seventh Asilomar Conference on Signals, Systems and Computers, pp. 1398–1402. IEEE (2003)

34. Heusel, M., Ramsauer, H., Unterthiner, T., Nessler, B., Hochreiter, S.: GANs trained by a two timescale update rule converge to a local nash equilibrium. In: Advances in Neural Information Processing Systems, pp. 6626–6637 (2017)

35. Liu, Z., Luo, P., Wang, X., Tang, X.: Deep learning face attributes in the wild. In: Proceedings of the IEEE International Conference on Computer Vision (ICCV) (2015)

Convolutional-Block-Attention Dual Path Networks for Slide Transition Detection in Lecture Videos

Minhuang Guan[1,2], Kai Li[1,2], Ran Ma[1,2(✉)], and Ping An[1,2]

[1] Shanghai Institute for Advanced Communication and Data Science,
Shanghai, China
maran@shu.edu.cn
[2] School of Communication and Information Engineering, Shanghai University,
Shanghai, China

Abstract. Slide transition detection is used to find the images where the slide content changes, which form a summary of the lecture video and save the time for watching the lecture videos. 3D Convolutional Networks (3D ConvNet) has been regarded as an efficient approach to learn spatio-temporal features in videos. However, 3D ConvNet gives the same weight to all features in the image, and can't focus on key feature information. We solve this problem by using the attention mechanism, which highlights more effective features information by suppressing invalid ones. Furthermore, 3D ConvNet usually costs much training time and needs lots of memory. Dual Path Network (DPN) combines the two network structures of ResNext and DenseNet and has the advantages of them. ResNext adds input directly to the convolved output, which takes advantage of extracted features from the previous hierarchy. DenseNet concatenates the output of each layer to the input of each layer, which extracts new features from the previous hierarchy. Based on the two networks, DPN not only saves training time and memory, but also extracts more effective features and improves training results. Consequently, we present a novel ConvNet architecture based on Convolutional Block Attention and DPN for slide transition detection in lecture videos. Experimental results show that the proposed novel ConvNet architecture achieves the better results than other slide detection approaches.

Keywords: Lecture video · Slide transition · 3D ConvNet · Convolutional Block Attention · DPN

1 Introduction

With the rapid development of the Internet, more and more people are watching online lecture video to gain knowledge. The original video records not only the course information but also the content that is not related to the course, such as the audience, wasting the viewer's time and reducing the enthusiasm of the

© Springer Nature Singapore Pte Ltd. 2020
G. Zhai et al. (Eds.): IFTC 2019, CCIS 1181, pp. 103–114, 2020.
https://doi.org/10.1007/978-981-15-3341-9_9

learner. So, automatically summarizing the lecture video makes sense for the learner. In this paper, we focus on the slide transition detection, which is significant in lecture video summarization. There are many types of lecture videos. For example, some lecture videos are composed of slide and some are composed of slide and speaker appearing on same screen, especially, some speaker blocks the slide. On the other hand, some lecture videos contain camera motion, and some contain sudden camera switch. Camera motion and camera switch have an effect on the result of the slide transition detection. Slide transition occurs in a very short period of time, with changes in the slide content. Slide transition detection is one of key researches in lecture video summarization.

The existing solution is to use the histograms of adjacent frames to detect slide transition. However, the histograms [1] are prone to the changes of viewing scale and angles and some images with different contents may have the same histogram, which greatly affect the accuracy of slide transition detection. In order to overcome these shortcomings, many improved methods have been proposed. For example, Li [2] tracked a set of corner points across all frames by using the standard Kanade-Lucas-Tomasi (KLT) feature tracking approach. And he analyzed the trajectories of these feature points. However, this method does not work well in the case of camera switch. Jaiswal [3] proposed an automatic method for aligning scripts of lecture videos with captions. Alignment is needed to extract time information from captions and insert the time information in the scripts, to create index of the videos. Yang [4] extracted textual metadata by applying video Optical Character Recognition (OCR) technology on key-frames and Automatic Speech Recognition (ASR) on lecture audio tracks. The OCR and ASR transcript as well as detected slide text line types were adopted for keyword extraction, by which both video- and segment-level keywords were extracted for content-based video browsing and search.

Based on the above analysis, the key to slide transition detection is how to get the internal representation of the feature. Deep learning is increasingly developed, and features extracted from convolutional neural networks can also be used for slide transition detection. Undoubtedly, 3D Convnet can work well. In this paper, we present a novel 3D ConvNet architecture based on Convolutional Block Attention and DPN for detecting slide transition in lecture videos, 3D ConvNet achieves good performance in shot boundary detection. Attention module effectly ignores the irrelevant information and focuses on the key information. DPN not only can extract new features from the previous hierarchy, but also can exploit the extracted feature from the previous hierarchy.

The rest of this paper is organized as follows. Section 2 reviews the development attention mechanism and DPN. Section 3 describes the proposed 3D ConvNet architecture based on Convolutional Block Attention and DPN. Section 4 discusses the experimental results. Section 5 concludes this paper.

2   Development of Related Technologies

2.1   Attention Mechanism

The basic idea of attention in computer vision is to let the system learn to ignore irrelevant information and focu on the key information. This helps to filter out unimportant information and improve the efficiency of information processing. Wang [5] proposed the residual attention network. The network utilizes the attention mechanism (which can be used in existing end-to-end convolutional networks), and the Residual attention network uses a method of stacking the attention structures to change the feature's attention. And as the network deepens, the attention module will make adaptive changes. In each attention module, the upsampling and downsampling structures are used. Zhu [6] proposed the Attention Couplenet. Attention CoupleNet incorporate the attention-related information and global and local information of objects to improve the performance. Fu [7] proposed the dual attention network. It captures context dependencies through the self attention mechanism. Such a structure can adaptively integrate local features and global dependencies. There are mainly two attention modules. The position attention module selectively aggregates the position of each feature by weighted summation of all locations. The channel attention module selectively emphasizes a feature map through features in the feature map of all channels. Finally, the outputs of the two modules are summed, and the two branches are connected in parallel to obtain the final feature expression.

2.2   DPN

In visual/image recognition, a key question is how to get the internal representation of the feature. In the traditional method, people get the feature through a hand crafted feature, then the feature is inputed into the classifier. The convolutional neural network can automatically learn the features, and the features it learns are hierarchical. As a consequence, convolutional neural networks (CNNs) significantly push the performance of vision tasks. At present, a variety of network architectures of convolutional neural network have been proposed. He [8] proposed the ResNet, and the idea of residual learning. Traditional convolutional networks or fully connected networks have information loss during information transmission, and also cause gradient disappear or gradient explosions. ResNet solves this problem to a certain extent. By directly bypassing the input information to the output and protecting the integrity of the information, the entire network only needs to learn the part of the input and output differences, simplifying the learning objectives and difficulty. He Kaiming consistently improved the residual network. He [9] proposed the ResNext and the concept of cardinatity. The cardinatity is used to measure the model complexity. Cardinatity refers to the number of identical branches in a block. Improving cardinatity can achieve better model performance than improving height or width. WideResNet [10] widens the residual network and inserts the dropout layer in the widened Residual block to effectively avoid overfitting. Huang [11] proposed the DenseNet,

Densenet directly connects all layers under the premise of ensuring maximum information transfer between layers in the network. The ultimate use of features for better results and fewer parameters. ResNet reuses the features extracted in the previous layer, and the feature "purity" extracted from the convolution is relatively high. The redundancy of features is relatively low. DenseNet exploits the features extracted by the front layer creating new features. The feature extracted by the convolution in the back layer of this structure is likely to have been extracted from the previous layer, DenseNet extracted features with high redundancy. One has a high reuse rate, but the redundancy is low; one can create new features, but the redundancy is high. So, Yan [12] combines these two structures, getting DPN.

Based on the Attention mechanism and DPN, we propose the Convolutional-Block-Attention Dual Path Networks for Slide Transition Detection in Lecture Videos.

3   Network Architecture

This section discusses in detail the Convolutional-Block-Attention Dual Path Networks by combining Convolutional-Block-Attention module and Dual Path Networks.

3.1   Convolutional Block Attention Module

Convolutional Block Attention module (CBAM) [13] is a simple and effective Attention Module designed for convolutional networks. For the feature map generated by the convolutional neural network, CBAM calculates the attention map of the feature map from the channel and space dimensions, and then multiplies the attention map with the input feature map to perform adaptive learning of the feature. We extend CBAM from 2D ConvNet to 3D ConvNet, so the attention map is calculated by the channel, space and time dimensions. To be specific, the feature map $F \in R^{C*T*H*W}$ as input, CBAM sequentially infers a 1D channel attention map $M_c \in R^{C*1*1*1}$, a 2D spatial attention map $M_s \in R^{1*1*H*W}$ and a 1D time attention map $M_t \in R^{1*T*1*1}$, getting F', F'', F''' in order. The overall attention process can be summarized as:

$$F' = M_c(F) \bigotimes F \tag{1}$$

$$F'' = M_s(F') \bigotimes F' \tag{2}$$

$$F''' = M_t(F'') \bigotimes F'' \tag{3}$$

where $\bigotimes$ denotes element-wise multiplication. Figure 1 depicts the attention module, consisting of three parts: channel attention module, spatial attention module, and time attention module.

Channel Attention Module M_c. Figure 2 depicts the computation process of channel attention module. To compute the channel attention efficiently, we

Channel Attention Module Spatial Attention Module Time Attention Module

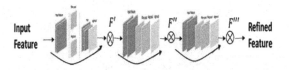

Fig. 1. Convolutional Block Attention module.

squeeze the spatial dimension and time dimension of the input feature map. Given a feature map $F \in R^{C*T*H*W}$ as input, we first squeeze the spatial dimension and time dimension by using average-pooling and max-pooling operations, generating two different descriptors. Then, the two descriptors seperately infer a multi-layer perceptron with one hidden layer, generating new descriptors. At the same time, the channel dimension is reduced to one-sixteenth of the original dimension and then is changed back to the original dimension. Add these two descriptors and then perform nonlinear operation by using sigmoid function on them. In short, the channel attention is computed as:

$$M_c(F) = \sigma(MLP(AvgPool(F)) + MLP(MaxPool(F))) \tag{4}$$

Finally, we get channel attention module M_c, $M_c \in R^{C*1*1*1}$.

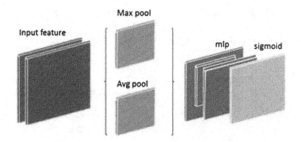

Fig. 2. Channel attention module.

Spatial Attention Module M_s. Fig. 3 depicts the computation process of spatial attention module. To compute the spatial attention efficiently, we squeeze the channel dimension and time dimension of the input feature map. Given a feature map $F \in R^{C*T*H*W}$ as input, We squeeze the channel dimension by using average-pooling and max-pooling operations, generating two different descriptors. We concatenate these two descriptors. Then We squeeze the time dimension by using convolution layer, the filter size of convolution kernel is set as $2*7*7$, producting new descriptor. Perform nonlinear operation on the new descriptor using sigmoid function. In short, the spatial attention is computed as:

$$M_s(F) = \sigma(f^{2*7*7}([AvgPool(F); MaxPool(F)])) \tag{5}$$

Finally, we get channel attention module M_s, $M_s \in R^{1*1*H*W}$.

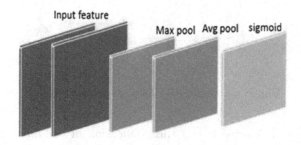

Fig. 3. Spatial attention module.

Time Attention Module M_t. Figure 4 depicts the computation process of time attention module. To compute the time attention efficiently, we squeeze the channel dimension and spatial dimension of the input feature map. Given a feature map $F \in R^{C*T*H*W}$ as input, we squeeze the channel dimension by using average-pooling and max-pooling operations, generating two different descriptors. We concatenate these two descriptors. Then we squeeze the time dimension by using convolution layer, producing new descriptor. Perform non-linear operation on the new descriptor using sigmoid function. In short, the time attention is computed as:

$$M_t(F) = \sigma(f([AvgPool(F); MaxPool(F)]))$$ (6)

Finally, we get time attention module M_t, $M_t \in R^{1*T*1*1}$.

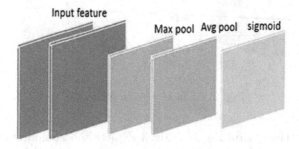

Fig. 4. Time attention module.

3.2 Dual Path Block

We formulate DPN [12] architecture as follows:

$$x^k = \sum_{t=1}^{k-1} f_t^k(h^t)$$ (7)

$$y^k = \sum_{t=1}^{k-1} v_t(h^t) = y^{k-1} + \emptyset^{k-1}(y^{k-1}) \tag{8}$$

$$r^k = x^k + y^k \tag{9}$$

$$h^k = g^k(r^k) \tag{10}$$

where k represents the current state is k, t represents t(th) state, $f_t^k(.)$ is a feature learning function, and Eq. (7) represents the DenseNet that enables exploring new features. $v_t(.)$ is a feature learning function, and Eq. (8) represents the ResNet that enables common features reuse. Equations (9) and (10) represents that the DPN adds the feature from DenseNet and the feature from ResNet, then sends them to the transformation function $g^k(.)$. DPN combines the ResNet with the DenseNet to take full advantage of the advantages of both networks. Figure 5 depicts the C3D dual path block. Given a feature map x as input, it infers a covlutional layer, performs BatchNorm operation, and performs nonlinear operation by using relu activation function, generating new descriptor. In practice, the $1 \times 3 \times 3$ convolutional layer uaually uses the group operation in ResNext to improve performance. Some descriptors are added to a part of x, and others are concatenated to the rest of x, the obtained result is used for making next mapping or prediction.

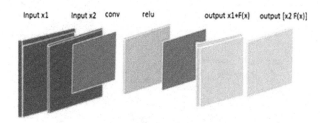

Fig. 5. Dual path block.

3.3 Convolutional-Block-Attention Dual Path Block

CBAM is a lightweight, universal module that can be integrated into a variety of convolutional neural networks for end-to-end training. The DPN structure has great advantages in image classification, target detection and semantic segmentation, and is an image recognition structure with excellent performance at present. In this paper, we integrate CBAM into dual path network. Figure 6 depicts the Convolutional block attention dual path block. We apply CBAM on the convolution outputs in each block.

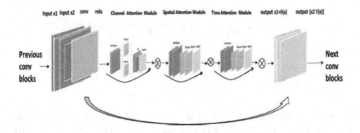

Fig. 6. Convolutional-Block-Attention Dual Path Block.

3.4 Convolutional-Block-Attention Dual Path Networks

Figure 7 depicts the network architecture of convolutional block attention dual path networks. The network architecture has eight convolutional layers and four Fully-Connected layers. The network architecture is similar to VGG. The number followed conv block is the size of channels.

We apply the ReLu activation function after each fully-connected layer. And dropout layer is behind the ReLu activation function. The dropout ratio is set to 0.5. The CrossEntropy Loss of the final layer is:

$$loss(x, class) = -log(e^{x[class]}/\sum_{j} e^{x[j]}) = -x[class] + log(\sum_{j} e^{x[j]}) \qquad (11)$$

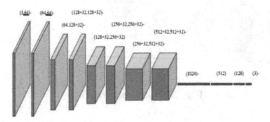

Fig. 7. Convolutional-Block-Attention Dual Path Networks.

4 Experiments

We use the own lecture videos library to do the experiment. The library has six types of lecture videos. Figure 8 depicts the all kinds of lecture videos. In Fig. 8 type 6 camera switch from speaker to slide. Table 1 depicts the characteristics of all kinds of lecture video.

Process the video into a frame by frame, every 2 frames feed into the convolutional block attention dual path network model. We divide the detection results into three classes. (1) slide transition; (2) the camera switches between the speaker and the slide; (3) no switching occurred.

Each video frame is resized to 112 * 112, and sent to the network. The batch size is set to 128. The optimization method we use is the Adam. The initial learning rate is 0.001. The epoch number is 100.

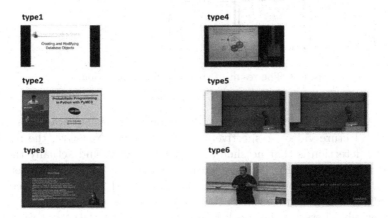

Fig. 8. The type of lecture video.

Table 1. The characteristics of all kinds of lecture video.

Type	Characteristics
Type 1	Be composed of slide
Type 2	Be composed of slide and speaker, they appear on same screen
Type 3	Speaker blocked the screen
Type 4	Speaker is near to the screen and don't block the screen
Type 5	Contain camera motion
Type 6	Lecture videos contain sudden camera switch

Equations (12) (13) (14) is used to judge the result of slide transition detection. F-score is the ultimate evaluation metric.

$$precision = s_c/s_t \tag{12}$$

$$recall = s_c/s_a \tag{13}$$

$$F_1 = (2 * precision * recall)/(precision + recall) \tag{14}$$

Where s_a is the total number of slide transitions in lecture videos library, s_c is the total number of slide transition correctly detected, and s_t is the total number of detected slide transitions, the detected result is not necessarily correct.

Figure 9 depicts the result of slide transition detection of the lecture videos. Our model can detect the slide transition.

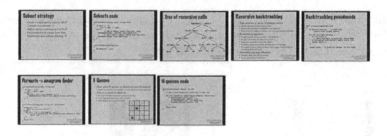

Fig. 9. The result of slide transition detection.

Figures 10 and 11 depicts the wrong detection result of slide transition detection of the lecture videos, respectively. In Fig. 10, false Negative, the result of the model detection is that no slide transition occurred, and actually the slide transition occurred. In Fig. 11, false Positive, the result of the model detection is that a slide transition occurred, and actually no slide transition occurred. In Fig. 10, subtle changes between slides effect the test results. In Fig. 11, the camera switches between the speaker and the speaker effect the test results.

Figure 12 depicts the precision, recall, and F-score of Convolutional-Block-Attention Dual Path Networks model on all kinds of lecture videos. Type 6 lecture videos have the lowest F-score. Type 1 lecture videos have the highest F-score. We compare our Convolutional-Block-Attention Dual Path Networks model with other three approaches on test dataset. These three methods are Singular Value Decomposition (SVD) [14], Frame Transition Parameters (FTP) [15] and the feature trajectories (SPD) [2]. Table 2 shows the average Precision, Recall, and F-score of four approaches on lecture videos library. The result of our method is better than other three methods. The F-score of our method is 23.9% higher than the SVD, 57.1% higher than the FTP, and 15.1% higher than the SPD.

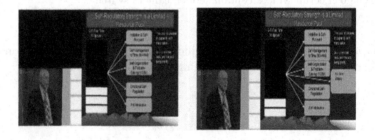

Fig. 10. False negative.

Fig. 11. False positive.

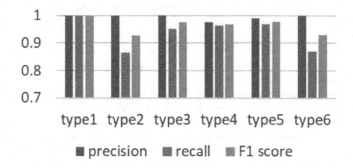

Fig. 12. The test result of model on all kinds of lecture videosthe.

Table 2. The test result of four approaches on lecture videos library.

	Precision	Recall	F1 score
SVD	0.700	0.745	0.722
FTP	0.748	0.264	0.390
SPD	0.773	0.848	0.810
Ours	0.993	0.935	0.961

5   Conclusion

In this paper, we present the convolutional block attention dual path network approach to detect slide transition in lecture videos. Our model extracts more effective features and improve training results. Experimental result shows that our system successfully detects slide transition invarious types of lectures videos and makes good result.

Acknowledgment. This work was supported by the Project of National Natural Science Foundation of (No. 61601278), "Chen Guang" project supported by Shanghai Municipal Education Commission and Shanghai Education Development Foundation (No. 17CG41).

References

1. Ma, D., Agam, G.: Lecture video segmentation and indexing. J. Proc. SPIE **8297**(1), 48 (2012)
2. Li, K., Wang, J., Wang, H., Dai, Q.: Structuring lecture videos by automatic projection screen localization and analysis. IEEE Trans. Pattern Anal. Mach. Intell. **37**(6), 1233–1246 (2015)
3. Jaiswal, S., Misra, M.: Automatic indexing of lecture videos using syntactic similarity measures. In: 2018 5th International Conference on Signal Processing and Integrated Networks (SPIN), pp. 164–169 (2018)
4. Yang, H., Meinel, C.: Content based lecture video retrieval using speech and video text information. IEEE Trans. Learn. Technol. **7**(2), 142–154 (2014)
5. Wang, F., et al.: Residual attention network for image classification. In: 2017 IEEE Conference on Computer Vision and Pattern Recognition (CVPR), pp. 6450–6458 (2017)
6. Zhu, Y., Zhao, C., Guo, H., Wang, J., Zhao, X., Lu, H.: Attention CoupleNet: fully convolutional attention coupling network for object detection. IEEE Trans. Image Process. **28**(1), 113–126 (2019)
7. Fu, J., et al.: Dual attention network for scene segmentation. In: IEEE Conference on Computer Vision and Pattern Recognition (CVPR) (2019)
8. He, K., Zhang, X., Ren, S., Sun, J.: Deep residual learning for image recognition. In: IEEE Conference on Computer Vision and Pattern Recognition (CVPR) (2016)
9. Xie, S., Girshick, R., Dollár, P., Tu, Z., He, K.: Aggregated residual transformations for deep neural networks. In: 2017 IEEE Conference on Computer Vision and Pattern Recognition (CVPR), pp. 5987–5995 (2017)
10. Zagoruyko, S., Komodakis, N.: Wide residual networks. In: BMVC (2016)
11. Huang, G., Liu, Z., Weinberger, K.Q., Maaten, L.: Densely connected convolutional networks. In: IEEE Conference on Computer Vision and Pattern Recognition (CVPR) (2017)
12. Chen, Y., Li, J., Xiao, H., et al.: Dual path networks. arXiv preprint arXiv:1707.01629 (2017)
13. Woo, S., Park, J., Lee, J.Y., et al.: CBAM: convolutional block attention module. arXiv preprint arXiv:1807.06521 (2018)
14. Gong, Y., Liu, X.: Video summarization using singular value decomposition. In: IEEE Conference on Computer Vision and Pattern Recognition (CVPR), pp. 174–180 (2000)
15. Mohanta, P.P., Saha, S.K., Chanda, B.: A model-based shot boundary detection technique using frame transition parameter. IEEE Trans. Multimed. **14**(1), 223–233 (2012)

Attention-Based Top-Down Single-Task Action Recognition in Still Images

Jinhai Yang[1], Xiao Zhou[2], and Hua Yang[1(✉)]

[1] Institution of Image Communication and Network Engineering,
Shanghai Jiao Tong University, Shanghai, China
{youngjh,hyang}@sjtu.edu.cn
[2] Suzhou Keensense Technology Co., Ltd., Suzhou, China
zhoux@1000video.com.cn

Abstract. Human action recognition via deep learning methods in still images has been an active research topic in computer vision recently . Different from the traditional action recognition based on videos or image sequences, a single image contains no temporal information or motion features for action characterization. In this study, we utilize a top-down action recognition strategy to analyze person instances in a scene respectively, on the task of detecting determine persons playing a cellphone. A YOLOv3 detector is applied to predict the human bounding boxes, and the HRNet (High Resolution Network) is used to regress the attention map centered on the area of playing a cellphone, taking the region of given human bounding box as the input. Experimental results on a custom dataset show that HRNet can reliably represent a person image to a heatmap where the region of interest (ROI) is highlighted. The accuracy of the proposed framework exceeds the performance of all the evaluated naive classification models, i.e., Densenet, inception_v3 and shufflenet_v2.

Keywords: Still image action recognition · Attention mechanism · High-resolution representation · Behavior analysis

1 Introduction

With the intelligent image processing drawing increasing interest, there have been urgent needs for researches on human action recognition recently. Action recognition can be applied in human behavior analysis, anomaly action alert, and human-machine interaction, etc. Recent progress of profound researches in the recognition of action in visual information is driven primarily by deep learning [3–7,17,22,24]. Although video-based action recognition [10,16–18,22,23] is still a hot topic, in recent years scholars have begun to turn their attention to the behavior analysis of static images. Some kinds of human actions, such as playing a cellphone (as shown in Fig. 1), can usually be inferred from a single frame. Compared to video-based action recognition, which highly relies on body

© Springer Nature Singapore Pte Ltd. 2020
G. Zhai et al. (Eds.): IFTC 2019, CCIS 1181, pp. 115–125, 2020.
https://doi.org/10.1007/978-981-15-3341-9_10

movement and temporal information, action recognition from single still image mainly depends on statics features.

Current approaches mining action knowledge from still images can be roughly grouped into two categories: holistic approaches and part-based approaches [12]. Holistic approaches extract human characteristics from the human bounding boxes and combine them with contextual characteristics from the entire picture to predict human behavior. Paper [14] suggested a straightforward fusion network for action prediction in 2016, which stacks features obtained from a bounding box with that from the background context. Some researchers [2,27] also consider inferring actions by a graphical model on the human body. Holistic approaches are considered in-efficient resulting from respective multi-level feature extraction and tend to be vulnerable to background interference.

Most part-based methods primarily make use of spatial information, such as human parts-poses-attributes [3,11], to acknowledge human actions in still pictures. These methods concern the unique pose features and relative-locations of different behaviors, emphasizing there is no need to extract features of the whole target human image. Certain body parts may be more linked to specified action execution than the complete human body. For instance, the behavior of playing a phone can generally be decomposed to actions and poses of certain body parts: eyes focused on the cellphone and hand(s) holding it or hearing a phone call. Partial features contain rich information for characterizing certain behaviors but can be incapable to discriminate analogous actions, e.g. playing a cellphone and holding a remote. What's more, the part-based approach takes into account no interaction between body parts, which is not conducive to the learning of advanced semantic information.

In this study, we propose a top-down action recognition architecture, which is composed of a YOLOv3-based human detector and an HRNet-based recognition model. YOLOv3 [15] has shown extraordinary performance in object detection, hence our work pays major attention to the performance of HRNet as the recognition model. To address the former-mentioned problems, we take advantage of HRNet to extract high-resolution attentive features from a bounding box. The repeated multi-scale information exchange in HRNet promotes knowledge integration between partial features and global features, thus overcoming the drawbacks of holistic methods and part-based methods. The HRNet is utilized to regress an attention map focused on the action area, which is proved to operate effectively. The attention mechanism has been widely applied in computer vision, including object detection [1], visual question answering [25], and image caption generating [26].

The major contributions of this work are three-fold: (1) proposal of a top-down action recognition framework, combining person detection and behavior characterization. (2) evaluate the proposed framework on a custom dataset for detecting persons who are playing a cellphone. (3) empirically demonstrate the effectiveness of HRNet to learn attentive characterization.

Fig. 1. Example scenarios captured on the street.

2   Method

2.1   Overall Framework

As illustrated in Fig. 2, the proposed attention-based top-down action recognition framework mainly contains two stages, i.e. (1) person detection and (2) action classification. We first obtain the bounding boxes of humans from a single frame using YOLOv3, and then accordingly crop and resize the regions of each person into 224×224 resolution. An attentive heatmap is generated by the attention-based action recognition model of HRNet, which is centered on the region of playing a cellphone. Experimental results indicate that the latter stage works efficiently to extract and learn and fuse multi-level characteristics, e.g. the texture of a cellphone (low-level feature), the pose of looking down on the cellphone (mid-level feature) and the semantics of action (high-level feature).

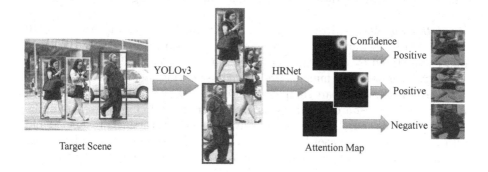

Fig. 2. Illustration of the overall framework.

2.2 Human Detector Based on YOLOv3

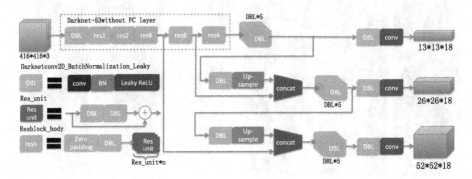

Fig. 3. Modified YOLOv3 structure for human detection. Output channels of the three YOLO layers are modified to 18 due to category reduction.

The modified YOLOv3 model is ported from our previous work, which is fine-tuned to serve as a human detector on a custom pedestrian dataset. This human detector achieves 88.8 AP (Average Precision) on this dataset and works efficiently and effectively in practical application scenarios. To be noticed, the YOLOv3 detector is only used during testing, and the human proposal is manually annotated for training the action recognition model. For more details on data preparation, please refer to Sect. 2.3.

2.3 Attention-Based Action Recognition Model

Dataset Construction. The process of data preparation is divided into two steps. Firstly, artificially annotate the bounding box of the attentive region. In the second stage, calculate the coordinates of the midpoint of the attentive region, and thereby synthesize an attention map for training HRNet.

Annotations of Action Attention Area. This dataset contains a total of 13110 images, which are human pictures manually cropped from video surveillance frames and stored by the original size. 5669 images among them are considered as positive samples, i.e., humans are judged as playing mobile phones in these instances. These pictures are sampled by interval, 80% of which are divided as the train set, and the rest are considered as the validation set. In positive samples, the bounding box of the ROI of the phone-playing behavior is artificially labeled and described by the corner coordinates. As the Fig. 4 shows, this manual bounding box mostly contains the mobile phone and a tight surrounding background.

Attention Map Synthesis. To suppress the subjectivity of manual annotation attention box and make full use of the HRNet's representation capability, we construct an attention map at the resolution of 56×56 for each positive sample resized to 224×224. Inspired by the method adopted in most pose estimation researches, we define the ground truth as a 2D Gaussian kernel centered on the midpoint of the bounding box, which can be denoted as:

$$G(x,y) = \frac{1}{2\pi\sigma^2} \exp(-\frac{(x-x_1)^2 + (y-y_1)^2}{2\sigma^2}) \qquad (1)$$

To best emphasize the ROI, the standard deviation σ is set to be 6 pixels after some tests on the human visual system. The criterion is to keep the image masked by the attention map action-recognizable while reserving the least unrelated surroundings.

The attention map for negative samples is a zero-filled matrix since these samples contain no attentive region of phone-playing.

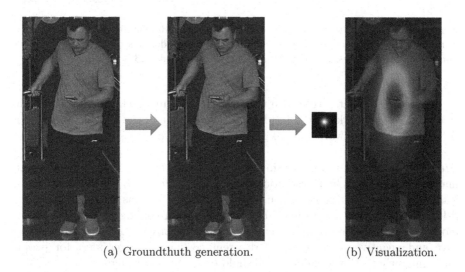

(a) Groundthuth generation.   (b) Visualization.

Fig. 4. An example of groundtruth construction. The bounding box of the phone-playing region is artificially annotated, and the grayscale heatmap centers on the ROI.

Deep High Resolution Network

Model Structure. An overview of the High Resolution Network (HRNet) proposed by paper [19] is shown in Fig. 5. The original intention of this innovative work was to promote the advancement of human pose estimation, but the HRNet used as an attentive sub-network in this study.

Most earlier techniques recover representations of high resolution from low-resolution representations generated by a network of high-to-low resolution.

Throughout the entire process, the HRNet maintains high-resolution representations and performs repeated multi-scale fusions. This unique mechanism promotes the fusion of multi-level characteristics, such as phone texture and looking-
at-phone behavior, enabling a better understanding of the action by this model.
The multi-level characterizing capacity, which is discussed in detail in Sect. 3.1,
can be indicated by some examples.

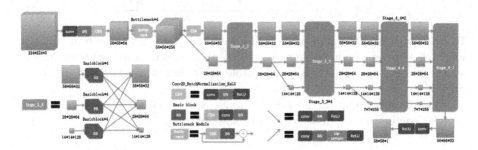

Fig. 5. Illustration of the architecture of HRNet-W32. Repeated multi-scale fusion is
completed in *StageModules*, where *m*, *n* in *stage_m_n* denotes the input and output
number of branches respectively. An instantiation of *stage_3_3* is also included.

In our experiments, we utilize the HRNet-W32, in which the width of high-
resolution subnetworks in the last three stages is 32 and the widths of the other
three parallel subnetworks are 64, 128, 256 respectively.

Attention Map Regression. We simply regress the attention maps from the ultimate exchange unit's high-resolution representations output, which works empirically well. The peak value of the attention map is seen as **recognition confidence**. By setting a threshold for the recognition confidence, we can obtain
models with diverse recognition sensitivities. Especially in some cases when positive and negative samples are of different cost, e.g. danger alert for people
infringing others, the confidence threshold ought to be set higher.

3   Experimental Result

3.1   Performance of the Attention-Based Model

Phone-Playing Action Recognition

Training. The boxes of human instances resized to 224×224 are fed into HRNet-
W32, thus the high-resolution representation output is 56×56 pixels. The images
are normalized the same as those of ImageNet, using $mean = [0.485, 0.456, 0.406]$
and $std = [0.229, 0.224, 0.225]$, and random horizontal flipping at the rate of 0.5
is taken as simple data augmentation.

The HRNet-W32 is trained using SGD optimizer, using batchsize 64 for 180 epochs. The initial learning rate is set to 0.1 and is lowered by 10 times at epoch 20, 60, 100, 130 and 160. We use a weight decay of 10^{-4} and a Nesterov momentum [20] of 0.9 and adopt the Kaiming method [8] for weight initialization. The loss function used to compare the anticipated heatmaps and groundtruth heatmaps is defined as the mean squared error.

Testing. The two-stage top-down paradigm introduced in Sect. 2.1 is used: detect the person instance using a human detector, and then determine each person instance is playing phone or not. This top-down procedure not only reduces the adverse influence from unrelated backgrounds but also restricts the feature scales, making it easier for automatic recognition.

Validating. We report the performance of the attentive HRNet-W32 on the validation set, as shown in Fig. 6. It's worth noting that the input images mainly contains a single person instance.

As indicated by the ROC (Receiver Operating Characteristic) curve, the AUC (Area under Curve) score achieves 0.9861. The attention-based action recognition model performs well on the task of identifying persons playing a cellphone. It is reasonable to believe that this model can easily be extended to other similar single-task action recognition, such as eating or exercising.

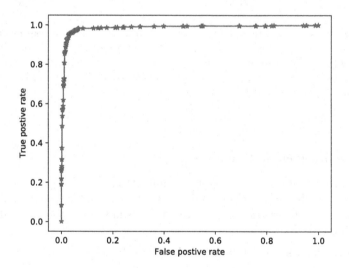

Fig. 6. The ROC curve of HRNet-W32 on the validation split of the custom dataset.

Multi-level Feature Representation. Back to our motivation to introduce HRNet to synthesize the attention map. One of the most challenging problems of image-based action recognition is that the network is required to learn multi-level features in a single image. The repeated multi-scale feature fusion in HRNet may provide an ideal solution for this difficulty. Some instances, as Fig. 7 shows, could be evidence that the attention-based action recognition doesn't rely on mono-level characteristic.

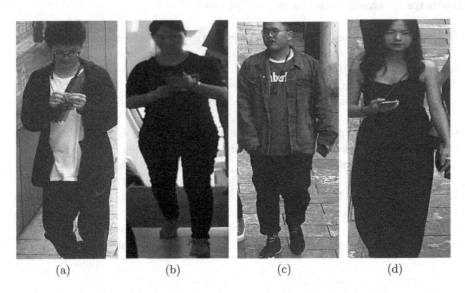

(a)   (b)   (c)   (d)

Fig. 7. Negative samples filtered by the proposed method. Figure (a) contains no phone textures, while these textures are confused by the clothes in Fig. (b), which is wrongly judged to be a negative sample. In Fig. (c), the man grasps the phone in hand without elbow being bent (strong evidence of not playing cellphone). The final picture shows a girl handing a cellphone, but looking into the distance.

3.2   Visualizations of Attention Map

Although the attention map consists of multi-level knowledge, we still expect it's centered on the midpoint of ROI. Both positive and negative samples are included in Fig. 8, where the attention map is visualized on the original image.

3.3   Comparisons with Classification Baselines

Classification Baselines. We model cellphone-playing recognition as a binary classification task in the baselines, and conduct transfer learning from pre-trained weights on ImageNet, by simply replace the final layers. The SGD optimizer is adopted with the initial learning rate of 0.001, and the learning rate is multiplied by 0.1 at epoch 100, 180 and 250 in the whole 300 epochs. We use cross entropy as the loss function in the baselines. Data preprocessing and other trivial configurations are identical to Sect. 3.1.

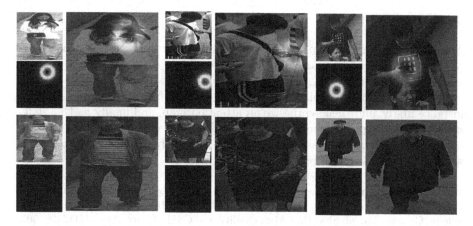

Fig. 8. Visualizations of samples randomly chosen. The figures on the top are positive samples while the figures on the bottom are negative samples. In each group, the top-left one is the original image, while the bottom-left one is the attention map, and the rest is the mixed figure for visualization. For better display, the grayscale attention map is colored, where cold color indicates a low value. (Color figure online)

Results on the Validation Set. As a comprehensive metric, the F1-score on the validation set is taken as the criterion to select the best weights for each model. We also conduct speed tests on each model, and the reported in Table 1, together with the results evaluated by other pivotal metrics. The HRNet-W32 is superior to all the other models in precision, F1-score, and accuracy. Its recall rate is only next to Densenet161 but performs better in speed. Due to frequent cross-scale information exchange, the HRNet-W32 has certain defects in processing speed compared to some classification baselines but still can complete the task in real-time. To conclude, the attention-based HRNet works effectively and efficiently in the task of cellphone-playing recognition.

Table 1. Comparisons of the performance between HRNet-W32 and the classification baselines. Tests are completed on 2 pieces of Tela-P4 with 11G memory.

Model	Precision	Recall	F1-score	Accuracy	FPS
Densenet121 [9]	95.03	95.77	95.40	95.26	246
Densenet169 [9]	94.72	96.03	95.37	95.21	217
Densenet201 [9]	93.70	96.29	94.97	94.77	186
Densenet161 [9]	94.45	**97.06**	95.74	95.57	144
Inception_v3 [21]	92.26	90.59	91.42	91.27	341
Shufflenet_v2 [13]	90.15	95.60	92.79	92.38	**465**
HRNet-W32	**96.61**	96.03	**96.32**	**96.23**	178

4 Conclusion

The absence of temporal information and motion features makes action recognition more challenging than video-based recognition. In this paper, we propose an attention-based top-down action recognition framework on the task of detecting persons playing a cellphone in still images. This framework detects human instance firstly, and then synthesize an attention map, and finally make a judgment by the recognition confidence. On a custom dataset, our method's performance surpasses all the classification baselines. We are planning to extend this work to multi-task action recognition in the future.

Acknowledgement. This work was supported in part by National Natural Science Foundation of China (NSFC, Grant No. 61771303 and 61671289), Science and Technology Commission of Shanghai Municipality (STCSM, Grant Nos. 17DZ1205602, 18DZ1200-102, 18DZ2270700), and SJTUYitu/Thinkforce Joint laboratory for visual computing and application. Director is funded by National Engineering Laboratory for Public Safety Risk Perception and Control by Big Data PSRPC.

References

1. Ba, J., Mnih, V., Kavukcuoglu, K.: Multiple object recognition with visual attention. arXiv preprint arXiv:1412.7755 (2014)
2. Desai, C., Ramanan, D.: Detecting actions, poses, and objects with relational phraselets. In: Fitzgibbon, A., Lazebnik, S., Perona, P., Sato, Y., Schmid, C. (eds.) ECCV 2012. LNCS, vol. 7575, pp. 158–172. Springer, Heidelberg (2012). https://doi.org/10.1007/978-3-642-33765-9_12
3. Diba, A., Mohammad Pazandeh, A., Pirsiavash, H., Van Gool, L.: DeepCamp: deep convolutional action & attribute mid-level patterns. In: Proceedings of the IEEE Conference on Computer Vision and Pattern Recognition, pp. 3557–3565 (2016)
4. Du, W., Wang, Y., Qiao, Y.: Recurrent spatial-temporal attention network for action recognition in videos. IEEE Trans. Image Process. **27**(3), 1347–1360 (2017)
5. Du, W., Wang, Y., Qiao, Y.: RPAN: an end-to-end recurrent pose-attention network for action recognition in videos. In: Proceedings of the IEEE International Conference on Computer Vision, pp. 3725–3734 (2017)
6. Gkioxari, G., Girshick, R., Malik, J.: Contextual action recognition with r* CNN. In: Proceedings of the IEEE International Conference on Computer Vision, pp. 1080–1088 (2015)
7. Guo, G., Lai, A.: A survey on still image based human action recognition. Pattern Recogn. **47**(10), 3343–3361 (2014)
8. He, K., Zhang, X., Ren, S., Sun, J.: Delving deep into rectifiers: surpassing human-level performance on ImageNet classification. In: Proceedings of the IEEE International Conference on Computer Vision, pp. 1026–1034 (2015)
9. Huang, G., Liu, Z., Van Der Maaten, L., Weinberger, K.Q.: Densely connected convolutional networks. In: Proceedings of the IEEE Conference on Computer Vision and Pattern Recognition, pp. 4700–4708 (2017)
10. Karpathy, A., Toderici, G., Shetty, S., Leung, T., Sukthankar, R., Fei-Fei, L.: Large-scale video classification with convolutional neural networks. In: Proceedings of the IEEE Conference on Computer Vision and Pattern Recognition, pp. 1725–1732 (2014)

11. Kwak, S., Cho, M., Laptev, I.: Thin-slicing for pose: learning to understand pose without explicit pose estimation. In: Proceedings of the IEEE Conference on Computer Vision and Pattern Recognition, pp. 4938–4947 (2016)
12. Liu, L., Tan, R.T., You, S.: Loss guided activation for action recognition in still images. In: Jawahar, C.V., Li, H., Mori, G., Schindler, K. (eds.) ACCV 2018. LNCS, vol. 11365, pp. 152–167. Springer, Cham (2019). https://doi.org/10.1007/978-3-030-20873-8_10
13. Ma, N., Zhang, X., Zheng, H.T., Sun, J.: ShuffleNet v2: practical guidelines for efficient CNN architecture design. In: Proceedings of the European Conference on Computer Vision (ECCV), pp. 116–131 (2018)
14. Mallya, A., Lazebnik, S.: Learning models for actions and person-object interactions with transfer to question answering. In: Leibe, B., Matas, J., Sebe, N., Welling, M. (eds.) ECCV 2016. LNCS, vol. 9905. Springer, Cham (2016). https://doi.org/10.1007/978-3-319-46448-0_25
15. Redmon, J., Farhadi, A.: YOLOv3: an incremental improvement. arXiv preprint arXiv:1804.02767 (2018)
16. Rodríguez, N.D., Cuéllar, M.P., Lilius, J., Calvo-Flores, M.D.: A survey on ontologies for human behavior recognition. ACM Comput. Surv. (CSUR) 46(4), 43 (2014)
17. Simonyan, K., Zisserman, A.: Two-stream convolutional networks for action recognition in videos. In: Advances in Neural Information Processing Systems, pp. 568–576 (2014)
18. Srivastava, N., Mansimov, E., Salakhudinov, R.: Unsupervised learning of video representations using LSTMs. In: International Conference on Machine Learning, pp. 843–852 (2015)
19. Sun, K., Xiao, B., Liu, D., Wang, J.: Deep high-resolution representation learning for human pose estimation. arXiv preprint arXiv:1902.09212 (2019)
20. Sutskever, I., Martens, J., Dahl, G., Hinton, G.: On the importance of initialization and momentum in deep learning. In: International Conference on Machine Learning, pp. 1139–1147 (2013)
21. Szegedy, C., Vanhoucke, V., Ioffe, S., Shlens, J., Wojna, Z.: Rethinking the inception architecture for computer vision. In: Proceedings of the IEEE Conference on Computer Vision and Pattern Recognition, pp. 2818–2826 (2016)
22. Tran, D., Bourdev, L., Fergus, R., Torresani, L., Paluri, M.: Learning spatiotemporal features with 3D convolutional networks. In: Proceedings of the IEEE International Conference on Computer Vision, pp. 4489–4497 (2015)
23. Wang, L., Xiong, Y., Wang, Z., Qiao, Y.: Towards good practices for very deep two-stream convnets. arXiv preprint arXiv:1507.02159 (2015)
24. Wang, Y., Zhou, L., Qiao, Y.: Temporal hallucinating for action recognition with few still images. In: Proceedings of the IEEE Conference on Computer Vision and Pattern Recognition, pp. 5314–5322 (2018)
25. Xu, H., Saenko, K.: Ask, attend and answer: exploring question-guided spatial attention for visual question answering. In: Leibe, B., Matas, J., Sebe, N., Welling, M. (eds.) ECCV 2016. LNCS, vol. 9911, pp. 451–466. Springer, Cham (2016). https://doi.org/10.1007/978-3-319-46478-7_28
26. Xu, K., et al.: Show, attend and tell: neural image caption generation with visual attention. In: International Conference on Machine Learning, pp. 2048–2057 (2015)
27. Yao, B., Fei-Fei, L.: Action recognition with exemplar based 2.5D graph matching. In: Fitzgibbon, A., Lazebnik, S., Perona, P., Sato, Y., Schmid, C. (eds.) ECCV 2012. LNCS, vol. 7575, pp. 173–186. Springer, Heidelberg (2012). https://doi.org/10.1007/978-3-642-33765-9_13

Adaptive Person-Specific Appearance-Based Gaze Estimation

Chuanyang Zheng[1,2], Jun Zhou[1,2(✉)], Jun Sun[1,2], and Lihua Zhao[3]

[1] Institute of Image Communication and Network Engineering,
Shanghai Jiao Tong University, Shanghai 200240, China
zhoujun@sjtu.edu.cn
[2] Shanghai Key Lab of Digital Media Processing and Transmissions,
Shanghai Jiao Tong University, Shanghai 200240, China
[3] Children's Hospital of Shanghai, Shanghai 200062, China

Abstract. Non-invasive gaze estimation from only eye images captured by camera is a challenging problem due to various eye shapes, eye structures and image qualities. Recently, CNN network has been applied to directly regress eye image to gaze direction and obtains good performance. However, generic approaches are susceptible to bias and variance highly relating to different individuals. In this paper, we study the person-specific bias when applying generic methods on new person. And we introduce a novel appearance-based deep neural network integrating meta-learning to reduce the person-specific bias. Given only a few person-specific calibration images collected in normal calibration process, our model adapts quickly to test person and predicts more accurate gaze directions. Experiments on public MPIIGaze dataset and Eyediap dataset show our approach has achieved competitive accuracy to current state-of-the-art methods and are able to alleviate person-specific bias problem.

Keywords: Gaze estimation · Meta-learning · Deep learning · Person-specific bias

1 Introduction

As eye is a pivotal organ for perceiving the world around us, eye gaze carries abundant important information that shows human attention, interest, sense, physical condition, etc. And it has been applied in numerous field such as human-computer interaction [16], medical analysis [17], Visual Reality industry [18,19], automotive driver assistance system [6,15] and is expected to extend to broader range of applications in the near future.

To obtain the eye gaze direction, earlier methods using invasive apparatus not only obstruct users seriously, but also obtain poor performance. Later, non-invasive methods, which keep attracting gaze objection of users with least interference, emerge and have dominated this field up to now. Recent approaches on

© Springer Nature Singapore Pte Ltd. 2020
G. Zhai et al. (Eds.): IFTC 2019, CCIS 1181, pp. 126–139, 2020.
https://doi.org/10.1007/978-981-15-3341-9_11

gaze estimation can be divided into two general categories: model-based gaze estimation and appearance-based gaze estimation [8]. Model-based approaches rely on an eye model of high complexity and the extracted local features by the model such as reflections from the eye images to estimate the gaze. While model-based approaches usually get high accuracy, they typically need to construct a sophisticated eye model according to anatomical eye knowledge and are sensitive to the variation of personal eye parameters such as eyeball radius. Moreover, they generally require additional hardware for illumination and caption, this leads to narrow application range. On the contrary, the latter methods estimate gaze direction directly from eye appearance or extracted eye features. That is to say that they straightly map the image contents to screen coordinates or gaze vectors. Recently, deep learning and convolution neural network have shown powerful strength in appearance-based gaze estimation [27]. With increasing deep learning approaches emerging, the angular error of gaze estimation continues to drop. Since deep learning methods trained on large amount of data and they don't need explicit manual feature, so they can get better result than model-based methods in low-resolution images which is difficult for model-based methods. However, it is laborious to collect sufficient amount of data with ground-truth labels for deep learning.

Apart from fewer datasets compared to other vision tasks like image classification, person-specific difference is another challenge. Most CNN approaches train a person-independent generic estimator on main public datasets. But their mean angular error is limited to around 5° [12] on data. Although datasets consist of different subjects images can reduce estimation error, person-specific bias still plays a crucial part in gaze error. [7] points out that the visual axis (the line joining the center of pupil and the center of the fovea) is not aligned with the optical axis (the line joining the centers of curvature of all the optical surfaces), such alignment differences are person-specific. In addition, each user has various eye ball structure, appearance, and so on. All these contribute to ambiguity in estimator.

In order to reduce the person-specific bias, one simple and straight way is to train a person-specific estimator for the current user. In practice, It is hard to collect enough data from a specific user. Even if we have large amount of specific user data, training for each user still costs large amount of time. So we propose a novel person-specific neural network to improve gaze estimation by alleviating person-specific bias using a little specific user data. Figure 2 illustrates our framework. Since meta-learning has been proved the ability of training a model that can quickly adapt to a new task using only a few data points and training iterations [2], we apply it to train a CNN estimator on the public datasets. At testing time, the trained estimator, with the help of meta-learning method, will take only 10 or less person-specific eye image samples as input which can be easily obtained through calibration process to learn person-specific features. Base on these features, the trained estimator will update our network and adapt quickly to the test user.

We evaluate the proposed framework on two main public MPIIGaze and Eyediap datasets. Our experiments show that our adaptive person-specific gaze estimator has significantly reduced the person-specific bias problem. Meanwhile, without the person-specific bias disturbance, our model achieved stat-of-the-art gaze estimation performance.

2   Related Work

In this paper, we study appearance-based gaze estimation methods and person-specific calibration detailing below.

2.1   Appearance-Based Gaze Estimation Methods

The appearance-based approaches take image as input to estimate gaze. They typically use a camera to capture the user eye image for input. Up to now, a variety of approaches have been proposed including adaptive linear regression [14], visual salience [23], support vector machine [24] and recently emerging deep learning networks. Because deep learning networks gain excellent performance, they have dominated the gaze estimation research. Zhang et al. firstly propose a deep CNN network for gaze estimation, which outperform previous methods a lot and help gaze estimation research turn to deep learning. Moreover, they present the first large MPIIGaze gaze dataset collected in everyday life rather than lab setting [27]. Followed by Krafka et al., they propose iTracker with four input channels processing face image, left eye, right eye and face grid separately. Then combine these multi-channel features obtained by four CNN network in full-connected layers to improve gaze accuracy [10]. As iTracker indicates that human face contains important information, Sugano et al. propose a simple but effective network, which deploys a spatial weight on top of convolution network to emphasize important area of interest in face and uses full face image as only input [22]. Both [27] and [22] have shown other regions rather than eye area also contain helpful information for gaze estimation.

While previous methods adopt both eye image and head pose as independent inputs, some work try to explore the relationship between head pose and gaze direction. Zhu et al. note that the network may suffer from head-gaze correlation overfitting and ambiguous landmark. To alleviate these problems, they apply two CNN network to find gaze in head coordinate and head pose in camera coordinate respectively, following that a matrix transformation is used to combine these two result to get final result [29]. Another network proposed by Ranjan et al., based on observation that under different head pose, eye appearance changes a lot and gaze has different distribution, apply a novel branched CNN architecture to improve the robustness of gaze classifiers to variable head pose [20].

Furthermore, other work tries different possible ways to improve gaze estimation. Instead of directly regressing gaze from eye images, Park et al. propose to regress eye images to an intermediate pictorial representation first, then regress the pictorial representation to gaze direction. Their CNN network is partly inspired by human pose, Cheng et al. take advantage of the two eye asymmetry characteristics observed during gaze estimation for the left and right eyes separately and propose a asymmetric regression-evaluation network, which teaches the network to find the eye easier to estimate [1]. Recently, Fischer et al. propose a new dataset collected in large camera-to-subject distances and high variations in head pose environments. The RT-GENE network in their paper also improve gaze estimation performance [3].

2.2 Person-Specific Calibration

As it is mentioned above, the problem of person-specific difference will cause bad performance when applying a generic estimator trained on public datasets to unseen test subjects. Currently, person-specific bias is hardly considered in appearance-based methods when doing cross subject gaze estimation. Zhang et al. mention the problem. They train and test data come from the same person that means training a person-specific gaze estimator. The pretty good gain in performance suggests that person-specific calibration is necessary [28] and generic estimator can not adapt to unseen person well. Accompanying researches focus on the person-specific problem. There are two main ways to solve that. The first way is to use the pre-trained network for further adjustment. Krafka et al. use feature maps from the last layer of the pre-trained model then apply them to train a SVR person-specific gaze prediction model on calibration samples data points [10]. Lindén et al. assign 2N calibration parameters to each person as specific features and connect them with feature map in the full-connected layers [11]. Yu et al. fine-tune a generic gaze estimator using many gaze redirected samples. In practice, we can only collect a few person-specific gaze calibration samples. So Yu applies gaze redirection network to synthesize many person-specific gaze data to overcome poor performance when testing with a few eye images. Different from the first way, another way aims to overcome the problem using novel and specially designed models which only require a few calibration samples [26]. Liu et al. propose a differential approach. Instead of directly regressing gaze, they use siamese network to obtain difference between input image and images of reference set. A weighting scheme following that is applied to compute the final gaze, obtaining fairly good result [13].

3 Method

In this section, we first introduce some notations for 3D gaze estimation. Then we will show the person-specific bias problem in previous methods. After that, we will present our network along with its train and test process.

3.1 3D Appearance-Based Gaze Estimation

For any gaze direction, it can be easily denoted by a 3D vector in 3D space. Meanwhile, eye appearance in a given eye image, such as the pupil location, provides rich information about the gaze direction. Moreover, other factors including

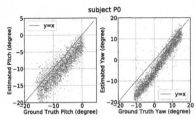
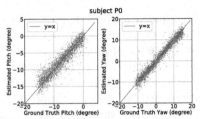

(a) Estimated by generic network versus ground truth angles of subject0

(b) Estimated by specific network versus ground truth angles of subject 0

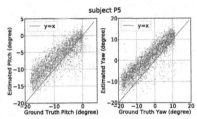
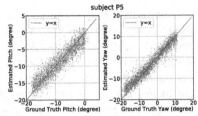

(c) Estimated by generic network versus ground truth angles of subject 5

(d) Estimated by specific network versus ground truth angles of subject 5

Fig. 1. Example of person-specific bias problem on MPIIGaze dataset above. All images plot estimated angles by trained network [28] versus ground truth angles on new subject 0 and subject 5. Figure a and c show results of trained generic network testing on these two subjects, while Fig. b and d show results of person-specific network.

major head pose, will also affect the gaze direction. So the problem of estimating the 3D gaze direction g from a given eye image I can be formulated as a complex non-linear regression problem as following. Generally, we consider head pose $h \in R^3$, and captured eye image $I^{H \times W \times 3}$ as major factors for gaze direction estimation.

$$g^e = f(I, h) \tag{1}$$

where f is the regression function.

3.2   Person-Specific Problem Analysis

In order to explain the person-specific problem in generic gaze estimator, we first train a standard convolution neural network following [28] for cross subject gaze estimation on MPIIGaze dataset and name it generic network. For better comparing, we also train a person-specific convolution neural network using training data from specific person on the same dataset and name it person-specific network. For example, take subject 0 as test. The generic network trains on data of other 14 subject, while person-specific network trains directly on data of this subject 0. On the basis of above network, we analyze the person-specific bias introduced by person-specific structure and appearance. Testing results are illustrated in Fig. 1. From results in Fig. 1a and c obtained by generic estimator,

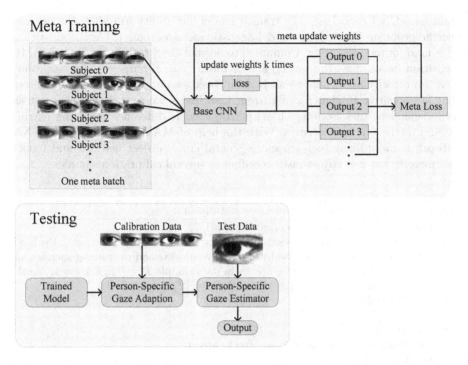

Fig. 2. Our proposed architecture overview. Meta-learning is applied to train on data consisting of many subjects. Following that, the trained model implement person-specific gaze adaption to obtain a person-specific CNN network using a few calibration samples. Thus, the final network adapts to current test subject and gives more accurate estimation.

it is apparent to see heavy and various data bias which changes according to different subject. This is because generic network fails to adapt to unseen person with certain different features from training subject, which includes extrinsic and intrinsic individual difference such as eyeball anatomical structure, the difference between optical axis and visual axis, and others [8]. On the contrary, using a person-specific estimator to test subject can avoid data bias. Results in Fig. 1b and d suggest that if the model can adapt well to the new subject, it will significantly improve gaze performance.

3.3   Meta-learning Gaze Estimation Network

In order to overcome the person-specific problem above, we propose a novel network illustrated in Fig. 2 which integrates meta-learning into normal CNN model. For our network, with even less than 10 calibration samples collected in the calibration process, it can still learn lots of person-specific characteristics, adapting quickly into current person, then becomes person-specific gaze estimator, which estimates more accurate gaze than generic estimator. As meta-learning works well for few-shot problem [25], we use meta-learning algorithm to train base CNN model and the trained model has ability to reduce the person-specific problem. Specifically, model-agnostic meta-learning (MAML) algorithm [2] is used for meta-learning. Compared to normal CNN training process, MAML algorithm encourages the emergence of more general-purpose representations that are broadly applicable to all tasks in task set and finds model parameters that are sensitive to changes in current task. Here, we regard every subject as an independent task because of the person-specific difference including extrinsic and intrinsic characteristics. With the help of MAML algorithm, our CNN network is more likely to learn more general cross-subject features and model parameters that can adjust fast according to several calibration samples.

Algorithm 1. Meta Learning of Gaze Estimator

Our algorithm is adapted from original MAML [2]

Assume dataset D consists of N subjects, the network takes the eye image I, the head pose h as inputs. We sample 1 subject from N subjects and m training samples for each subject in D. Thus we have one meta data sample $T = \{(I_i, h_i), i = 1, 2...m\}$. And we set one meta training batch as $B_{meta} = \{T_j, j = 1, 2...n\}$.

Weight θ initialization

while not done **do**
    sample meta training batch B_{meta} from dataset.
    for all T_j in B_{meta} **do**
        normal CNN training: get output $g^e = f(I_i, h_i)$ for $i = 1, 2...m$.
        compute L2 loss and do gradient descent:
        $\mathcal{L} = \sum_{i=1}^{m} \left(g_i^e - g_i^{gt} \right)^2$
        $\theta_j' = \theta - \alpha \bigtriangledown_\theta \mathcal{L}$
    end for
    meta training: update θ:
    $\theta = \theta - \beta \bigtriangledown_\theta \sum_{all j} Loss(\mathcal{L}_j')$
end while

Network Architecture. As shown in Fig. 2, we first use meta-learning to train our base model and attain a generic gaze estimator. Later, the trained model needs to do person-specific gaze adaption using only a few calibration samples for testing. Once these have been done, the trained generic model adapts to current test person and becomes person-specific gaze estimator.

Since transfer learning from CNNs training for similar computer vision tasks on huge dataset is common. In our architecture, our base CNN model adopts VGG network [21] pre-trained on huge ImageNet dataset, which is good at extracting complex image features. As loss function, we use the same loss as in [3] i.e. the sum of the L2 losses of every sample between the estimated and ground truth gaze vectors using 2D polar coordinate representations. And the performance assessment index of model is defined as the mean of angles of every sample between estimated vectors and ground truth gaze vectors using 3D Cartesian coordinates representation.

Train and Test Process. The whole train process can be divided into two train phase i.e. normal CNN train phase and meta train phase. The latter can be seen as more abstract learning on basis of former train. The detailed process of the train can be seen in Algorithm 1. Both the two phase will update weights. Apart from the update in normal CNN train process, which helps the network to find best result under the current subject, meta learning will update the weights according to meta loss consisting of various subject losses. In this way, not only can our network find more transferable features across subjects, but also will make our model easier to adapt quickly into new person with several calibration samples.

During test on new person, a small amounts of calibration data collected in the calibration process before test, is used for model weights to update fast to fit the characteristics of current person by meta-learning. After adjustment, the person-specific model can reduce the person-specific bias greatly. In this way, our proposed model obtains better performance. We will show it in our experiments.

Train Details. The weights of our base CNN model except full-connected layers are initialized from a pre-trained model trained on huge ImageNet dataset [21]. The weights of the full-connected layers are initialized using the Xavier initialization [5]. And We train our model on a desktop PC with an Intel Xeon E5-2630 and Nvidia Quadro M5000 GPU, applying the Adam optimizer [9] with initial meta learning rate 0.01 and meta learning batch size of 16. As for choosing the number of calibration size, we get our best results using 20 calibration samples.

4   Experiment

In our experiments, we compare our performance with that of several baselines, including published person-specific methods. Further, we evaluate our approach on reducing the person-specific bias and the impact of different number of calibration samples. Here are several methods used for comparison.

- iTracker [10]: One earlier classical deep gaze estimation network. It comprises four input channels include left eye, right eye, face image and face grid. It achieves a significant reduction in gaze error.
- GazeNet [28]: Another classical and robust gaze estimation network architecture modified from [27]. It outperforms previous methods.
- Lin-Ad [12]: Based on GazeNet, as there is usually a linear relationship between estimated gaze and ground truth, Lin-Ad tries to learn this relation and obtain an adapted gaze by Least Mean Square Error (LMSE).
- ARE-Net [1]: One multi-streams CNN founding on the core of "two eye asymmetry" observed during gaze estimation for the left and right eyes.

- Spatial weight CNN [22]: One full-face appearance-based gaze estimation with spatial weight CNN. Virtually, the spatial weight CNN is a kind of attention models which can be viewed as a tool to bias the allocation of available processing resources toward the most informative components.
- Diff-NN [12]: Differential network learns the gaze difference between input image and reference images. Then estimates gaze by add weighted these gaze difference to the corresponding reference gaze.

4.1 Dataset

MPIIGaze Dataset [28]. This dataset consists of 213659 images from 15 subjects. It is the first dataset collected in the wild and covers various head pose, distance, and illuminations. The MPIIGaze dataset contains an "Evaluation" set which provides 1500 left and right eye images of each participant along with 2D gaze point in the screen and 3D gaze vector. All the images have been normalized to 36×60 pixels to eliminate the error caused by face alignment following the normalization approach in [24]. In our experiment, we perform a leave-one-person-out cross-validation for all $N = 15$ participants within this dataset D and calculate the final average angular error of all folds for performance comparison. During test, in order to do specific adaption, we randomly select $m = 10$ calibration samples from test set for 512 times rather than using directly the whole test set.

Eyediap Dataset [4]. This dataset contains 94 sessions. Each session comprises recording videos, camera parameters, eye tracking data, screen targets, etc. In our experiment, we use continuous screen target (CS) sessions from 14 subjects because they have sufficient manual annotations while Discrete screen target sessions (DS) does not. We sample images at 10 fps from video lasting 2.5 min of each session. All the images are normalized as described in [22], choosing the center between two eyes as the reference point. Accordingly, all gaze ground truth with respect to world coordinate system are converted to gaze vectors with respect to camera coordinate system. After normalization, we conduct 5-fold cross-validation following [22].

4.2 Performance Comparison

In this section, we compared with several baselines described above on MPIIGaze dataset and Eyediap dataset under the same leave-one-out protocol. As MPIIGaze dataset is the most recognized gaze estimation dataset, all the proposed methods evaluate their performance on it. However, gaze estimation research still lacks a second mostly recognized dataset due to various problems such as collection setting, image quality and coverage of head pose. Commonly, the proposed method will choose another dataset differently. That is why we see the number of approaches evaluating on Eyediap dataset decreases. The results are shown in Fig. 3. The proposed method has achieved the lowest error on both datasets. For the results on MPIIGaze, our model outperforms the classical GazeNet architecture [28] 21%. This suggests that our network adapts well to test subjects,

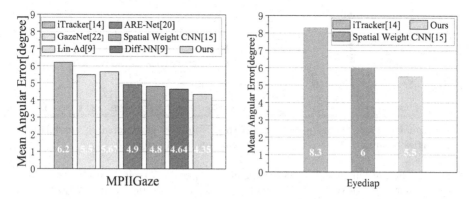

Fig. 3. Mean angular error in degree of baselines methods and ours on MPIIGaze dataset and Eyediap dataset.

becoming more robust to person-specific bias of test subjects such as various eye appearance. It is important to note that Lin-Ad [13] has an error 23% higher than ours. This is because Lin-Ad fails to find good linear function fitting the current subject characteristics with very few calibration samples and does not really account for the specificity of subject's eyes. And it even performs worse than GazeNet demonstrating that simple linear fitting can not solve the person-specific bias problem well. Meanwhile, for results on Eyediap, our model has obtained as low as 5.5° of mean angular error, which is 8% lower than Spatial Weight CNN [22]. The low-quality eye images in Eyediap leads to higher error of the mentioned methods than that of results on MPIIGaze.

4.3   Results on Personal-Specific Bias

As our model aims to reduce person-specific bias, it is necessary to discuss the ability of our model to reduce the person-specific bias detailed above. From results illustrated in Fig. 4, the person-specific bias shown in Fig. 1 has reduced a lot. When testing on arbitrary new subject, the estimated gaze angle of the proposed model versus ground truth angle approximates the same linear distribution $y = x$. While our model has largely reduce the person-specific bias, the dispersion becomes larger than that of person-specific network in Fig. 1b and d. This happens because person-specific network in Fig. 1 trains and tests on the same person, while our meta learning model trains on other persons and tests on the new person with several calibration samples. Although the dispersion appears to be larger than that of generic network in Fig. 1a and c, the mean angular error shows lower. The reason for that is that trained generic network is more robust to individual variation, while our model becomes more sensitive to person-specific variation, that is also why our model can alleviate person-specific bias. Further, how to decrease the large dispersion will remain as our future work.

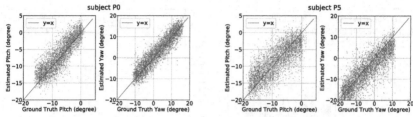

(a) Estimated gaze versus ground truth gaze angles of subject 0

(b) Estimated gaze versus ground truth gaze angles of subject 5

Fig. 4. Results of our model when testing on new person. The images plot estimated angles by our network versus ground truth angles on new subject 0 and subject 5.

4.4   Performance of Different Calibration Set Size

Since calibration data is required to estimate gaze in our method, in this session, we study the impact of different calibration set size. We follow the same train and test protocol above in the experiments and do leave-one-person-out cross validation for all subjects on MPIIGaze dataset. The trained model will do person-specific gaze adaption given different calibration set size. And the performance changes with it as illustrated in Fig. 5. Even with only one sample in the calibration set, the model performs still better than baseline GazeNet [28]. This is because our model extracts more general-purpose features, it adapts to test person better. As the calibration samples increases, the mean angular error continues to drop. The more samples, the more person-specific features our model learns, which assist our model to adjust fast and well to person-specific network. The lowest error emerges when the number of calibration samples reaches 20.

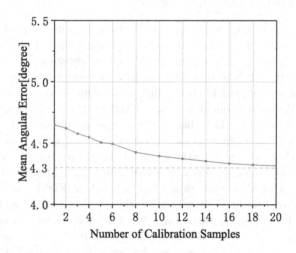

Fig. 5. Comparison of mean angular error in degree for our method with different number of calibration samples on MPIIGaze dataset.

5   Conclusion and Future Work

Although deep-learning-based gaze estimation has gained large improvement, person-specific calibration in gaze estimation still lacks enough attention. We study the person-specific bias problem when applying generic gaze estimator to new person. To address this, we present a novel network adopting MAML for training, which encourages our network to learn more general-purpose features across persons and gives our model the ability to adapt quickly to new person with only a few calibration samples. Through experiments on public datasets, we show our model can alleviate the person-specific bias and obtain lower mean angular error than previous methods. However, there are still future works to do along this way. Since our model becomes sensitive to the testing person, large dispersion emerges. It is possible to further improve performance by reducing the dispersion. In the future, we will aim to design more robust network with more general features.

Acknowledgment. The paper was supported by Multidisciplinary Development Project of Shanghai Jiao Tong University under Grant YG2017MS33, Science and Technology Commission of Shanghai Municipality (STCSM) under Grant 12DZ1200102, and NSFC under Grant 61471234, 61771303.

References

1. Cheng, Y., Lu, F., Zhang, X.: Appearance-based gaze estimation via evaluation-guided asymmetric regression. In: Proceedings of the European Conference on Computer Vision (ECCV), pp. 100–115 (2018)
2. Finn, C., Abbeel, P., Levine, S.: Model-agnostic meta-learning for fast adaptation of deep networks. In: Proceedings of the 34th International Conference on Machine Learning, vol. 70, pp. 1126–1135. JMLR. org (2017)
3. Fischer, T., Jin Chang, H., Demiris, Y.: Rt-gene: real-time eye gaze estimation in natural environments. In: Proceedings of the European Conference on Computer Vision (ECCV), pp. 334–352 (2018)
4. Funes Mora, K.A., Monay, F., Odobez, J.M.: EYEDIAP: a database for the development and evaluation of gaze estimation algorithms from RGB and RGB-D cameras. In: Proceedings of the ACM Symposium on Eye Tracking Research and Applications. ACM, March 2014. https://doi.org/10.1145/2578153.2578190
5. Glorot, X., Bengio, Y.: Understanding the difficulty of training deep feedforward neural networks. In: Proceedings of the Thirteenth International Conference on Artificial Intelligence and Statistics, pp. 249–256 (2010)
6. Guasconi, S., Porta, M., Resta, C., Rottenbacher, C.: A low-cost implementation of an eye tracking system for driver's gaze analysis. In: 2017 10th International Conference on Human System Interactions (HSI), pp. 264–269. IEEE (2017)
7. Guestrin, E.D., Eizenman, M.: General theory of remote gaze estimation using the pupil center and corneal reflections. IEEE Trans. Biomed. Eng. **53**(6), 1124–1133 (2006)
8. Hansen, D.W., Ji, Q.: In the eye of the beholder: a survey of models for eyes and gaze. IEEE Trans. Pattern Anal. Mach. Intell. **32**(3), 478–500 (2009)

9. Kingma, D.P., Ba, J.: Adam: a method for stochastic optimization. arXiv preprint arXiv:1412.6980 (2014)
10. Krafka, K., et al.: Eye tracking for everyone. In: Proceedings of the IEEE Conference on Computer Vision and Pattern Recognition, pp. 2176–2184 (2016)
11. Lindén, E., Sjöstrand, J., Proutiere, A.: Learning to personalize in appearance-based gaze tracking. arXiv e-prints arXiv:1807.00664, July 2018
12. Liu, G., Yu, Y., Funes Mora, K.A., Odobez, J.M.: A differential approach for gaze estimation. arXiv e-prints arXiv:1904.09459, April 2019
13. Liu, G., Yu, Y., Funes Mora, K.A., Odobez, J.M.: A differential approach for gaze estimation with calibration. In: 29th British Machine Vision Conference (2018)
14. Lu, F., Sugano, Y., Okabe, T., Sato, Y.: Adaptive linear regression for appearance-based gaze estimation. IEEE Trans. Pattern Anal. Mach. Intell. **36**(10), 2033–2046 (2014)
15. Mavely, A.G., Judith, J., Sahal, P., Kuruvilla, S.A.: Eye gaze tracking based driver monitoring system. In: 2017 IEEE International Conference on Circuits and Systems (ICCS), pp. 364–367. IEEE (2017)
16. Mutlu, B., Shiwa, T., Kanda, T., Ishiguro, H., Hagita, N.: Footing in human-robot conversations: how robots might shape participant roles using gaze cues. In: Proceedings of the 4th ACM/IEEE International Conference on Human Robot Interaction, pp. 61–68. ACM (2009)
17. Nakano, T., et al.: Atypical gaze patterns in children and adults with autism spectrum disorders dissociated from developmental changes in gaze behaviour. Proc. R. Soc. B: Biol. Sci. **277**(1696), 2935–2943 (2010)
18. Padmanaban, N., Konrad, R., Stramer, T., Cooper, E.A., Wetzstein, G.: Optimizing virtual reality for all users through gaze-contingent and adaptive focus displays. Proc. Natl. Acad. Sci. **114**(9), 2183–2188 (2017)
19. Patney, A., et al.: Towards foveated rendering for gaze-tracked virtual reality. ACM Trans. Graph. (TOG) **35**(6), 179 (2016)
20. Ranjan, R., De Mello, S., Kautz, J.: Light-weight head pose invariant gaze tracking. In: Proceedings of the IEEE Conference on Computer Vision and Pattern Recognition Workshops, pp. 2156–2164 (2018)
21. Simonyan, K., Zisserman, A.: Very deep convolutional networks for large-scale image recognition. arXiv preprint arXiv:1409.1556 (2014)
22. Sugano, Y., Fritz, M., Andreas Bulling, X., et al.: It's written all over your face: full-face appearance-based gaze estimation. In: Proceedings of the IEEE Conference on Computer Vision and Pattern Recognition Workshops, pp. 51–60 (2017)
23. Sugano, Y., Matsushita, Y., Sato, Y.: Appearance-based gaze estimation using visual saliency. IEEE Trans. Pattern Anal. Mach. Intell. **35**(2), 329–341 (2012)
24. Sugano, Y., Matsushita, Y., Sato, Y.: Learning-by-synthesis for appearance-based 3D gaze estimation. In: The IEEE Conference on Computer Vision and Pattern Recognition (CVPR), June 2014
25. Wang, Y., Yao, Q.: Few-shot learning: a survey. CoRR abs/1904.05046, http://arxiv.org/abs/1904.05046 (2019)
26. Yu, Y., Liu, G., Odobez, J.M.: Improving few-shot user-specific gaze adaptation via gaze redirection synthesis. In: The IEEE Conference on Computer Vision and Pattern Recognition (CVPR), June 2019
27. Zhang, X., Sugano, Y., Fritz, M., Bulling, A.: Appearance-based gaze estimation in the wild. In: Proceedings of the IEEE Conference on Computer Vision and Pattern Recognition, pp. 4511–4520 (2015)

28. Zhang, X., Sugano, Y., Fritz, M., Bulling, A.: MPIIGaze: real-world dataset and deep appearance-based gaze estimation. IEEE Trans. Pattern Anal. Mach. Intell. **41**(1), 162–175 (2017)
29. Zhu, W., Deng, H.: Monocular free-head 3D gaze tracking with deep learning and geometry constraints. In: Proceedings of the IEEE International Conference on Computer Vision, pp. 3143–3152 (2017)

Preliminary Study on Visual Attention Maps of Experts and Nonexperts When Examining Pathological Microscopic Images

Wangyang Yu, Menghan Hu$^{(\boxtimes)}$, Shuning Xu, and Qingli Li

Shanghai Key Laboratory of Multidimensional Information Processing,
East China Normal University, Shanghai 200241, China
mhhu@ce.ecnu.edu.cn

Abstract. Pathological microscopic image is regarded as a gold standard for the diagnosis of disease, and eye tracking technology is considered as a very effective tool for medical education. It will be very interesting if we use the eye tracking to predict where pathologists or doctors and persons with no or little experience look at the pathological microscopic image. In the current work, we first establish a pathological microscopic image database with the eye movement data of experts and nonexperts (PMIDE), including a total of 425 pathological microscopic images. The statistical analysis is afterwards conducted on PMIDE to analyze the difference in eye movement behavior between experts and nonexperts. The results show that although there is no significant difference in general, the experts focus on a broader scope than nonexperts. This inspires us to respectively develop saliency models for experts and nonexperts. Furthermore, the existing 10 saliency models are tested on PMIDE, and the performance of these models are all unacceptable with AUC, CC, NSS and SAUC below 0.73, 0.47, 0.78 and 0.52, respectively. This study indicates that the saliency models specific to pathological microscopic images urgent need to be developed using our database—PMIDE or the other related databases.

Keywords: Pathological microscopic images · Visual attention map · Saliency model · Database

1 Introduction

In medical diagnostic analysis, the pathological microscopic image is often considered as a gold standard for diagnosing disease [1]. Pathological microscopic image is a special kind of image. Pathologists or doctors and persons often zoom in pathological microscopic image to see more details. As the magnification increases, the depth of field decreases [2]. This phenomenon inevitably leads to the co-existence of focusing and defocusing regions in the same field of view, which in turn causes the distortion of pathological microscopic image. In addition, some other factors including but not limited to the performance of camera and lighting conditions will deteriorate the quality of pathological microscopic image. Undoubtedly, the poor quality of image will significantly affect the results of diagnosis.

© Springer Nature Singapore Pte Ltd. 2020
G. Zhai et al. (Eds.): IFTC 2019, CCIS 1181, pp. 140–149, 2020.
https://doi.org/10.1007/978-981-15-3341-9_12

When experts look at images, they will pay more attention to certain areas visually, and the visual attention is tightly interplayed with eye movements [3]. Eye tracking technique has been verified to accurately obtain the eye movements' information of image readers [4] which can be used as ground truth when training a saliency model. We can get a person's gaze pattern by analyzing these gaze data.

Eye tracking technology has been applied in many medical fields such as radiology diagnosis to improve the effect of clinical radiology training programs as well as users' speed and accuracy in diagnostic reading [5]. The eye tracking is also used to predict where the children with autistic spectrum disorder (ASD) look [6] and to evaluate the perceptual skill of pathologists or doctors in medical education [7, 8]. The eye tracking technology has been widely used in many fields, however the analysis of pathological microscopic images with this technique is still lacking. Hence, it is critical to understand experts' gaze behavior of pathological microscopic image, and statistically analyze its difference with the gaze behavior of nonexperts.

In this paper, we establish a pathological microscopic image database with the eye movement data of experts and nonexperts (PMIDE), which contains a total of 425 pathological microscopic images. PMIDE is capability of providing the strong statistic evidence for analyzing of gaze behavior of experts and nonexperts. Moreover, based on PMIDE, we respectively predict the fixation points of experts and nonexperts using the traditional saliency models. We also evaluate the performance of these models and check the plausibility of predicting fixation points by it.

2 Experiment and Analysis

2.1 Experiment

The PMIDE consists of 425 images. The images in PMIDE were collected from various medical centers to ensure the diversity of samples. Figure 1 shows some image samples in PMIDE. The contents of images include but are not limited to retina and tissue cells. The quality of these images are significantly different, including blur and clear, cool color cast and warm color cast, with and without bright blur, taking by high magnification and low magnification. The rich diversity of pathological microscopic images can well simulate the complex and various viewing environment in reality, and provides researchers a deeper understanding of the visual attention difference of people. The resolution of each image is 2304 × 1728.

Fig. 1. Some image samples in PMIDE.

A total of 15 participants ranging from nineteen to twenty-nine years old were recruited as subjects. Two of them are experts on pathological microscopic images processing and acquisition while the other participants are nonexperts. Especially, both experts have more than 2 years researching experience on their own fields. In this experiment, we use the eye tracker Tobii T120 and a 19 inch screen with resolution of 1280 × 1024. The distance between the participants and the eye tracker is approximately 65 cm. Considering the limited attention and the fatigue level of participants, the experiment is divided into five sections while 85 images are randomly selected in each section. We will calibrate the eye tracker and inform the participants of the announcements when using the eye tracker before the beginning of the experiment. Each image will be displayed for 5 s, with a 1 s gray image between two images.

All the displayed images are adjusted into 1/4 (1152 × 864) and shown in the middle of the screen.

2.2   Analysis

In this subsection, we focus on the data of eye movement and try to study the difference between the fixation points of experts and nonexperts. The fixation points' minimum fixation duration is set to be 100 ms. To get a continuous visual attention map of one pathological microscopic image from eye tracking data, we overlay all fixation points of one image fixated by all experts and nonexperts into one map respectively. The map is further smoothed with a Gaussian kernel (visual angle: 2 degree) [9] and normalized to unit. After overlapping the shaded map with the original sample, we obtain the fixation heat map. By comparing the fixation heat map of experts and nonexperts, we can analyze and get a lot of characteristics. Through intuitive analysis, we observe four significant differences in eye movement behavior between experts and nonexperts. We choose the several representational images containing these four kinds of difference, and show their corresponding heat map in Fig. 2. There are four subfigures in Fig. 2. The three columns of each subfigure represent the sample image, the experts' visual attention map and the nonexperts' visual attention map respectively. These three lines of each subfigure represent three different samples.

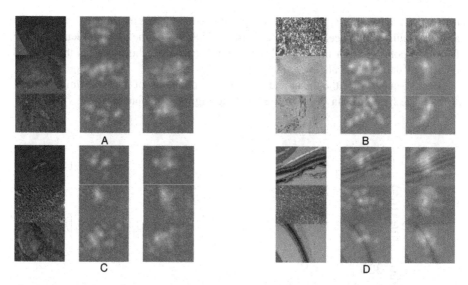

Fig. 2. Comparison between experts' visual attention map and nonexperts' visual attention map (three columns of each subfigure from left to right are sample image, heat map of experts and heat map of nonexperts respectively). ((a) Spatial difference of gaze point between experts and nonexperts; (b) experts gaze points are more scattering and they have more gaze points; (c) same gaze points; and (d) experts' gaze points tend to be more concentrated.)

Figure 2(a) shows the spatial difference of gaze point between experts and non-experts. This feature is quite common in our database. For complex images, experts and nonexperts have totally different fixation points. In some cases, experts even do not look at the point that the nonexperts often focus. In Fig. 2(b), it can be obviously observed that the experts' gaze points are more scattering and they have more fixation points. This feature is a commonplace in PMIDE. The finding may reveal that the experts will focus on a broader scope than nonexperts. Figure 2(c) shows the same position of the fixation points of experts and nonexperts. These images tend to have distinct image features. Figure 2(d) shows that the characteristics of expert gaze points tend to be more concentrated than those of nonexpert gaze points.

As shown in Fig. 2, although the gaze points of experts and nonexperts are the roughly same, the scope of the expert's attention is much smaller. In reality, seeing key points accurately and viewing images in specific fixation modes can help experts diagnose quickly and correctly. These differences of fixation may be useful for further study like creating the new saliency algorithms and make some contributions in diagnosis.

To further analyze the fixation mode's differences between the gaze points of experts and nonexperts, we perform an ANOVA (Analysis of Variance) analysis of the average fixation time. Although the mean gaze time is not significant overall

($p = 0.077$), it is significant on some images (there are 33 images whose $p < 0.05$). To analyze spatial difference, we also perform an ANOVA of the diagonal length of the first gaze point of each image (with the top left corner of the image as the original point) and the result is also not significant.

For the analysis of the mean and the standard deviation of the gaze time, we deal with all the fixation data of all images and the Fig. 3 is obtained.

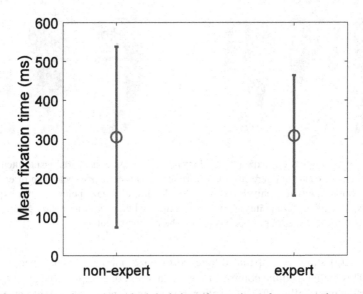

Fig. 3. Average value and standard deviation of gaze time of experts and nonexperts.

The mean fixation duration of experts is a little longer, and their standard deviation is much smaller which means they use more average time to fix on one point. Figure 4 shows the gaze time's average value and standard deviation of experts and nonexperts of each image with the red line refers to the nonexperts and the blue line refers to the experts. Figure 4(a) shows the average fixation time of each image, and Fig. 4(b) shows the standard deviation of fixation time of each image. We can tell from the figure that expert's average fixation time is higher than nonexpert's in general, and their standard deviation is lower than nonexpert's in most cases.

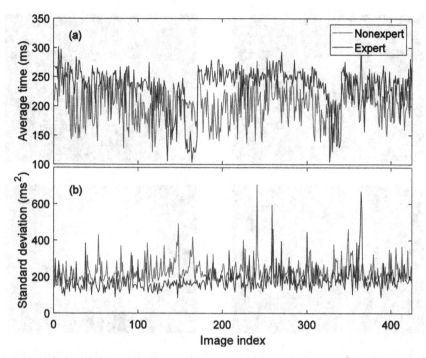

Fig. 4. Gaze time's average number and standard deviation of experts and nonexperts of each image. The red line refers to the nonexperts and the blue line refers to the experts. (a) Gaze time's average number of experts and nonexperts for each image; (b) gaze time's standard deviation of experts and nonexperts for each image. (Color figure online)

In order to further study the spatial position difference between the gaze points of experts and nonexperts, we collect all the gaze points of specific person in the same image and therefore get the fixation collecting point map, as shown in Fig. 5. Figure 5 (a) and (b) show one particular expert's fixation collecting point map and one particular nonexpert's fixation collecting point map in PMIDE, respectively. Figure 5(c) and (d) show all experts' fixation collecting point map and all nonexpert's fixation collecting point map in PMIDE, respectively. By comparing Fig. 5(a) and (b), we can conclude that while experts do have an accumulation point in the mass, their fixation points' distribution are more scattered. This suggests that experts focus on a larger scale. By comparing Fig. 5(c) and (d), we can conclude that both experts and nonexperts tend to have an accumulation point, which is at the top left corner in this experiment.

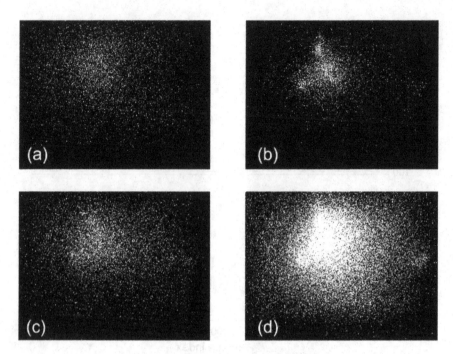

Fig. 5. Experts' and nonexperts' fixation collecting point map. (a) A particular expert's fixation collecting point map; (b) a particular nonexpert's fixation collecting point map; (c) all expert's fixation collecting point map; and (d) all nonexpert's fixation collecting point map.

3   Check the Plausibility of Predicting Fixation Points by Traditional Saliency Model

For the purpose of predicting the gaze points where people look at the pathological microscopic image, we perform the algorithmic processing on various saliency models [10–12]. A total of 10 algorithms are selected in this study, they are, Tobrralba [13, 14], IT [15], AIM [16], SeR [17], CA [18], SUN [19], GBVS [20], HouNIPS [21], AWS [22], SR [23]. Taking the views of experts and nonexperts collected by eye tracking as ground truth, the saliency maps estimated by the aforementioned salient algorithms are scored by four evaluation criteria viz. AUC, CC, NSS, and SAUC [24–26]. In general, when the absolute value of criteria CC is close to 1, the similarity is high and it is low when criteria CC is close to 0. When criteria NSS>0 or criteria AUC>0.5, the measure of similarity is significantly better than chance, and the higher the value is the more similar the models are.

Figure 6 shows the shaded result of some well-performed models.

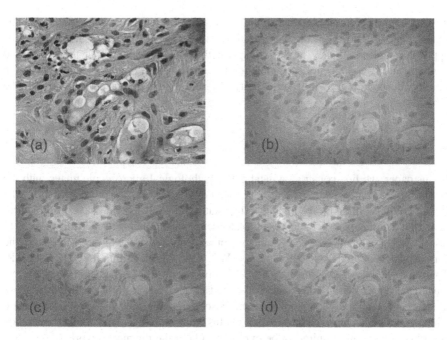

Fig. 6. The results of different saliency models. (a) Sample image; (b) GBVS model image; (c) IT model image;

Figure 6(a) depicts the sample image. The subfigure Fig. 6(b), (c) and (d) depict the predicted heat maps of GBVS model, IT model and HouNIPS model, respectively.

As shown in Table 1, we compare the scores of the top 4 best performing models. It can be seen from Table 1 that the GBVS and HouNIPS models give the best performance among these 4 algorithms.

Table 1. Four evaluation criteria of the top 4 performing models.

	GBVS	HouNIPS	IT	SR
Exp-AUC	**0.7293**	0.7042	0.6355	0.6043
Exp-CC	**0.4620**	0.3932	0.2716	0.1920
Exp-NSS	**0.7798**	0.6833	0.4633	0.3455
Exp-SAUC	0.5120	**0.5170**	0.5143	0.5160
None-AUC	**0.7256**	0.7041	0.6377	0.6018
None-CC	**0.3543**	0.3025	0.2109	0.1395
None-NSS	**0.7146**	0.6212	0.4271	0.2870
None-SAUC	0.5063	**0.5121**	0.5091	0.5106

In addition to the SAUC evaluation metric, HouNIPS yields the relatively good performance. The GBVS model outperforms the other models when considering the

AUC, CC and NSS. Notice that all CC values are less than 0.5. This means that the saliency model and the ground truth are dissimilar. Therefore, it is unreasonable to apply the current saliency models for predicting the gaze point that the expert and nonexpert look, and the more accurate, robust and specialized models are required to be developed according to the particularity of pathological microscopic image in the further study.

4 Conclusion

In this paper, we first construct a database of pathological microscopic image with the eye movement data of experts and nonexperts (PMIDE), which can be applied for the further study on this kind of images. After statistical analysis, we get several features including that the experts focus on a broader scope than nonexperts. This inspires us to respectively design the saliency models for experts and nonexperts. Moreover, the current saliency models namely AIM, AWS, CA, GBVS, HouNIPS, IT, SeR, SR, Tobrralba, SUN are examined on PMIDE, and the results demonstrate that the eye movement data of both expert and non-expert are not successfully predicted by these models. In future study, a pathological microscopic image specialized saliency model, which is trained by the characteristics of this kind of images and its fixation mode described above, are required to be explored to help the diagnostic work.

Acknowledgement. This work is sponsored by the Shanghai Sailing Program (No. 19YF1414100), the National Natural Science Foundation of China (No. 61831015, No. 61901172), the STCSM (No. 18DZ2270700), and the China Postdoctoral Science Foundation funded project (No. 2016M600315).

References

1. Glaser, A.K., et al.: Light-sheet microscopy for slide-free non-destructive pathology of large clinical specimens. Nat. Biomed. Eng. **1**(7), 0084 (2017)
2. Mohapatra, S., et al.: Blood microscopic image segmentation using rough sets. In: 2011 International Conference on Image Information Processing. IEEE (2011)
3. Itti, L., et al.: Computational modelling of visual attention. Nat. Rev. Neurosci. **2**(3), 194 (2001)
4. Cornish, L., et al.: Eye-tracking reveals how observation chart design features affect the detection of patient deterioration: an experimental study. Appl. Ergon. **75**, 230–242 (2019)
5. Lévêque, L., et al.: State of the art: eye-tracking studies in medical imaging. IEEE Access **6**, 37023–37034 (2018)
6. Duan, H., et al.: Learning to predict where the children with ASD look. In: 2018 25th IEEE International Conference on Image Processing (ICIP). IEEE, pp. 704–708 (2018)
7. Li, R., et al.: Modeling eye movement patterns to characterize perceptual skill in image-based diagnostic reasoning processes. Comput. Vis. Image Underst. **151**, 138–152 (2016)
8. Van der Gijp, A., et al.: How visual search relates to visual diagnostic performance: a narrative systematic review of eye-tracking research in radiology. Adv. Health Sci. Educ. **22** (3), 765–787 (2017)

9. Liu, H., et al.: Visual attention in objective image quality assessment: based on eye-tracking data. IEEE Trans. Circuits Syst. Video Technol. **21**(7), 971–982 (2011)
10. Min, X., et al.: Fixation prediction through multimodal analysis. ACM Trans. Multimed. Comput. Commun. Appl. (TOMM) **13**(1), 6 (2017)
11. Min, X., et al.: Visual attention analysis and prediction on human faces. Inf. Sci. **420**, 417–430 (2017)
12. Gu, K., et al.: Visual saliency detection with free energy theory. IEEE Signal Process. Lett. **22**(10), 1552–1555 (2015)
13. Bylinskii, Z., et al.: What do different evaluation metrics tell us about saliency models? IEEE Trans. Pattern Anal. Mach. Intell. **41**(3), 740–757 (2019)
14. Bylinskii, Z., et al.: Mit saliency benchmark, vol. 12, p. 13 (2014/2015). http://saliency.mit. edu/resultsmit300.html
15. Walther, D., et al.: Modeling attention to salient proto-objects. Neural Netw. **19**, 1395–1407 (2006)
16. Bruce, N.D.B., et al.: Saliency based on information maximization. In: Proceedings of the Advances in Neural Information Processing Systems (NIPS), pp. 155–162 (2005)
17. Seo, H.J., et al.: Static and space-time visual saliency detection by self-resemblance. J. Vis. **9** (12), 15, 1–27 (2009)
18. Goferman, S., et al.: Context-aware saliency detection. IEEE Trans. Pattern Anal. Mach. Intell. **34**(10), 1915–1926 (2012)
19. Zhang, L., et al.: SUN: a Bayesian framework for saliency using natural statistics. J. Vis. **8** (7), 1–20 (2008)
20. Harel, J., et al.: Graph-based visual saliency. In: Advances in Neural Information Processing Systems (2007)
21. Hou, X., et al.: Dynamic attention: searching for coding length increments. In: Proceedings of the Advances in Neural Information Processing Systems (NIPS), pp. 681–688 (2008)
22. Garcia-Diaz, A., Fdez-Vidal, X.R., Pardo, X.M., Dosil, R.: Decorrelation and distinctiveness provide with human-like saliency. In: Blanc-Talon, J., Philips, W., Popescu, D., Scheunders, P. (eds.) ACIVS 2009. LNCS, vol. 5807, pp. 343–354. Springer, Heidelberg (2009). https://doi.org/10.1007/978-3-642-04697-1_32
23. Hou, et al.:Saliency detection: a spectral residual approach. In 2007 IEEE Conference on Computer Vision and Pattern Recognition. Ieee (2007)
24. Judd, T., et al.: Learning to predict where humans look. In: 2009 IEEE 12th International Conference on Computer Vision. IEEE, pp. 2106–2113 (2009)
25. Harel, J., et al.: Graph-based visual saliency. In: Advances in Neural Information Processing Systems, pp. 545–552 (2007)
26. Goferman, S., et al.: Context-aware saliency detection. IEEE Trans. Pattern Anal. Mach. Intell. **34**(10), 1915–1926 (2011)

Few-Shot Learning for Crossing-Sentence Relation Classification

Wen Wen, Yongbin Liu$^{(\boxtimes)}$, and Chunping Ouyang

School of Computer, University of South China,
28 West Changsheng Road, Hengyang 421001, Hunan, China
yongbinliu03@gmail.com

Abstract. There is heavy dependence on the large amount of annotated data in most existing methods of relation classification, which is a serious problem. Besides, we cannot learn by leveraging past learned knowledge in most situation, which means it can only train from scratch to learn new tasks. Motivated from humans' ability of learning effectively from few samples and learning quickly by utilizing learned knowledge, we use both meta network based on co-reference resolution and prototypical network based on co-reference resolution to resolve the problem of few-shot relation classification for crossing-sentence task. Both of the two network aim to learn a transferrable deep distance metric to recognize new relation categories given very few labelled samples. Instead of single sentence, paragraphs containing multi-sentence is a major concern in the experiment. The results demonstrate that our approach performs well and achieves high precision.

Keywords: Few-shot · Crossing-sentence · Relation classification

1 Introduction

Relation classification is an important task in information extraction, which aims to recognize the correct relation between two entities in a given paragraph. Supervised models suffer from the limitation of the amount of annotated data and have to train new tasks from scratch at most situations. Distant supervision [1,20] becomes a relatively common method in this task, however, the lack of diversity and inevitable noise is a serious problem.

As we all know, humans are good at recognizing new objects from prior knowledge and able to achieve high accuracy through generalization and inference. Motivated from the few-shot learning and quickly learning abilities of humans, how to make few-shot learning come true has become a hot topic to deal with the problem of few-shot relation classification.

Few-shot learning (FSL) aims to label new relations from small group samples per class by incorporating prior knowledge. The kernel of FSL is to learn transferrable distance metric, which makes it easy to generalize new classes.

In recent year, many constructive methods have been devoted to FSL. In 2015, Koch et al. approach the problem of few-shot learning by utilizing deep

© Springer Nature Singapore Pte Ltd. 2020
G. Zhai et al. (Eds.): IFTC 2019, CCIS 1181, pp. 150–161, 2020.
https://doi.org/10.1007/978-981-15-3341-9_13

siamese network [4], which trains on multiple convolutional layers to classify by ranking similarity between inputs. Vinyals et al. has some work on matching network [6], which requires the match of test and train conditions and predicts classes by calculating the cosine similarities. Prototypical network, proposed [7] by Snell et al., assumes that a prototype must exist for each class, and the model classifier works by comparing the distance between query and every prototype. Then, in 2017, Sachin et al. put forward meta network [9,23], which learns how to extract transferrable distance metric and then acquires knowledge quickly from few examples by mimicking humans. However, most of these works pay attention to image domain, seldom applying it on NLP tasks.

Besides, there are lots of crossing-sentence relations in our daily life. Crossing-sentence relations could only be extracted by considering over multi-sentences. Just as Table 1 illustrates, for the relation of institution, "Julia Bell" is the head entity, which is mentioned in sentence 1 at first time, while "Girton College" is the tail entity mentioned in sentence 2.

Table 1. Examples from dataset without co-reference resolution. Different color represents different entities, Red for head entity, Pink for tail entity.

Sentence	Relation
Julia Bell (January 28, 1879 – April 26, 1979) was a pioneering English human geneticist. She attended Girton College in Cambridge and took the Mathematical Tripos exam in 1901. But because women could not officially receive degrees from Oxford or Cambridge, she was awarded a master's degree at Trinity College, Dublin for her work investigating solar parallax at Cambridge Observatory. In 1908, She moved to University College London and obtained a position there as an assistant in statistics	Institution

To address these issues, we propose a novel method which utilizes meta network [2] and prototypical network [3] for few-shot learning based on crossing-sentence relation classification.

Similar to other crossing-sentence tasks, our approach is supposed to resolve co-referential problem, first at all. Co-reference resolution plays an important role in process of NLP tasks. Co-reference aims to find as full as possible expressions, which refer to the same entity in a paragraph, and cluster them to the real entity based on syntactic analysis. Stanford CoreNLP [4,5] we used is a famous NLP tool and able to handle co-reference resolution. Our experimental results on crossing-sentence show that our method achieves significant and consistent improvement as compared to baselines.

2 Related Work

Relation classification has been given widespread attention abroad all the time. Supervised models heavy depends on large-scared and high-quality annotated data, which not readily available. Then, few-shot learning is proposed to alleviate the burden of annotated data by mimicking human's talent. It retains prior knowledge, therefore, it can learn quickly to solve the new tasks [22].

In recent years, many approaches have been come up with to solve the problem of few-shot, and make few-shot learning come true. Distant supervision mechanism [1,19] is proposed by Mintz et al. for extracting relations between entities through aligning existing knowledge graphs (KGs) with unstructured data like text. However, Distant supervision suffers from lots of wrong labeling caused by its hypothesis, which says that two entity having a relationship in a known KGs will be labelled with the same relation as any sentence containing the pair of entities in corpus.

Contemporary method for Few-shot learning based on metric-learning or based on meta-learning. For metric learning, matching network and prototypical network are two typical representations. Matching network [6] means that test and train conditions are required matching, and it can be regard as a weighted nearest-neighbor classifier measured by the cosine similarities of embeddings. Its classifier is able to predict classes by using an attention mechanism. Prototypical network [7] assumes that a prototype must exist for each class, and the model classifier works by comparing the distance between query and every prototype. Meta network [8,18,23] is a typical meta-learning model, which plays more and more significant role nowadays as a key to achieve human-level intelligence in the future. Meta network utilizes meta learning algorithm based on learning-to-learn strategy, which is a significant presence transferrable distance distributions to achieve better performance on classifying the relation of testing set. Meta-learning can split into two levels: the first is to quickly acquire knowledge, which is guided by the second level which aims to extract information. Then, in 2017, Finn et al. focus on model-agnostic meta-learning (MAML) [18], which is an algorithm for meta-learning. MAML receives extensive concern after publishing, due to the characteristic of its applicable to many different learning problems [11, 21]. Besides, Han et al. have some recent work which provides a Large-Scale Supervised Few-Shot Relation Classification Dataset, named FewRel [2]. FewRel consists of 100 relations and has 700 instance for each class, which is derived from Wikipedia and annotated by crowdworkers and then adapts various few-shot learning methods for relation classification which achieves state-of-art result. FewRel boosts the velocity of research of few-shot relation classification, however, it is sentence-level, so it is not available on the task of extracting crossing-sentence relation, while a large amount of crossing-sentence relations exists in real life. Due to the lack of research for crossing-sentence relation classification, we focus on utilizing meta network [2] and prototypical network [3] to the task of crossing-sentence relation classification based on few-shot learning [17].

3 Methodology

3.1 Task Definition

Approaches for few-shot learning aim to train a classifier given only a small amount of samples for each of classes. The few-shot relation classifier is supposed to predict the unknown relation class label y between two entity mentioned, named head entities $e1$ and tail entities $e2$ respectively, in a query instance x [2,8,18]. However, most researches for few-shot problem pay attention to the sentence-level relation classification, ignoring the ubiquity of crossing-sentence relation in our daily life. Thus, our method tends to solve the problem of crossing-sentence relation classification for few-shot learning, utilizing meta network [2] and prototypical network [3]. Besides, to improve the performance of crossing-sentence relation extraction, we apply co-reference resolution on our dataset, as Table 2 shows.

Table 2. Examples from dataset being co-reference resolution. Different color represents different entities, Red for head entity, Pink for tail entity.

Sentence	Relation
Julia Bell -LRB- January 28, 1879 – April 26, 1979 -RRB- was a pioneering English human geneticist. Julia Bell attended Girton College in Cambridge and took the Mathematical Tripos exam in 1901. But because women could not officially receive degrees from Oxford or Cambridge, Julia Bell was awarded a master's degree at Trinity College, Dublin for Julia Bell work investigating solar parallax at Cambridge Observatory. In 1908, Julia Bell moved to University College London and obtained a position there as an assistant in statistics	Institution

3.2 Model

As mentioned in Sect. 1, we use meta network [2] and prototypical network [3] which both are few-shot learning networks, to solve the problem of crossing-sentences relation classification. The network usually consists two main modules: embedding module f and relation module g. Embedding module f solves the problem of how to represent the query and supporting instances, while relation module g determines whether these embeddings are from matching categories or not [10,12].

As Fig. 1 shows, in each episode, we first sample C classes from the list of classes in training set and randomly sample K labelled instances from each of those classes to yield support set S, and then, the query set Q is produced by sampling an additional fixed amount from the rest of the examples. Then, support set S and query set Q are fed to the embedding module f, where each

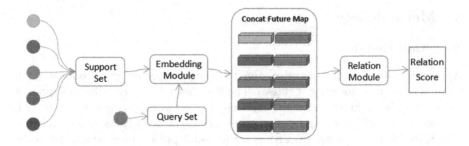

Fig. 1. Model architecture for 5-way 1-shot with one query example [8].

sample x_i in support set S and each sample $f(x_j)$ in query set Q produce feature maps $f(x_i)$ and $f(x_j)$ respectively. Then combined each feature maps as $c(f(x_i), f(x_j))$. Finally, fed the relation module g with the combined feature maps $c(f(x_i), f(x_j))$ to get C relation scores $r_{i,j}$, each of which represents the similarity between support sample x_i and query input x_j. Then acquire the predict relation label y by getting the highest relation scores [2,8].

$$r_{i,j} = g(c(f(x_i), f(x_j))) \tag{1}$$

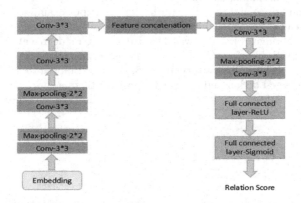

Fig. 2. Meta network architecture [8].

Meta Network. In meta network, meta-learning performs on training set to extract transferrable knowledge through learn to learn distance metric, thus, meta network can retain learned knowledge from past. It extracts the key features to recognize classes based on the hypothesis, which believes similar classes are adjacent in the distance space. Meanwhile, meta network learns to learn a transferrable distance metric, instead of pre-defined, to make rapidly learning come true. Besides, as Fig. 2 shows, the network architecture mainly contains two modules: embedding module and relation module. Embedding module consists of four convolutional blocks, each of which is a 64-filter 3 * 3 convolution, then concatenate the output feature maps from support set and query set. Finally, the concatenation feeds to the relation module containing two 3 * 3 convolutional blocks and two full connected layers, which aims to get relation scores [8, 18, 20].

Prototypical Network. Prototypical Network usually use Euclidean distance as the standard of distance metric, because its outperforms others like cosine similarity in matching network. Based on the assumption that every classes must have a prototype and the distance between query sample and relation is equal to the distance between sample and the prototype, the network classifies relations by calculating the distances to every class prototypes produced by averaging over all the output embeddings. All instances will map to an embedding space, and items in one class are adjacent to the class's prototype, while items in different classes are far away from other class's prototype points. Then, classifier can simply find the relation label for an embedded query point. Prototypical network architecture utilizes attention and memory to enable rapid learning.

3.3   Co-reference Resolution

Co-reference resolution is the process of clustering mentions in a paragraph and referring to the same entities in the real-world by adding a co-reference link between two entity-mention in a same paragraph. As Table 3 shows that, "Budrys" and He refer to the same person, meanwhile, the pen name and his Lithuanian name having the same meaning refer to the same entity, too [15].

Table 3. Co-reference resolution. Different color represents different entity corefChains.

Co-reference resolution	
Orial paragraph	Budrys was educated at the University of Miami, and later at Columbia University in New York. Beginning in 1952 Budrys worked as editor and manager for such sci ence fiction publishers as Gnome Press and Galaxy Sci ence Fiction. Some of his science fiction in the 1950s was published under the pen name "John A. Sentry", a re configured Anglification of his Lithuanian name. He also wrote several stories under the names "Ivan Janvier" or "Paul Janvier". He also used the pen name "Alger Rome" in his collaborations with Jerome Bixby.

We would like to apply co-reference Resolution to our method, which is an essential step of crossing-sentence tasks. After comparing the advantages of current popular tools, we utilize Stanford Coreference Resolution by sending requests to Stanford CoreNLP server [4, 5], which generates both pronominal and nominal co-reference resolution. It not only helps us find co-referential entity, but also divides one paragraph into tokens. Result of the CorefAnnotator is pretty enough to satisfy our experiment, and can be saved as various file types such as json, which is easy to save and access data [12,16].

4   Experiment

4.1   Datasets

The dataset we utilized, called FewSP, is collected from online open resource and then performs data cleaning. A part of FewSP comes from Google Code Relation-extraction-corpus[1] having over 10,000 sentences with five kinds of relationships: institution, place of birth, place of death, date of birth and graduate degree, and the remainder part is extracted from TAC containing around 1000 sentences with five kinds of relationships: personal social, general affiliation, physical, origin affiliation, part whole. The two parts both from Wikipedia. Thus, just as Table 4 shows that, our dataset consists of 10 relations, and each relationship has over 1,500 instances on average, which including over 800 crossing-sentence instances. Then, we divide our dataset into three parts: training set, validation set, testing set, besides, training set has its own label space which is disjoint with the others.

For validation and testing, we produce enough datasets to evaluate each method. Accuracy is a standard measure to evaluate the performance of both meta network and prototypical network.

Table 4. Dataset.

Dataset		Relation	Instance	Crossing instances
FewSP	Google Code Relation-extraction-corpus	Institution	5,626	3,491
		Place of birth	2,999	849
		Place of death	864	425
		Date of birth	317	33
		Graduate degree	322	145
	TAC	Personal social	1,243	800
		General affiliation	609	434
		Physical	1,926	1,212
		Origin affiliation	1,209	793
		Part whole	650	450

---

[1] https://code.google.com/archive/p/relation-extraction-corpus/downloads.

4.2   Experimental Settings

While the validation set is able to get C unique classes and K labelled examples per class, then, we called it C-way K-shot. In our experiments, we only consider 5-way 1-shot classification task for both meta network and prototypical network. Besides, both of the two networks sample mini-batch for each episode to simulate few-shot during training.

During one training episode, we randomly choosing C classes and then randomly sample $K + Q$ examples from each of the C classes. The K examples serve as support set S, while the other Q examples serve as query set X. Support set S and query set X both are subsets of examples, but they are mutually exclusive. The development set and testing set share the same way to split sample/query set as training set does [2,8].

To tune all hyperparameters, we randomly choose different values on one hyperparameter while others remain unchanged, and then observe the result to find which ones affected the results most and choose initial hyperparameter values. In our experiment, we find that learning rate is a key metric to improve our accuracies on both meta network and prototypical network. For example, we sample various values of learning rate and fed into the prototypical network, and as Fig. 3 demonstrates that when learning Rate equals to 5e-2, the model achieves better performance.

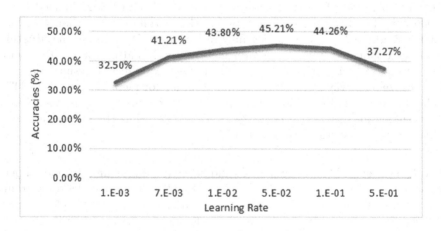

Fig. 3. How Learning Rate influence the accuracies on prototypical network.

All the hyperparameters are shown in Table 5. For both of meta network and prototypical network, we use step policy to decay the learning rate, that's to say, the learning rate will be changed every learning-rate step size as the degree of training improves. Due to the different convergence speed, we adopt different learning rate to different models. For meta network, we train 30000 iterations with learning-rate step size 20000, while training prototypical network needs 30000 iterations with learning-rate step size 10000. Both of the two models are trained on training set, then pick the best epochs on the validation set and test on testing set [16].

Table 5. Parameter setting.

Hyperparameter	Initial value (meta network)	Initial value (prototypical network)
Batch size	1	4
Query size	5	1
Learning rate	1e−3	5e−2
Learning-rate step size	20000	10000
Optimizer	ADAM	SGD
Training iterations	30000	30000

4.3   Experimental Result and Discussion

Our experiments are implemented under the parameter setting shown above. Both of the two networks uses Adam optimizer with 30000 Training iterations. For meta network, the initial learning rate is 10^{-3}, and being changed every 20000 episodes. For prototypical network, the learning rate is changed every 10000 episodes, while the initial learning rate $5 * 10^{-2}$. We consider the situation of 5-way 1-shot and 5-way 5-shot on meta network and prototypical network for meta network and prototypical network. Our experimental results shows in Table 6: Under 5-way 1-shot setting, the accuracy for meta network is 39.34%, while the number is 45.21% for prototypical network. Under 5-way 5-shot setting, the accuracy for meta network increases to 39.59%, while the number achieves 50.88% for prototypical network [10,14].

The Influence of Dataset Size. To find how dataset size influent model's performance, we compare the result of 10%, 50%, 100% datasets under our parameter setting shown above. In 5-Way 1-Shot task, the size of dataset increases doubled, the accuracy improves over 0.3% for meta network while the data is around 1.3% for prototypical network, and the size of dataset increases tenfold, the accuracy improves over 2.5% while the data achieves 5% (Table 7).

Table 6. Results under 5-way 1-shot and 5-way 5-shot. CR: having resolved co-reference.

Model	5-Way 1-Shot	5-Way 5-Shot
Meta network (CR)	39.34%	39.59%
Prototypical network (CR)	45.21%	50.88%

Table 7. The influence of dataset size under 5-Way 1-Shot setting for meta network and prototypical network.

Dataset size	Meta network accuracy (%)	Prototypical network accuracy (%)
10%	36.67%	39.93%
50%	39.03%	43.94%
100%	39.34%	45.21%

The Effective of Considering Crossing-Sentence Relation Classification and Co-reference Resolution for the Task. To prove training on document-level dataset and resolving co-reference can improve the performance of networks, we make a comparison to the results of training with sentence-level dataset (FewRel) [2] and none co-reference resolution document-level dataset on prototypical network. As Table 8 reports that, under 5-Way 1-Shot setting, the accuracy of prototypical network increases over 6.5% compared to the results of training on FewRel dataset, meanwhile, the accuracy of prototypical network having been resolved co-reference increases 14% compared to the result of none co-reference resolution [13].

Table 8. Accuracies(%) of models under 5-Way 1-Shot setting. OT: means training on FewRel(sentence-level) training set, while valuation set and testing set are document-level; NCR: means not to resolve co-reference; CR: means having resolved co-reference.

Model	5-Way 1-Shot
Prototypical network (OT)	38.68
Prototypical network (NCR)	31.00
Prototypical network (CR)	**45.21**

5   Conclusion

We identify the difficulties in few-shot problem caused by the limitation of annotated datasets and the importance of crossing-sentence relation extraction, and we present to utilize existing meta network and prototypical network to solve the few-shot problem of crossing-sentence relation classification. Our experiments prove that, for the tasks of crossing-sentence relation extraction, adding co-reference resolution on paragraph is a significant step, because it significantly outperforms none co-reference resolution paragraph on both meta network and prototypical network. We believe that this work is one key step toward a general-purpose few-shot learning. Besides, we propose a dataset, FewSP, collected from online resources for cross-sentence relation classification task. Through the implement of various few-shot learning methods, we find there are still huge gaps to human's ability. Future researches in this area may pay attention on the few shot learning of Chinese text relation classification.

Acknowledgments. This research work is supported by National Natural Science Foundation of China (No. 61402220, No. 61502221), the Philosophy and Social Science Foundation of Hunan Province (No. 16YBA323), Scientific Research Fund of Hunan Provincial Education Department for excellent talents (No.18B279).

References

1. Mintz, M., Bills, S., Snow, R., Jurafsky, D.: Distant supervision for relation extraction without labeled data. In: Proceedings of the Joint Conference of the 47th Annual Meeting of the ACL and the 4th International Joint Conference on Natural Language Processing of the AFNLP, vol. 2, pp. 1003–1011. Association for Computational Linguistics, Suntec, Singapore (2009)
2. Han, X., et al.: FewRel: a large-scale supervised few-shot relation classification dataset with state-of-the-art evaluation. In: Proceedings of the 2018 Conference on Empirical Methods in Natural Language Processing, pp. 4803–4809. Association for Computational Linguistics, Brussels, Belgium (2018)
3. Gao, T., Han, X., Liu, Z., Sun, M.: Hybrid attention-based prototypical networks for noisy few-shot relation classification. In: The Thirty-Third AAAI Conference on Artificial Intelligence. Association for the Advancement of Artificial Intelligence, Hawaii, USA (2019)
4. Koch, G., Zemel, R., Salakhutdinov, R.: Siamese neural networks for one-shot image recognition. In: Proceedings of the 32nd International Conference on Machine Learning, Lille, France, vol. 37. JMLR, July 2015
5. Clark, K., Manning, C.D.: Improving coreference resolution by learning entity-level distributed representations. In: Proceedings of the 54th Annual Meeting of the Association for Computational Linguistics, pp. 643–653, Berlin, Germany. Association for Computational Linguistics (2016)
6. Vinyals, O., Blundell, C., Lillicrap, T., Wierstra, D.: Matching networks for one shot learning. In: Neural Information Processing Systems, pp. 3630–3638. Curran Associates Inc., Barcelona, Spain (2016)
7. Snell, J., Swersky, K., Zemel, R.: Prototypical networks for few-shot learning. In: Neural Information Processing Systems, pp. 4077–4087. Curran Associates Inc., Long Beach, USA (2017)
8. Sung, F., Yang, Y., Zhang, L., Xiang, T., Torr, P.H., Hospedales, T.M.: Learning to compare: relation network for few-shot learning. In: Proceedings of the IEEE Conference on Computer Vision and Pattern Recognition, UT, USA, pp. 1199–1208. IEEE (2018)
9. Ravi, S., Larochelle, H.: Optimization as a model for few-shot learning. In: International Conference on Learning Representations. Toulon, France (2017)
10. Luo, H., Glass, J.: Learning word representations with cross-sentence dependency for end-to-end co-reference resolution. In: Proceedings of the 2018 Conference on Empirical Methods in Natural Language Processing, Brussels, Belgium, pp. 4829–4833. Association for Computational Linguistics (2018)
11. Cheng, J., Dong, L., Lapata, M.: Long short-term memory-networks for machine reading. In: Proceedings of the 2016 Conference on Empirical Methods in Natural Language Processing, pp. 551–561. Association for Computational Linguistics, Austin, Texas (2016)

12. Lee, K., He, L., Lewis, M., Zettlemoyer, L.: End-to-end neural coreference resolution. In: Proceedings of the 2017 Conference on Empirical Methods in Natural Language Processing, pp. 188–197. Association for Computational Linguistics, Copenhagen, Denmark (2017)
13. Durrett, G., Klein, D.: Easy victories and uphill battles in coreference resolution. In: Proceedings of the 2013 Conference on Empirical Methods in Natural Language Processing, Seattle, Washington, USA, pp. 1971–1982. Association for Computational Linguistics (2013)
14. Peters, M.E., et al.: Deep contextualized word representations. In: Proceedings of the 2018 Conference of the North American Chapter of the Association for Computational Linguistics: Human Language Technologies, New Orleans, Louisiana, vol. 1, pp. 2227–2237. Association for Computational Linguistics (2018)
15. Wiseman, S., Rush, A.M., Shieber, S.M.: Learning global features for coreference resolution. In: Proceedings of NAACL-HLT 2016, pp. 994–1004, San Diego, California. Association for Computational Linguistics (2016)
16. Bengio, Y.: Deep learning of representations for unsupervised and transfer learning. In: Proceedings of ICML Workshop on Unsupervised and Transfer Learning, Washington, USA, pp. 17–36. JMLR.org (2011)
17. Feng, J., Huang, M., Zhao, L., Yang, Y., Zhu, X.: Reinforcement learning for relation classification from noisy data. In: Thirty-Second AAAI Conference on Artificial Intelligence. Association for the Advancement of Artificial Intelligence, New Orleans, Louisiana, USA, April 2018
18. Finn, C., Abbeel, P., Levine, S.: Model-agnostic meta-learning for fast adaptation of deep networks. In: Proceedings of the 34th International Conference on Machine Learning, Sydney, Australia, vol. 70, pp. 1126–1135. JMLR.org, August 2017
19. Ji, G., Liu, K., He, S., Zhao, J.: Distant supervision for relation extraction with sentence-level attention and entity descriptions. In: AAAI 2017 Proceedings of the Thirty-First AAAI Conference on Artificial Intelligence, San Francisco, California, USA, pp. 3060–3066. Association for the Advancement of Artificial Intelligence, February 2017
20. Luo, B., et al.: Learning with noise: Enhance distantly supervised relation extraction with dynamic transition matrix. In: Proceedings of the 55th Annual Meeting of the Association for Computational Linguistics, Vancouver, Canada, pp. 430–439. Association for Computational Linguistics (2017)
21. Andrychowicz, M., et al.: Learning to learn by gradient descent by gradient descent. In: Neural Information Processing Systems, Barcelona, Spain, pp. 3981–3989. Curran Associates Inc. (2016)
22. Husken, M., Goerick, C.: Fast learning for problem classes using knowledge based network initialization. In: Proceedings of the IEEE-INNS-ENNS International Joint Conference on Neural Networks, IJCNN 2000. Neural Computing: New Challenges and Perspectives for the New Millennium, Como, Italy, vol. 6, pp. 619–624. IEEE (2000)
23. Munkhdalai, T., Yu, H.: Meta networks. In: Proceedings of the 34th International Conference on Machine Learning, vol. 70, pp. 2554–2563, Sydney, Australia. JMLR.org, August 2017

Image Classification of Submarine Volcanic Smog Map Based on Convolution Neural Network

Xiaoting Liu[ID], Li Liu[✉], and Yuhui Chen

University of South China, Hengyang 421001, China
1713698@qq.com

Abstract. In order to meet the problem of smoke image classification in submarine volcanic scene, in this paper, depth convolution neural network (Deep Convolutional Neural Networks, DCNN) is used to classify smoke seafloor map and smoke-free seafloor map under small-scale data set and limited computing power. Firstly, the data enhancement technology is used to expand the data set through angle rotation, horizontal flipping, random cutting and adding Gaussian noise, and then the depth convolution neural network is built for training. Finally, the recognition and classification is carried out according to the prediction image label of the classifier. The experimental results show that the classification accuracy of the proposed method is more than 91%.

Keywords: Submarine volcanic smoke · Image classification · Convolutional neural network

1 Preface

More than 70% of all volcanic activity on Earth actually occurs in the ocean, and undersea volcanoes are also unique ecosystem bases that tell us how seafloor geological activity works on the ocean and even on land. This is closely related to the earth's ecosystem. Detecting the existence of submarine volcanoes is of great significance for the development of science. However, the vast majority of undersea volcanoes are located in deep-sea areas, and the submarine earthquakes associated with volcanism are usually very small, and most of the instruments on land are far from the seismic area, resulting in most of the undersea volcanic activity still undetected. An important way to determine whether there are volcanoes on the seafloor is to detect smoke on the seafloor. At present, the deepest operational manned submersible in the world is the Jiaolong, with a maximum diving depth of 7000 m. But the maximum depth of the ocean is more than 10,000 m, resulting in the need for deep-sea smoke detectors to automatically detect the existence of undersea craters where the Jiaolong cannot dive.

Supported by University of South China.

The main studies on smoke recognition include: Zhou et al. [1] use the local extreme region segmentation method for smoke recognition for long-distance smoke images. However, this method will produce a false alarm of smoke in areas with thick fog. Dimitropoulos et al. [2] proposed a high-order linear dynamic system (h-LDS) description feature operator to analyze the dynamic texture of smoke and improve the recognition rate of smoke features. However, most of the smoke feature extraction operators are designed by hand, which may not be able to reflect the essential features of smoke, and the selection of threshold depends on experience, and the rationality of threshold will affect the effect of smoke recognition [3]. Chen et al. [4] fused the static and dynamic texture information of smoke, and proposed a framework of smoke texture recognition based on cascade convolution neural network, which can improve the accuracy of smoke recognition. However, the processing of static and dynamic texture information will increase the complexity of the algorithm and affect the real-time performance of smoke detection. Xu et al. [5] based on domain adaptability method, convolution neural network (CNN) model is trained by synthetic and real smoke images [6]. Although the false detection rate of smoke can be reduced, the use of synthetic smoke images will affect the smoke recognition performance of the training model in real scenes to a certain extent.

There are some problems in the above smoke recognition methods, and there is no smoke recognition for deep seabed scenes. In this paper, the deep convolution neural network training model is used to classify the seafloor images with and without smoke for the deep seafloor scene. The experimental results show that the model has obvious recognition effect for the smog seafloor map and the smoke-free seafloor map. It is suitable for smoke image classification in deep seafloor scene, which can greatly improve the efficiency and accuracy of deep-sea detector in detecting smoke, and detect the possibility of submarine volcanoes and eruptions as soon as possible.

2   Recognition of Submarine Volcanic Smog Map Based on Convolution Neural Network

One of the representative algorithms of convolution neural network depth learning is widely used in the field of image recognition.

2.1   Network Structure

The convolution neural network model used in this experiment consists of 4 convolution layers, 4 pooling layers, 2 fully connected layers and 1 output layer(see Fig. 1).

The input of this network is 208 * 208 RGB pictures processed by the mean value. The step size of the convolution layer is 1 (stride = 1) and the complement is 1 circle 0 (pad = 1). The size of convolution kernel is 3 * 3. The pooling layer is the maximum pooling (MaxPooling), pooling window is 2 * 2, and the step size is 2. The ReLU layer is added after each hidden layer, and the dropout layer is added after the full connection layer (Table 1).

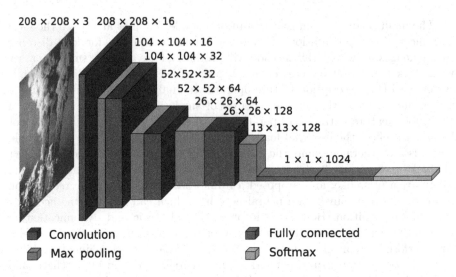

Fig. 1. Structure of CNN

2.2 Network Parameter Settings

In this experiment, the convolution neural network is designed and optimized. The network model includes convolution layer, pooling layer and full connection layer. To prevent network overfitting, dropout processing is used at the fully connected layer, and the probability is set to an empirical value of 0.5. The activation function is divided into two categories by using Softmax, output nodes. The learning rate is initially set to $1e - 4$, iterated 6000 times, and the learning speed is controlled by the optimizer AdamOptimizer. After bias correction, the parameters are relatively stable.

Table 1. Parameters of four-layer convolutional neural network model.

Number of layers	Name	Input	Nuclear	Step size	Fill	Output
0	Input	208 * 208 * 3	-	-	-	208 * 208 * 3
1	Conv1	208 * 208 * 3	3 * 3 * 16	1	1	208 * 208 * 16
2	MaxPool1	208 * 208 * 16	2 * 2	2	0	104 * 104 * 16
3	Conv2	104 * 104 * 16	3 * 3 * 32	1	1	104 * 104 * 32
4	MaxPool2	104 * 104 * 32	2 * 2	2	0	52 * 52 * 32
5	Conv3	52 * 52 * 32	3 * 3 * 64	1	1	52 * 52 * 64
6	MaxPool3	52 * 52 * 64	2 * 2	2	0	26 * 26 * 64
7	Conv4	26 * 26 * 64	3 * 3 * 128	1	1	26 * 26 * 128
8	MaxPool4	26 * 26 * 128	2 * 2	2	0	13 * 13 * 128
9	FC1	1 * 1 * 1024	-	-	-	1024
10	FC2	1024	-	-	-	1024

2.3 Activation Function Optimization

In order to improve the performance of the network model, the modified linear element (Rectified linear unit, ReLU) activation function is used (see Fig. 2).

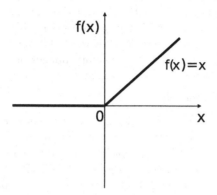

Fig. 2. Relu activation function representation

$$f(x) = x, x = 0 f(x) = x, x \leq 0 \tag{1}$$

As can be seen from Eq. (1), ReLU is a linear correction. If the calculated value is less than 0, make it equal to 0, otherwise leave the original value unchanged. A the network using ReLu can introduce sparsity by itself, which can not only accelerate the training speed of convolution neural network, but also effectively reduce the training error.

2.4 Cross Entropy Loss Function

Cross entropy is the distance between the actual output probability and the expected output probability. The smaller the value of cross entropy is, the closer the two probability distributions are, and the higher the accuracy of expected output is. After regression processing, the Softmax layer can transform the actual output of the neural network into probability distribution, and then use the cross entropy loss function to calculate the close distance coefficient between the actual output probability and the expected output probability. Suppose the original output of the neural network is a1, a2, ai, then the output after Softmax regression processing is shown in Eq. (2). Let the probability distribution p be the expected output and the probability distribution q be the actual output, then the cross entropy H (p, q) is shown in Eq. (3):

$$softmax(a_i) = \frac{e^{a_i}}{\sum_{j=1}^{n} e^{a_i}} \tag{2}$$

$$H(p, q) = -\sum_{x} p(x) log q(x) \tag{3}$$

3   Model Training and Verification

3.1   Experimental Environment

The CPU of the computer used in the experiment is Intel (R) Core (TM) i7 5500U, the main frequency is 2.40 GHz, and the memory is 8G. It is written in Python language and OpenCV visual processing library under Windows7 operating system. The TensorFlow depth learning framework is used to build the DCNN model and to train and test the smoke images of submarine volcanoes.

3.2   Image Data Processing

Data Set. The data set of this experiment is generated in batches by data enhancement technology from the images of submarine volcanic smoke collected on the Internet. The experimental pictures are divided into training set and test set, including smoke and no smoke, a total of 6696 training samples, 600 test samples, as shown in Table 2. Figure 3 is a partial sample of the experiment. Among them, Fig. 3(a) is a partial smoke sample, and Fig. 3(b) is a partial smoke-free sample.

Table 2. Category statistics of submarine volcanic smog maps

Picture type	Training sample (sheet)	Test sample (sheet)
There is smoke	3348	300
There is no smoke	3348	300

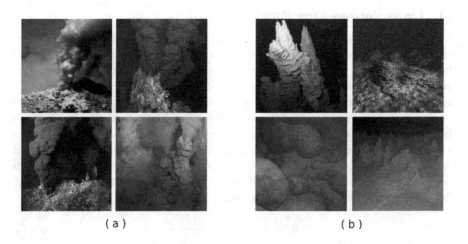

(a)   (b)

Fig. 3. Partial sample of the experiment

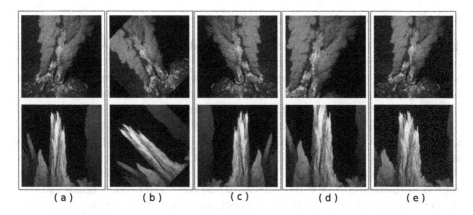

$$(a) \qquad (b) \qquad (c) \qquad (d) \qquad (e)$$

Fig. 4. Example of enhancement of submarine volcanic smoke data

Data Enhancement. Through data enhancement, the amount of training data and noise data are increased, the generalization ability of the model is improved, and the robustness of the model is improved. This experiment uses the data enhancement method shown in Fig. 3. In Fig. 4, (a) is the original image, (b) is angular rotation, (c) is horizontal flipping, (d) is random clipping, and (e) is adding Gaussian noise.

Using the above four data set enhancement methods, the training data is generated in batches from the original limited data set through the Python program, in order to achieve the amount of data required for the target number of iterative rounds. The enhanced data can effectively reduce the network overfitting and increase the ability of convolution neural network to recognize deep seafloor smoke images.

Data Preprocessing. The image needs to be preprocessed after data enhancement. Due to the unequal resolution of seafloor images, in order to complete the unification of network input, all images are zoomed to 208 * 208 pixels. Due to inconsistent image channels, some images need to be converted to channel 3. Because the image pixel is from 0 to 225, the input computation is more complex, in order to simplify the input of the network, the picture pixel is normalized, the range is [0, 1].

Table 3. Comparison of accuracy of convolution neural network model

Model	Total number	Number of correct identification	Number of misidentifications	Accuracy
Neural network with three layers of convolution	600	514	86	85.67%
Neural network with four layers of convolution	600	551	49	91.83%

3.3   Classification Results and Analysis

The training of convolution neural network model of submarine volcanic smoke consists of two processes, network training and network fine-tuning. The data sets after data enhancement are trained by convolution neural network. The training process consists of two steps: network training and network fine-tuning.

In the process of network training, the learning rate of convolution neural network is initially set to 1e/4, iterated 6000 times, and AdamOptimizer is used to control the learning speed. In the comparative experiment, the test set is used to test the correct rate of the model. When the network is fine-tuned, the dropout, is added to the full connection layer to improve the generalization ability, the recognition rate of the network and the speed of convergence, so that the performance of the convolution network is optimized. In the experiment, the changing trend of the parameters of the structural training process of the two different models is shown in Figs. 5 and 6.

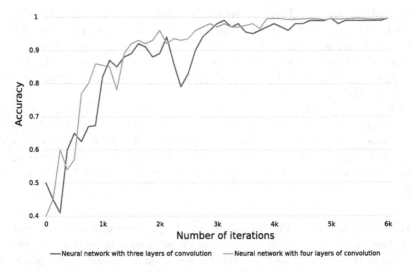

Fig. 5. Variation trend of model cross entropy loss in iterative process

It can be seen from the diagram that the convolution neural network with four convolution layers is better than the three convolution layers, its convergence speed is faster, and the recognition accuracy of the trained model is higher. The accuracy of different network models is compared in Table 3, in which the recognition results of four-layer convolution neural network for smoke and smoke-free are shown in Fig. 7.

The experimental results show that: Finally, four convolution layers, four maximum pooling layers, two fully connected layers with dropout and one Soft-Max classifier layer are more suitable for seafloor smoke recognition. Compared with the convolution neural networks of three convolution layers, the convolution

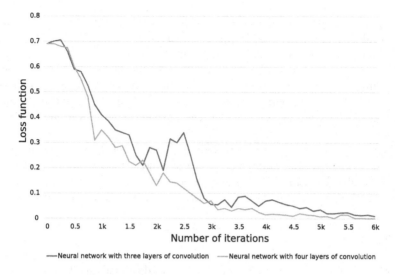

Fig. 6. Trend of model accuracy during iteration

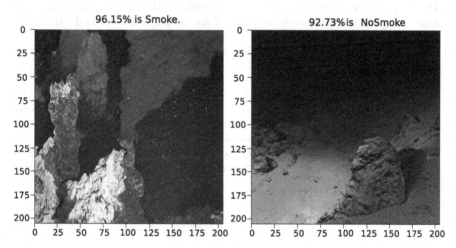

Fig. 7. Recognition result of four-layer CNN

neural networks of four convolution layers, four maximum pooling layers, two fully connected layers and one convolution neural network are more suitable for seafloor smoke recognition. 6.1% increase in accuracy.

4 Conclusion

In order to meet the needs of smoke recognition in undersea volcanic scenes, the previous smoke recognition algorithms are time-consuming and labor-consuming in feature extraction, and have not been applied to deep seafloor scenes. In this paper, a classification model of submarine volcanic smoke based on convolution neural network is proposed, which can effectively automatically have smoke seafloor image and smoke-free seafloor image.

In this paper, in view of the small number of deep seabed images, the method of data enhancement is used to expand the number of images. The neural network model can initialize the parameters randomly and then fine-tune the parameters according to the training data, which can save more time and get higher test accuracy.

In the later work, we can increase the diversity of samples (such as black smoke, gray smoke, etc., submarine volcanic smoke map combined with many types of smoke, bottom map in complex scenes with seafloor organisms, etc.). The accuracy of submarine volcanic smoke detection is further improved.

References

1. Zhou, Z., Shi, Y., Gao, Z., et al.: Wildfire smoke detection based on local extremal region segmentation and surveillance. Journal **85**, 50–58 (2016)
2. Dimitropoulos, K., Barmpoutis, P., Grammalidis, N.: Higher order linear dynamical systems for smoke detection in video surveillance applications. Journal **27**(5), 0303–0322 (2018)
3. Shi, J.T., Yuan, F.N., Xia, X.: Higher order linear dynamical systems for smoke detection in video surveillance applications. Journal **23**(3), 1143–1154 (2016). (in Chinese)
4. Chen, J.Z., Wang, Z.J., Chen, H.H., et al.: Video dynamic smoke detection based on cascade convolution neural network. Journal **46**(6), 992–996 (2016). (in Chinese)
5. Xu, G., Zhang, Y.M., Zhang, Q.X., et al.: Deep domain adaptation based video smoke detection using synthetic smoke images. Journal **93**, 53–59 (2017)
6. Zhou, F.Y., Jin, L.P., Dong, J.: Deep domain adaptation based video smoke detection using synthetic smoke images. Journal **40**(06), 1229–1251 (2017). (in Chinese)

Multi-Scale Depthwise Separable Convolutional Neural Network for Hyperspectral Image Classification

Jiliang Yan[1], Deming Zhai[1], Yi Niu[2], Xianming Liu[1], and Junjun Jiang[1(✉)]

[1] School of Computer Science and Technology, Harbin Institute of Technology, Harbin 150001, China
jiangjunjun@hit.edu.cn
[2] School of Artificial Intelligence, Xidian University, Xi'an 710071, China

Abstract. Hyperspectral images (HSIs) have far more spectral bands than conventional RGB images. The abundant spectral information provides very useful clues for the followup applications, such as classification and anomaly detection. How to extract discriminant features from HSIs is very important. In this work, we propose a novel spatial-spectral features extraction method for HSI classification by Multi-Scale Depthwise Separable Convolutional Neural Network (MDSCNN). This new model consists of a multi-scale atrous convolution module and two bottleneck residual units, which greatly increase the width and depth of the network. In addition, we use depthwise separable convolution instead of traditional 2D or 3D convolution to extract spatial and spectral features. Furthermore, considering classification accuracy can benifit from multi-scale information, we introduce atrous convolution with different dilation rates parallelly to extract more discriminant features of HSIs for classification. Experiments on three standard datasets show that the proposed MDSCNN has got the state-of-the-art accuracy among all compared methods.

Keywords: Hyperspectral images classification · Multi-scale · Depthwise separable convolution · Residual learning

1 Introduction

Recently, hyperspectral imaging technology has attracted widespread attention in the remote sensing society. The hyperspectral imager can capture accurate spectral response characteristics and spatial details of surface materials, which makes it possible to identify and classify the landcovers. HSI classification aims to assign a unique category to each pixel in the image, enabling automatic identification of categories and serving for following applications. However, due to the limit of labeled samples, the existence of mixed pixels, and the Houghes phenomenon, HSI classification is a very challenge problem.

© Springer Nature Singapore Pte Ltd. 2020
G. Zhai et al. (Eds.): IFTC 2019, CCIS 1181, pp. 171–185, 2020.
https://doi.org/10.1007/978-981-15-3341-9_15

Based on traditional machine learning methods, many HSI classification approaches such as support vector machine (SVM) [1], multiple logistic regression [2], decision trees [3], *etc.* are proposed for pixel-level classification of HSI. However, HSIs usually provide hundreds of spectral bands, which contain a large amount of redundant information. Therefore, using raw spectral information directly not only results in high computational cost, but also reduces classification performance. Consequently, there are some methods that focus on mitigating the redundancy of HSIs with principal component analysis (PCA) [4] or linear discriminant analysis (LDA) [5]. Furthermore, spatial information has been reported to be very helpful in improving the representation of HSI data [6]. Thus more and more classification frameworks based on spatial-spectral features have been presented [7,8]. Although these spatial-spectral classification methods have achieved some progress, they all need to perform feature extraction engineering through human prior knowledge, which limits these methods in different scenarios.

Deep learning has become an important tool for big data analysis, and has made great breakthroughs in many computer vision tasks, such as image classification, object detection and natural language processing. Recently, it has been introduced into the HSI classification as a powerful feature extraction tool and shows great performance. Compared with the traditional artificial feature extraction methods, deep convolutional neural network can extract rich features from the original data through a series of layers. Since the learning process is completely automatic, deep learning is more suitable for dealing with complex scenes. Chen *et al.* [9] first applied the Stacked Autoencoder (SAE) to the HSI classification which is composed of multiple sparse autoencoders. Mughees *et al.* [10] proposed a Spectral-Adaptive Segmentation DBN (SAS-DBS) for HSI classification that exploits the spatial-spectral features by segmenting the original spectral bands into small sets and processing each group separately by local DBNs. However, deep neural networks such as SAE are based on the fully connected layer. Although the above-mentioned deep neural network models can effectively extract deep features in HSIs, they may ignore the spatial information of HSIs. Unlike SAE, Convolutional Neural Networks (CNN) can directly extract spatial and spectral features of HSIs while keeping the input shape. For this reason, most of the current HSI classification networks with spatial-spectral features are based on CNN structure. They can be divided into two main categories. The first is to extract spatial and spectral features separately and then combine them and feed to the classifier [11]. Another strategy is to extract the spatial-spectral joint features of HSIs simultaneously by 3D convolution [12,13]. Although these methods can effectively extract the spatial spectral information of HSIs, they all ignore the multi-scale characteristics. Because of the complexity and diversity of HSI scenery, it is often difficult to extract spatial information from a single scale.

In this work, we propose a Multi-Scale Depthwise Separable Convolutional Neural Network (MDSCNN) for HSI classification which can effectively exploit spatial-spectral features and achieve competitive HSI classification performance.

This new model consists of a multi-scale atrous convolution module and two bottleneck residual units, which greatly increase the width and depth of the network. In addition, we use depthwise separable convolution instead of traditional 2D or 3D convolution, which leads to extract spectral features poorly or has high computational complexity. In contrast, the depthwise separable convolution can not only extract spatial-spectral features separately, but also greatly reduce the amount of training parameters. Furthermore, considering classification accuracy can benifit from multi-scale information, we introduce atrous convolution with different dilation rates parallelly to extract more discriminant features of HSIs. Experiments on three standard datasets show that the proposed MDSCNN has got the state-of-the-art accuracy among all compared methods.

The remainder of this paper is organized as follows: In Sect. 2 several related techniques are described. Section 2 introduces the proposed MDSCNN model. The experiments and results analysis are shown in Sect. 3, A conclusion is made in Sect. 4.

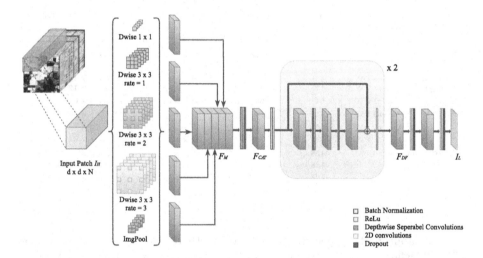

Fig. 1. Overview of the proposed Multi-Scale Depthwise Separable CNN (MDSCNN) model.

2 Method

In this section, we have discussed the overall architecture of the proposed MDSCNN firstly. Then we provide a detailed explanation about each module.

2.1 Overall Architecture

We have constructed a wide and deep network with a specially developed multi-scale atrous convolution module and two depthwise separable bottleneck residual units for HSI classification. As shown in Fig. 1, the proposed MDSCNN is a fully convolutional network (FCN) [19] without any fully connected layers, so that it can handle any input patches with arbitrary size and produce the same size output. Let's denote I_H and I_L as the input HSI patch and predicted labels of MDSCNN.

Table 1. The proposed MDSCNN topology. M is the number of bands.

Module	Layer	Input channel	Output channel	Kernel size	Padding	Parameters
Multi-scale atrous conv module	DW 1 × 1	M	M	1 × 1	0	1 × 1 × M
	PW 1 × 1	M	128	1 × 1	0	1 × 1 × M × 128
	DW 3 × 3 (r = 1)	M	M	3 × 3	1	3 × 3 × M
	PW 1 × 1	M	128	1 × 1	0	1 × 1 × M × 128
	DW 3 × 3 (r = 2)	M	M	3 × 3	2	3 × 3 × M
	PW 1 × 1	M	128	1 × 1	0	1 × 1 × M × 128
	DW 3 × 3 (r = 3)	M	M	3 × 3	3	3 × 3 × M
	PW 1 × 1	M	128	1 × 1	0	1 × 1 × M × 128
	ImgPool	M	M	2 × 2	0	0
		M	128	1 × 1	0	1 × 1 × M × 128
Concatenate Conv	Conv	640	128	1 × 1	0	1 × 1 × 640 × 128
Depthwise separable bottleneck residual unit	Layer1 DW 3 × 3	128	128	3 × 3	1	3 × 3 × 128
	Layer1 PW 1 × 1	128	64	1 × 1	0	1 × 1 × 128 × 64
	Layer2 DW 3 × 3	64	64	3 × 3	1	3 × 3 × 64
	Layer2 PW1 × 1	64	64	1 × 1	0	1 × 1 × 64 × 64
	Layer3 DW 3 × 3	64	64	3 × 3	1	3 × 3 × 64
	Layer3 PW 1 × 1	64	128	1 × 1	0	1 × 1 × 64 × 128
Classification module	Conv 1	128	128	3 × 3	1	3 × 3 × 128 × 128
	Conv 2	128	128	3 × 3	1	3 × 3 × 128 × 128
	Conv 3	128	C_Number	1 × 1	0	1 × 1 × 128 × C_Number

The input to the spectral pixel based methods usually is a pixel vector $x^{1 \times 1 \times M}$, where M is the number of spectral bands. In order to simultaneously exploit the spatial and spectral features, it is necessary to introduce a three-dimensional approach to incorporate the contextual information. In this method, we feed the network with a $d \times d$ patch P centered on x, where d is the width and height of the patch. In this way, the original spatial and spectral features can be considered simultaneously. Especially, the model is designed to predict the label of center pixel, whose position index is $[d/2+1, d/2+1, M]$. Meanwhile, we need to select the value of d carefully. It will result in lacking spatial information if d is too small. On the other hand, when we set d too large, it may introduce some pixels that are not belong to the same class. Furthermore, a multi-scale atrous convolution module is introduced to extract rich spatial and spectral features. It extracts multi-scale features F_M from I_H

$$F_M = H_{MAC}(I_H),\tag{1}$$

where $H_{MAC}(\cdot)$ denotes multi-scale atrous convolution operation. F_M is a joint maps with multi-scale features, then F_M is concatenated together via one 1×1 Conv layer

$$F_{CAT} = W_{CAT}(F_M),\tag{2}$$

where $W_{CAT}(\cdot)$ and F_{CAT} denote the weight set to the Conv layer and joint features respectively. The following are backbone of the network, two specially designed Depthwise Separable bottleneck Residual (DSR) units, implemented with depthwise separable convolution

$$F_{DF} = H_{DSR}(H_{DSR}(F_{CAT})),\tag{3}$$

where $H_{DSR}(\cdot)$ denotes our residual unit, F_{DF} is the obtained deep discriminative feature. The end of the model are three convolutional layers for classification, and we insert the dropout layer (p = 0.5) during training to prevent overfitting

$$I_L = H_{CLS}(F_{DF}),\tag{4}$$

where $H_{CLS}(\cdot)$ and I_L denate classification module and label map predicted by MDSCNN.

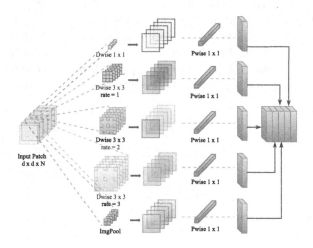

Fig. 2. The developed multi-scale atrous convolution module.

In this paper, we select the cross-entropy as the loss function to train the network, which can be formulated as:

$$E = -\sum_{i=1}^{T} y_i \cdot \log h(x_i),\tag{5}$$

where T denotes the total number of training samples, y_i is the ground truth of x_i, and $h(\cdot)$ denotes the softmax function which is computed as:

$$h_{x_i} = \frac{e_i^z}{\sum_{j=1}^{C} e_j^z}, \tag{6}$$

where z_i is the features learned from sample x_i, and C is the number of label categories.

We have summarized the proposed MDSCNN in Table 1, which includes the number of channels, kernel size, padding value and parameters for each convolution or pooling layer of each module.

2.2   Multi-scale Feature Extraction with Atrous Convolution

It has been proved that classification can benifit from abundant contextual features [20]. Inspired by the Atrous Spatial Pyramid Pooling (ASPP) module [15] which is commonly used in semantic segmentation, we design a multi-scale atrous convolution module based on depthwise separable convolution. As shown in Fig. 2, it consists of four filters: 1×1, 3×3 $(r = 1)$, 3×3 $(r = 2)$, 3×3 $(r = 3)$, and an $ImagePooling$ branch. Atrous convolution can enlarge the receptive field of the filter while maintaining the amount of parameters. These convolutions are extracted in parallel with different dilation rates, and then the generated feature maps are concatenated together. Therefore, we pad the input patches to ensure the shape of generated feature maps are same. Early studies have shown that a 3×3 atrous convolution with an extremely large rate will degenerate into a simple 1×1 convolution. In this way, it will not be able to capture long range information due to image boundary effects [15]. Therefore, considering the spatial size d of input patch is generally between 9–25, we set the maximum dilation

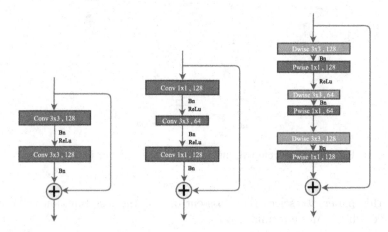

Fig. 3. Different residual unit architectures. (Left) Traditional residual units, (Middle) Bottleneck residual units, (Right) The developed Depthwise Separable bottleneck Residual (DSR) units.

rate to 3. In addition, The pooling filter preserves the image-level features of the original HSIs and enriches the diversity of features.

2.3 BottleNeck Residual Block with Depthwise Separable Convolution

Deep convolutional neural networks often appear degradation phenomenon due to inadequate training, therefore, He *et al.* [21] constructed an identity mapping to ease the training process. The basic idea is that if E is a perfect network with best performance, the T is a deeper network with some redundant layers, so the goal is to make redundant layer become an identical transformation. That is to say, T's performance is the same as E. Therefore, the network needs to learn a residual $F(x) = H(x) - x$, where x is original feature, $H(x)$ denotes the features learned from x. $H(x)$ will be equivalent with x if the network learns nothing, *i.e.* $F(x) = 0$. Since fitting the residual $F(x)$ is easier than fitting the original $H(x)$, residual network can effectively avoid degradation of network. The design of the residual units becomes a point worth exploring, as we can see, there are three different residual units showed in Fig. 3. Basic residual unit (Left) contains two convolutional layers. Bottleneck residual unit (Middle) [22] is more economical than the conventional residual block, and its input and output feature maps dimension is first reduced and then restored, which reduces the calculation amount of the middle layer and allows a faster execution.

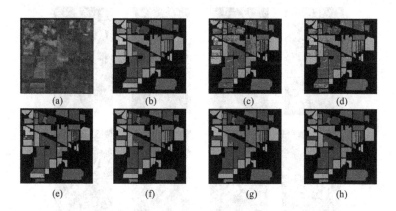

Fig. 4. Classification maps for IP dataset. (a) Simulated RGB composition of the scene. (b) Ground-Truth classification map. Classification maps obtained by (c) MLP, (d) SVM, (e) 2D-CNN, (f) 3D-CNN, (g) HybridSN, (h) MDSCNN

As we know, these traditional residual units mainly focus on the spatial features of RGB images, the spectral features are not well extracted. Inspired by bottleneck, here we have specially designed a DSR unit for HSI classification as shown in Fig. 3 (Right). It mainly consists of three convolutional layers, which extract spatial features using depthwise convolution firstly, and then convolute point by point to extract spectral features. After the first convolutional layer, the feature map dimension is reduced (from the number of channels point of view). A nonlinear activation function is introduced between the first and second convolutional layers

$$F_{MID} = W_{pw}W_{dw}\sigma(W_{pw}W_{dw} \cdot x), \tag{7}$$

where W_{pw} and W_{dw} denate the weights of pointwise convolution and depthwise convolution respectively. $\sigma(\cdot)$ denotes the ReLu activation function. F_{MID} is the feature map after the second Conv layer in DSR. Since the depthwise convolution's output is shallow, in order to retain as much information as possible, a linear output is put between the second and third convolutional layers without adding any nonlinear activation function. Thus a output F_{DF} is obtained via the shortcut connection:

$$F_{DF} = W_{pw}W_{dw} \cdot F_{MID} + x, \tag{8}$$

where $+$ is an elementwise addition that does not change the size of the feature map.

The experimental results show that adding two depthwise separable convolutional units improves the classification accuracy while using limited training samples.

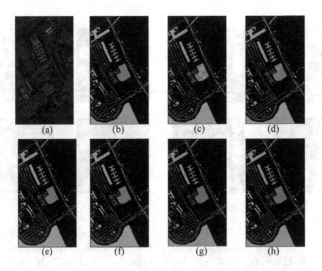

Fig. 5. Classification maps for UP dataset. (a) Simulated RGB composition of the scene. (b) Ground-Truth classification map. Classification maps obtained by (c) MLP, (d) SVM, (e) 2D-CNN, (f) 3D-CNN, (g) HybridSN, (h) MDSCNN

3   Experiments

3.1   Experimental Datasets

We have evaluated our model on three well-know HSI datasets, which are widely used for HSI classification, Indian Pines (IP), University of Pavia (UP) and Salinas Valley (SV).

IP: This scene was gathered by AVIRIS sensor in North-western Indiana, which consists of 145×145 pixels and 224 spectral reflectance bands in the wavelength range from 400 nm to 2500 nm. We have also reserved the number of bands to 200 by removing 24 damaged bands.

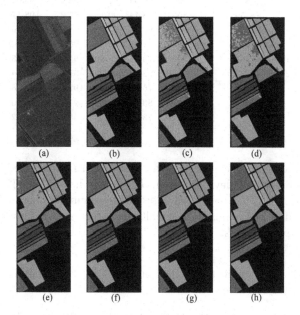

Fig. 6. Classification maps for UP dataset. (a) Simulated RGB composition of the scene. (b) Ground-Truth classification map. Classification maps obtained by (c) MLP, (d) SVM, (e) 2D-CNN, (f) 3D-CNN, (g) HybridSN, (h) MDSCNN

UP: This dataset captured the urban area around the University of Pavia, Italy. The spatial resolution of the image is 1.3 m per pixel, the spectral coverage ranges from 0.43 m to 0.86 m. After 12 bands is removed due to noise, there are 103 bands left. The image consists of 610×340 pixels, but it contains many background pixels.

Table 2. Classification results for IP dataset using 15% of the labeled data for training and 15 × 15 input spatial size.

Class	Training/Test	SVM	MLP	2D-CNN	3D-CNN	HybirdSN	MDSCNN
Alfalfa	6/46	72.50	63.63	0.00	75.86	95.65	**95.65**
Corn-notill	214/1428	81.24	72.15	92.45	93.10	98.47	**99.72**
Corn-mintill	124/830	81.14	77.02	**99.21**	94.65	98.67	97.87
Corn	35/237	73.45	64.71	73.96	87.83	99.58	**100.00**
Grass-pasture	72/483	90.56	81.42	93.36	97.38	98.74	**100.00**
Grass-trees	109730	93.08	91.44	99.17	98.38	99.05	**100.00**
Grass-pasture-mowed	4/28	100.00	100.00	0.00	99.04	89.28	**100.00**
Hay-windrowed	71/478	95.36	89.33	90.87	**99.58**	99.17	98.97
Oats	3/20	70.00	100.00	100.00	78.26	95.24	**100.00**
Soybean-nottill	145/972	78.63	69.91	94.92	97.64	99.37	**99.49**
Soybean-mintill	368/2455	79.52	70.04	92.95	98.61	99.02	**99.79**
Soybean-clean	88/593	80.42	58.74	90.18	97.22	98.17	**99.32**
Wheats	30/205	94.74	89.03	97.15	98.55	98.56	**100.00**
Woods	189/1265	93.36	91.35	93.71	95.01	**99.92**	98.21
Building-Grass-Trees	57/386	79.61	73.74	93.80	95.97	96.98	**98.92**
Stone-Steel-Towers	13/93	97.78	98.90	98.93	97.85	98.94	**98.94**
OA		81.96	76.62	96.27	97.84	98.88	**99.33**
AA		85.09	80.71	81.92	94.12	97.80	**99.45**
Kappa		81.96	73.11	92.52	95.75	98.72	**99.23**

SV: The SV dataset was captured by an onboard visible/infrared imaging spectrometer over Salinas Valley,California. The image has 512 × 217 pixels with a spatial resolution of 3.7 m per pixel. The image originally contained 224 bands, but the remaining 204 bands were usually used for experiments after removing 20 water absorption bands.

Table 3. Classification results for UP dataset using 15% of the labeled data for training and 15 × 15 input spatial size.

Class	Training/Test	SVM	MLP	2D-CNN	3D-CNN	HybirdSN	MDSCNN
Asphalt	994/6631	94.32	88.72	96.61	99.14	**99.89**	99.77
Meadows	2797/18649	95.28	86.02	98.03	98.45	**99.97**	98.47
Gravel	314/2099	84.91	78.88	**100.00**	90.19	99.43	99.81
Trees	459/3064	97.95	87.39	98.49	98.89	99.19	**99.71**
Sheets	201/1345	99.48	99.48	100.00	100.00	100.00	100.00
Bare soils	754/5029	93.40	91.90	100.00	100.00	98.28	100.00
Bitumen	199/1330	89.65	84.70	99.32	86.24	100.00	100.00
Bricks	552/3682	85.19	76.27	94.16	99.53	95.00	**99.95**
Shadows	142/947	100.00	100.00	100.00	99.79	99.58	100.00
OA		93.82	86.38	97.98	98.14	98.52	**99.26**
AA		93.35	88.15	98.51	96.92	99.04	**99.75**
Kappa		91.76	81.48	97.34	97.56	98.04	**99.03**

In this paper, we implement the proposed method with Pytorch framework. Before training, we have enhanced the data by randomly flipping and adding noise. We use the Adam optimizer to train the network with a batch size of 64 and initially set a base learning rate as 0.001 then reduce it with poly (0.9).

3.2 Experimental Results

We compare the proposed MDSCNN with several classical and state-of-the-art HSI classification methods. (1) SVM; (2) MLP; (3) 2D-CNN [23]; (4) 3D-CNN [12]; (5) HybidSN [13]. SVM and MLP both are spectral-based methods. 2D-CNN is based on spatial features. 3D-CNN utilizes spatial-spectral features of HSIs with 3D convolution, which consists of two 3D convolutional layers and one fully connected layer. The HybridSN firstly get a low-dimensional data with PCA as input, and it contains three 3D convolutional layers, one 2D convolutional layer, and three fully connected layers in the end of model. We evaluate all of these methods on three standard datasets described above. In order to evaluate the proposed MDSCNN and demonstrate the effectiveness of the multi-scale strategy, we have conducted the following three experiments.

(1) In our first experiment: the first step is to randomly divide the original IP,UP and SV dataset respectively into two subsets: the training set and testing set, whose sample numbers are shown in the first column of Tables 2, 3 and 4. We train all methods mentioned above with some optimal parameters. In addition, for our model, the input patch size is set to $15 \times 15 \times M$.
(2) In our second experiment: intuitively, different spatial size of patch has significant effect on the classification performance of model. We takes three different sizes of patch as input: $9 \times 9, 15 \times 15, 21 \times 21$ and 15% of the available training data for experiment.
(3) In our third experiment: to verify the effectiveness of the multi-scale atrous convolution module used to jointly extract the mutli-scale spatial-spectral features, we compare the proposed MDSCNN to the network without the multi-scale module. To verify the effectiveness of the DSR units, we also compare the performance of the proposed MDSCNN to a similar network with the DSR unit replaced with traditional two convolutional layers residual unit.

To evaluate the performance of different methods, three objective metrics: overall accuracy (OA), average accuracy (AA), and the Kappa coefficient, are adopted in these experiments.

Experiment 1: Tables 2, 3 and 4 show the quantitative results, moreover, the best result is highlighted in bold font. As shown, the results of traditional spectral-based pixel-level classification methods are not satisfactory, like SVM and MLP, which are far worse than spatial-spectral based methods like 2D-CNN, 3D-CNN, HybridSN. The proposed MDSCNN achieves the best classification accuracy on each dataset. Furthermore, the proposed MDSCNN has improved

Table 4. Classification results for SV dataset using 15% of the labeled data for training and 15 × 15 input spatial size.

Class	Training/Test	SVM	MLP	2D-CNN	3D-CNN	HybirdSN	MDSCNN
Brocoli-green-weeds-1	302/2009	100.00	100.00	99.87	100.00	100.00	**100.00**
Brocoli-green-weeds-2	559/3726	99.71	98.59	99.97	99.84	99.97	**100.00**
Fallow	297/1976	98.64	95.40	99.83	99.92	**99.92**	99.85
Fallow-rough-plow	210/1394	98.58	98.86	99.21	99.78	**99.85**	99.50
Fallow-smooth	402/2678	98.92	91.49	99.22	**99.78**	99.18	99.51
Stubble	594/3959	99.98	99.98	99.94	99.92	99.95	**100.00**
Celery	537/3579	99.78	99.02	99.66	**100.00**	99.80	99.92
Graphes-untrained	1691/11271	79.21	68.92	90.04	98.86	94.68	**99.34**
Soil-vinyard-develop	931/6203	99.18	97.14	99.69	99.96	99.66	**100.00**
Corn-senesced-green-weeds	492/3278	95.46	87.82	99.33	**99.96**	99.69	99.91
Lettuce-romaine-4wk	161/1068	97.71	90.77	94.19	**100.00**	99.98	99.91
Lettuce-romaine-5wk	290/1927	98.32	97.27	99.43	99.84	99.94	**100.00**
Lettuce-romaine-6wk	138/916	99.12	94.23	99.89	99.93	100.00	**100.00**
Lettuce-romaine-7wk	161/1070	98.03	95.64	97.43	99.90	100.00	**100.00**
Vinyard-untrined	1091/7268	83.69	80.02	78.22	82.95	**99.98**	99.01
Vinyard-vertical-trellis	272/1807	99.89	97.51	99.83	**100.00**	99.93	99.63
OA		92.65	87.69	94.45	97.18	98.70	**99.87**
AA		96.64	93.29	97.53	98.86	99.55	**99.78**
Kappa		91.80	86.21	95.14	96.86	98.57	**99.86**

the OA value 0.5% at least compared to the suboptimal method on all testing sets, and there are surprising improvements in AA and Kappa. In addition to the qualitative results, the Figs. 4, 5 and 6 show three visual classification maps of the different methods on three datasets respectively. It can be observed that the traditional single-pixel-based methods have a lot of noise due to the lack of spatial information. Meanwhile, the classification maps obtained by the spatial-spectral based methods are smoother, and the most of the ewrong classified pixels exist around the boundaries of some categories. Taking all these observations into account, it is possible to state that the MDSCNN provides a more accurate and robust classification result than all of the other tested methods.

Table 5. Results on the proposed MDSCNN when considering different spatial size input patches.

Spatial size	IP			UP			SV		
	OA	AA	Kappa	OA	AA	Kappa	OA	AA	Kappa
9 × 9	98.41	97.11	98.19	98.79	99.15	98.40	98.57	99.27	98.42
11 × 11	98.91	99.08	98.75	99.23	99.94	98.98	98.73	99.57	98.58
15 × 15	**99.33**	**99.45**	**99.23**	**99.27**	**99.96**	**99.03**	**99.87**	**99.93**	**99.86**
21 × 21	98.62	97.71	98.43	99.24	99.95	98.99	97.17	98.86	96.86

Experiment 2: Table 5 shows the classification results of the proposed MDSCNN when using different spatial size patches as input. The OA, AA, and Kappa values all increase firstly and then decrease on the three datasets. The highest score was reached as the spatial size is set to 15 × 15. It is not difficult to understand that increasing the spatial size of the patch will introduce a certain amount of spatial information at first. But as the spatial size continues to increase, a lot of noise or pixels with different classes will be also introduced.

Table 6. Classification performance comparison of the proposed MDSCNN and the network without multi-scale atrous convolution module and the network with traditional two layers residual unit.

Method	UP			IP		
	OA	AA	Kappa	OA	AA	Kappa
w/o MS	94.71	93.80	93.02	89.65	88.23	88.92
w/o DRS	96.34	96.05	95.29	92.51	92.64	92.36
MDSCNN	**99.26**	**99.75**	**99.03**	**99.33**	**99.45**	**99.23**

Experiment 3: As shown in Table 6, the multi-scale atrous convolution module outperforms the network without it (by 4.55% for the UP dataset, 9.68% for the IP dataset in OA classification performance). Beyond that, our developed DRS units achieve better performance than traditional residual units.

4 Conclusion

In this paper, a novel multi-scale separable convolutional network for HSI classification is proposed, the model leverages a multi-scale atrous convolutional module to extract spatial-spectral features from a HSI patch. In addition, a specially designed depthwise separable bottleneck residual unit is applyed to increase the depth of the network and improve classification performance. The proposed MDSCNN is deep while it doesn't introduce large quantities of training parameters because of the depthwise separable convolution. The final experimental results show that our method achieves outstanding classification performance with a relative small number of training samples. Although multi-scale features fusion has been adopted in our MDSCNN, the features at different stages of the network is not considered. In the future, we will continue to explore some new multi-stage information fusion ways for HSI classification.

Acknowledgements. This work is supported by the National Science Foundation under Grant Nos. 61922027, 61672193, and 61971165.

References

1. Mercier, G., Lennon, M.: Support vector machines for hyperspectral image classification with spectral-based kernels. In: 2003 IEEE International Geoscience and Remote Sensing Symposium. Proceedings (IEEE Cat. No. 03CH37477), IGARSS 2003, vol. 1, pp. 288–290, July 2003

2. Li, J., Bioucas-Dias, J.M., Plaza, A.: Semisupervised hyperspectral image segmentation using multinomial logistic regression with active learning. IEEE Trans. Geosci. Remote Sens. **48**(11), 4085–4098 (2010)

3. Li, S., Zhang, B., Gao, L., Zhang, L.: Classification of coastal zone based on decision tree and PPI. In: 2009 IEEE International Geoscience and Remote Sensing Symposium, vol. 4, pp. IV-188–IV-191, July 2009

4. Chen, H., Chen, C.H.: Hyperspectral image data unsupervised classification using Gauss-Markov random fields and PCA principle. In: IEEE International Geoscience and Remote Sensing Symposium, vol. 3, pp. 1431–1433, June 2002

5. Bandos, T.V., Bruzzone, L., Camps-Valls, G.: Classification of hyperspectral images with regularized linear discriminant analysis. IEEE Trans. Geosci. Remote Sens. **47**(3), 862–873 (2009)

6. Demir, B., Ertürk, S.: Improving SVM classification accuracy using a hierarchical approach for hyperspectral images. In: 2009 16th IEEE International Conference on Image Processing (ICIP), pp. 2849–2852, November 2009

7. Fauvel, M., Tarabalka, Y., Benediktsson, J.A., Chanussot, J., Tilton, J.C.: Advances in spectral-spatial classification of hyperspectral images. Proc. IEEE **101**(3), 652–675 (2013)

8. Wang, J., Jiao, L., Wang, S., Hou, B., Liu, F.: Adaptive nonlocal spatial-spectral kernel for hyperspectral imagery classification. IEEE J. Sel. Top. Appl. Earth Obs. Remote Sens. **9**(9), 4086–4101 (2016)

9. Chen, Y., Lin, Z., Zhao, X., Wang, G., Gu, Y.: Deep learning-based classification of hyperspectral data. IEEE J. Sel. Top. Appl. Earth Obs. Remote Sens. **7**(6), 2094–2107 (2014)

10. Mughees, A., Tao, L.: Multiple deep-belief-network-based spectral-spatial classification of hyperspectral images. Tsinghua Sci. Technol. **24**(2), 183–194 (2019)

11. Yang, G., Gewali, U.B., Ientilucci, E., Gartley, M., Monteiro, S.T.: Dual-channel DenseNet for hyperspectral image classification. In: 2018 IEEE International Geoscience and Remote Sensing Symposium, IGARSS 2018, pp. 2595–2598, July 2018

12. Hamida, A.B., Benoit, A., Lambert, P., Amar, C.B.: 3-D deep learning approach for remote sensing image classification. IEEE Trans. Geosci. Remote Sens. **56**(8), 4420–4434 (2018)

13. Roy, S.K., Krishna, G., Dubey, S.R., Chaudhuri, B.B.: HybridSN: exploring 3D–2D CNN feature hierarchy for hyperspectral image classification. ArXiv, abs/1902.06701 (2019)

14. He, K., Zhang, X., Ren, S., Sun, J.: Spatial pyramid pooling in deep convolutional networks for visual recognition. CoRR, abs/1406.4729 (2014)

15. Chen, L.C., Papandreou, G., Kokkinos, I., Murphy, K., Yuille, A.L.: DeepLab: semantic image segmentation with deep convolutional nets, atrous convolution, and fully connected CRFs. CoRR, abs/1606.00915 (2016)

16. LeCun, Y., et al.: Backpropagation applied to handwritten zip code recognition. Neural Comput. **1**(4), 541–551 (1989)

17. Krizhevsky, A., Sutskever, I., Hinton, G.E.: ImageNet classification with deep convolutional neural networks. In: Proceedings of the 25th International Conference on

Neural Information Processing Systems, NIPS 2012, vol. 1, pp. 1097–1105. Curran Associates Inc., USA (2012)

18. Szegedy, C., et al.: Going deeper with convolutions. In: 2015 IEEE Conference on Computer Vision and Pattern Recognition (CVPR), pp. 1–9, June 2015

19. Long, J., Shelhamer, E., Darrell, T.: Fully convolutional networks for semantic segmentation. CoRR, abs/1411.4038 (2014)

20. Huang, G., Chen, D., Li, T., Wu, F., van der Maaten, L., Weinberger, K.Q.: Multi-scale dense convolutional networks for efficient prediction. CoRR, abs/1703.09844 (2017)

21. He, K., Zhang, X., Ren, S., Sun, J.: Deep residual learning for image recognition. CoRR, abs/1512.03385 (2015)

22. Tishby, N., Zaslavsky, N.: Deep learning and the information bottleneck principle. CoRR, abs/1503.02406 (2015)

23. Liu, B., Yu, X., Zhang, P., Tan, X., Yu, A., Xue, Z.: A semi-supervised convolutional neural network for hyperspectral image classification (2017)

Joint SPSL and CCWR for Chinese Short Text Entity Recognition and Linking

Zhiqiang Chong, Zhiming Liu$^{(\boxtimes)}$, Zhi Tang, Lingyun Luo,
and Yaping Wan

School of Computer, University of South China, Hengyang 421001, China
nhdxlzm@foxmail.com

Abstract. Entity Recognition Linking (ERL) is a basic task of Natural Language Processing (NLP), which is an extension of the Named Entity Recognition (NER) task. The purpose of the ERL is to detect the entity from a given Chinese short text, and the detected entity is linking to the corresponding entity in the given knowledge library. ERL's task include two subtasks: Entity Recognition (ER) and Entity Link (EL). Due to the lack of rich context information in Chinese short text, the accuracy of ER is not high. In different fields, the meaning of the entity is different and the entity cannot be accurately linking. These two problems have brought a big challenge to the Chinese ERL task. In order to solve these two problems, this paper proposes based on neural network model joint semi-point semi-label (SPSL) and Combine character-based and word-based representations (CCWR) embedding. The structure of this model enhances the representation of entity features and improve the performance of ER. The structure of this model enhances the representation of contextual semantic information and improve the performance of EL. In summary, this model has a good performance in ERL. In the ccks2019 Chinese short text ERL task, the F1 value of this model can reach 0.7463.

Keywords: Entity Recognition · Entity linking · CCWR · SPSL

1 Introduction

In the era of big data, the Internet has accumulated a large amount of Chinese short texts (such as search Query, Weibo, user dialogue content, article title), and these short texts have great value in the NLP field. In recent years, with the re-ignition of deep learning and the support of massive data, the NLP field is booming. As event knowledge and entity knowledge become more and more important, NER has a new wave of development, which has been an important research foundation of NLP. As an extension of NER, ERL has gradually emerged in the field of NLP.

In 1995, the concept and definition of NER was presented at the MUC-6[1] meeting, MUC-7's MET-2, IEER-99, Co NLL-2002, Co NLL-2003, IREX, LREC and other international conferences also took NER as a task of evaluation, and determine the category of entities in the NER evaluation task. In a broad sense, there are three classes

---

[1] https://cs.nyu.edu/cs/faculty/grishman/muc6.html.

© Springer Nature Singapore Pte Ltd. 2020
G. Zhai et al. (Eds.): IFTC 2019, CCIS 1181, pp. 186–199, 2020.
https://doi.org/10.1007/978-981-15-3341-9_16

of NER entities (entity class, time class, and number class). And seven subclasses (name, institution name, place name, time, date, currency and percentage) of named entities in the text. In practical applications, the category of the entity also includes the movie name, music name, dish name, game name, etc.

With the development of machine learning technology and the renaissance of deep learning technology, NER as the basic task of information extraction, it is a hot topic in the deep learning technology. NER is widely used in EL, machine translation, information retrieval, relationship extraction, question and answer systems, etc. In fact, this paper can be seen as a NER application in ERL. Traditional NEL task refers to for long documents, and long documents have rich context information, which can assist entities to complete the task. In contrast, Chinese short text ERL has great challenges. Chinese colloquial is more serious, Chinese short text context is not rich, Chinese has its own linguistic characteristics (word disorder and the diversity of constructed words) relative to English. In order to solve the above challenges, we propose a structure that combines CCWR embedding and SPSL decoding based on Long Short-Term Memory (LSTM) and Convolutional Neural (CNN). In the ER model, our experiment uses three methods of embedding: character embedding, word embedding, CCWR embedding. Experiments show that CCWR embedding performance is higher than the other two embeddings 0.032 and 0.0553, and it can be seen that CCWR embedding is more suitable for ERL tasks than character embedding and word embedding. In the EL model, we used SPSL annotation and CCWR embedding to improve the accuracy of NEL. In the ccks2019ERL evaluation task, the F1 value of this paper reaches 0.7463.

2   Related Work

The essence of Chinese ER is to detect the boundary of entity. The easiest way is to use word-based representation methods: First, the Chinese short text is divided into words, then find the entity from the entity library. However, if word segmentation error occurs in the process of word segmentation, this error will encounter error propagation, it will lead to ERL error. Out of vocabulary (OOV) may be encountered in the process of word segmentation, which is a serious problem in open areas such as knowledge maps. Zhang et al. [1] experiments show that the character-based representation method is more effective than the word-based representation method.

Solving the NER task is mainly summarized as three types of methods. (1) Rule-based approach: After the Chinese short text segmentation, then collect feature words into a dictionary and study the pattern of the dictionary, summarizing the rules of the entity. However, such methods have serious shortcomings, such as poor portability, high cost, long system cycle time, need to build knowledge library in different fields to improve the ability of the NER system. (2) Statistical methods: Hidden Markov Model (HMM) [2], Maximum Entropy (ME) [3], Support Vector Machine (SVM) [4], Conditional Random Field (CRF) [5] and other sequence labeling models. The disadvantages of such methods are long training time and high system overhead. (3) Based on deep learning methods: Turian et al. [6] used pre-trained word vectors as additional features, they used traditional CRF and LSTM grid structures compared to SVM and ME traditional methods have a better promotion.

Collobert et al. [7] used neural networks for NER research, they using window methods and sentence methods based on neural networks. The main difference between the two methods is that the window method simply uses the context window of the current word as input and then uses the traditional neural network. The sentence method uses the entire sentence as the input of the whole neural network, and add the relative position information of the sentence, it further uses a layer of convolutional neural network (CNN) structure [8]. Turian et al. [6] use pre-trained word vectors to represent word information, then use the CRF method to solve the problem of NER [6]. In 2015, Lample et al. [9] used RNN structures combined with CRF for NER study. According to their research, RNN-CRF is more effective than traditional neural network methods and rule-based methods and statistical-based methods. Peters et al. [10] used a massive unlabeled corpus to train a bidirectional neural network language model, then use this trained language model to get the language model vector of the current annotation, finally, the vector is entered as a feature into the original RNN-CRF. This study shows that adding pre-training language model can greatly improve the effect of NER, even if the language vector is added to a large number of unlabeled training data, still can improve the effect of NER.

In recent years, ERL has attracted the attention of academic and business circles. The purpose of ERL is to identify the entities in the text and link the entities to the knowledge library. ER has already been explained in the first part of this article. When using traditional statistical methods and regular methods to solve ER, the selection of features is strict, and these methods solve ER problems in different fields, it needs to re-dig entity rule and context feature rules, stop words feature and semantic features. Since different languages have different characteristics, traditional methods need to re-dig the rules when faced with different languages, therefore, the traditional method to solve the problem of ER is not strong in generalization. Although the attention mechanism can be added to the LSTM-CRF to dynamically learn word vectors and character vectors, however it does not fully learn the semantic information of context information, such as "中国科技大学" is detected as "中国" and "科技大学". Although the bag of words model or similar with bag of word model can solve the problem of entity disambiguation, but these models only capturing shallow feature information instead of capturing deep semantic information. The essence of EL is entity disambiguation. Bagga et al. [11] used the bag of words model to solve the problem of name disambiguation. In addition, Fleichman et al. [12] used the network information feature to train ME model to solve the problem of entity disambiguation. Miller et al. [13] found that similar words appear in similar contexts, then we can think of the approach of entity disambiguation, that is to perform text context similarity analysis on candidate entities and detected entity.

In order to solve these problems that text context semantics are not rich and the feature representation is insufficient, this paper proposes a model based on BiLSTM [15] (Bidirectional LSTM)-CNN joint SPSL and CCWR embedding. BILSTM can not only get text information from front to back, but also get text information from the back to the front, it combines CCWR embedding to enhance contextual semantic information, it has better performance in ERL than traditional methods. CNN combined with

SPSL has better performance in entity decoding than traditional methods. CCWR embedding not only learns the relationship between characters, but also learns the relationship between words and words, and effectively solves the problem of OOV [16]. In this paper, CCWR embedding uses a matrix to fuse word embedding and character embedding, the advantages of CCWR embedding enhance embedding representation capability and contain rich contextual semantic information. SPSL is more excellent in decoding entities than traditional methods. In the ccks2019 Chinese ERL evaluation task, the model f1 value of this paper reaches 0.7463.

3 Model

The model proposed in this paper refers to Chiu et al. [14]. The model is based on BiLSTM-CNN and joint CCWR embedding and SPSL annotation to solve the problems of ER and EL. The ERL task can be divided into two steps, the ER step and the EL step. The ER step is divided into two steps, predict the entity boundary sequence and extract the entity. ERL structure of this paper (see Fig. 1). In NEL, we use CCWR to generate the vector S of the Chinese short text as input into BILSTM-CNN to get the entity set $E = \{e_1, e_2, ..., e_n\}$, then we find the set of candidate entities M of e_i from the knowledge library, finally S, E, M input into BiLSTM-CNN model to gets the final result.

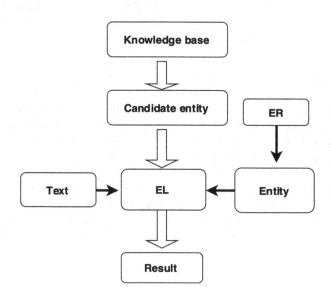

Fig. 1. ERL task model entire structure.

3.1   Entity Recognition (ER)

In the ER model, BiLSTM-CNN is used as the main neural network structure, then combined with CCWR embedding to complete the ER task.

Combine Character-Based and Word-Based Representations (CCWR). The CCWR embedding mode (see Fig. 2).

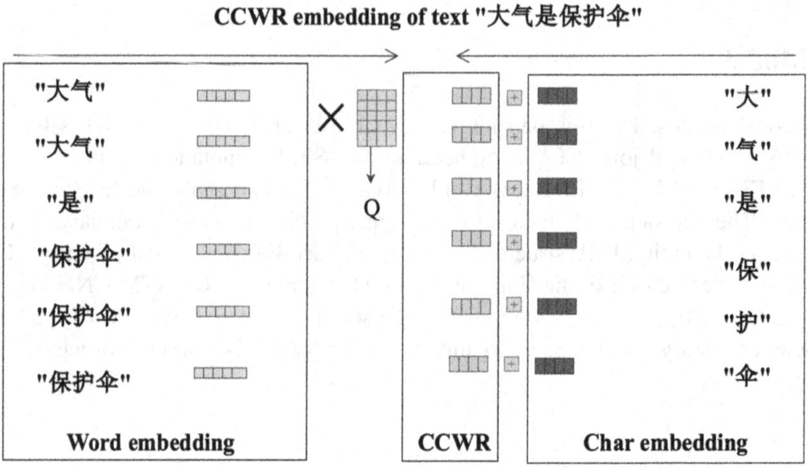

CCWR embedding of text "大气是保护伞"

Fig. 2. CCWR embedding mechanism.

Character embedding: First we get a Chinese character set $L = \{c_1, c_2, ..., c_n\}$ from knowledge library and training datasets, $c_i! = c_j \, \forall i! = j$. We get the vector $S = \{w_1, w_2, ..., w_n\}$ from L, $w_i \in R$, w_i indicates the position of the i-th character of Chinese short text in the character library. Each of the elements in vector S is randomly initialized into vector C, $C \in R^{1*n}$, C is n-dimensional [17]. Word embedding step: Use Google's open source gensim[2] library and *Word2vec*[3] *skip Gram negative sampling* technology to train the Wikipedia-provided Latest Chinese Article corpus[4] and training datasets, then use the opencc[5] library to convert traditional Chinese into simplified, unification of different punctuation marks, space processing. Finally, after the word segmentation, the word vector set W is obtained, $W = \{w_1, w_2, ..., w_m\}$, each word vector $w_i \in R^{1*m}$ (W is m-dimensional) [18]. CCWR embedding step: Training a transform matrix

---

[2] https://radimrehurek.com/gensim/.

[3] https://radimrehurek.com/gensim/models/word2vec.html.

[4] https://dumps.wikimedia.org/zhwiki/latest/.

[5] https://www.byvoid.com/en/project/opencc.

$Q \in R^{m*n}$ with the character vector C and the word vector W, W is n-dimensional, then use the following formula to get a fused word vector $W* \in R^{1 \times n}$ [19, 20].

$$W^* = W*Q + C \tag{1}$$

Entity Recognition (ER). The ER model (see Fig. 3). The vector W^* is generated in Eq. 1 input to the BILSTM model to extract the local feature representation, an enhanced embedded representation feature vector is obtained. In the ER model, Chinese short text input to the LSTM model after CCWR embedding, then we obtained two vectors V1, V2, concatenate V1 vector and V2 vector become V3 vector.

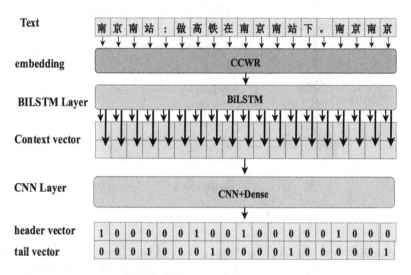

Fig. 3. ER task model structure.

The entity position annotation vectors L and V3 are input as feature vectors into the CNN Dense model, then we got two vectors B and E. Vector B_i represents the probability that the i-th character in the text is the beginning of the entity. Vector E_i represents the probability that the i-th character in the text is the end of the entity. ER decoding uses two thresholds (*theta1* and *theta2*) to determine the begin of an entity and the end of an entity, *theta1* value is 0.35, *theta2* value is 0.35, completed the ER task.

3.2 Entity Linking (EL)

The EL model (see Fig. 4). The essence of entity linking is entity disambiguation. According to the above ER model, the entity *e* can be detected from Chinese short text.

Perform similarity calculation for each element in the set of candidate entities with detected entity e, finally get the most similar entity with e.

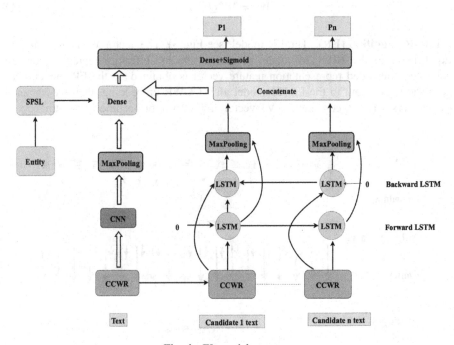

Fig. 4. EL model structure

Candidate Entities. The set of candidate entities M built from the knowledge library, $M = \{m_n:\{e_1, e_2, ..., e_d\}\}$, $m_i! = m_j$ $\forall i! = j$, n is the number of entities in the knowledge library, d is the number of candidate entities of detected entity m_i. When generating a candidate entity, deleting the candidate entity that does not match with the length of the detected entity, can improve the accuracy and speed of the model.

SPSL Annotation. Traditional sequence tagging tasks use BIOES[6] (begin, inside, outside, end, simple) and BIO (begin, inside, other) to mark entities, their disadvantage is that they require a lot of manual feature engineering. The SPSL method proposed in this paper use three one-dimensional vectors B, E, T (The dimension of the three vectors is as same as the length of the given text) [21]. Vector B: The element value equals to 1 indicate that the corresponding position in the text is the beginning of an entity. Vector E: The element value equals 1 to indicate that the corresponding position

---

[6] https://en.wikipedia.org/wiki/Inside–outside–beginning_(tagging).

in the text is the end of an entity. Vector T: The element value equal to 1 indicate that the corresponding position in the text is a part of an entity. The SPSL annotation method can mark text sequence without considering the type of the entity, the most prominent advantage of SPSL is that SPSL annotation is hierarchical, and the entity marker vector is divided into an entity start vector and entity end vector and entity position vector.

In the EL model. Firstly, generating a set of candidate entities from the detected entity of ER model, then the information of each candidate entity is concatenate into a sentence, and the sentence is embedded by using CCWR embedding to obtain $x \in R^d$, d represents the dimension of the word vector. x is used as a feature input into the BiLSTM model to obtain Y1 vector and Y2 vector, extracting the text semantic vector H1 from front to back and the text semantic vector H2 from back to front, Concatenate H1 and H2 get vector H3.

$$H1 = maxpooling(Y1) \tag{2}$$

$$H2 = maxpooling(Y2) \tag{3}$$

According to the above method, the H3 vectors of all candidate entities are obtained and concatenate into H vector. The original text $s = \{s1, s2, ..., sn\}$ of the entity e is embedded by CCWR, extracting local feature vectors through CNN, and the text entity feature vector of s is obtained through the maximum pooling $W = [w1, w2, ..., wn]$. For entity e, using the CCWR embedding and SPSL methods to obtain text entity position vector F. Concatenating H, W, and F into a vector input to the Dense layer can obtain the similarity Pi between i-th candidate entity and the detected entity e.

$$S_i = \frac{1}{(1 + e^{-P_i})} \tag{4}$$

$$l = argmax(S_i) \tag{5}$$

Tracing the candidate entity of the detected entity e to search the maximum similarity, refer to Eqs. 4 and 5, finally get an index l, l is the location of the entity that is most similar with entity e in the candidate entity library. (e, l) is the result of the EL and ERL tasks.

4   Experiments

In this paper, the experimental platform is Ubuntu 16.04, Tesla P100 GPU, memory is 16G, based on Python using the open source artificial neural network framework Keras[7] to design program.

---

[7] https://keras.io/.

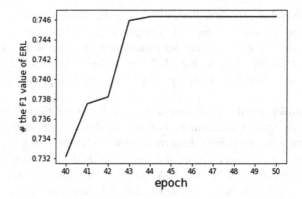

Fig. 5. Experiment of NEL F1 value and epoch number.

The epoch value has a slight rise after 40 (see Fig. 5), the F1 value increases by about 0.0141 when the epoch is 45 times. This is because each epoch randomly selects part of the dataset s as test data, and s is not used for train, after 40 epochs, the model will training data that previous epoch has not trained before, so the model shows a slight improvement. After 45 epochs, almost all the data has been fitted, F1 tends to stabilize, so the model epoch parameter is set to 45.

4.1   Metrics

The evaluation standard of this experiment is F1 value and accuracy rate P and recall rate R, the evaluation standard for NEL is mainly F1 value, and the F1 value can judge the ability of ERL model. The relevant definitions are as follows:

$$P = \frac{a}{b} \tag{6}$$

$$R = \frac{a}{c} \tag{7}$$

$$F1 = \frac{2*P*R}{(P+R)} = \frac{2*a}{(b+c)} \tag{8}$$

a represents the number of correct detected entities, b represents the number of detected entities, c represents the number of sample entities, P, R, and F1 are between 0 and 1, and their value is from 0 to 1, their value is close to 1, indicating the higher the accuracy or recall rate.

4.2   Datasets

The experiment is based on CCKS2019 Chinese ERL task. The training datasets and knowledge library[8] used in this article are all from Baidu Encyclopedia. The data set contains 90000 Chinese short texts, and 399252 entity information. Table 1 shows the knowledge library, each entity form is as follows:

$$kb = \{alias, subject\_id, subject, type, data\} \tag{9}$$

$$data = \{\{predicate : p_1, object : o_1\}...\{predicate : p_n, object : o_n\}\} \tag{10}$$

kb represents the knowledge library, $data$ is the attribute of the knowledge library kb, p_i represents the abstract, and o_i represents the description sentence.

Table 1. Knowledge library attribute

Attribute	Interpretation
alias	Other alias
subject_id	Knowledge library uniqueness identifier
subject	Entity
type	Type of entity (movie, music)
data	Descriptive information about the entity

The training dataset is described in Table 2, it is derived from real Internet page title data, generating raw data based on user retrieval and generated by Baidu crowdsourcing annotation (manual evaluation, average accuracy is greater than 0.95).

Table 2. Standard datasets attribute

Attribute	Interpretation
text_id	Standard data text uniqueness identifier
text	Chinese short text
mention_data	Entity information collection

The mention_data attribute in Table 2 is shown in Table 3.

Table 3. mention_data attribute

Attribute	Interpretation
kb_id	Knowledge library uniqueness identifier
Mention	Entity
Offset	The offset position of the entity in the text

---

[8] https://www.biendata.com/competition/ccks_2019_el/data/.

4.3   Hyperparameters

The parameters of the model proposed in this paper are described in Table 4.

Table 4. The parameters of the model in this paper

Attribute	Interpretation
Char-embedding size	128
Batch-size	64
Dropout	0.5
Learning rate	0.001
LSTM hidden	64
Adam beta_1	0.9
Adam beta_2	0.999
Decay	0.0

The model in this paper uses sigmoid as the activation function, which can get the similarity between the entity and the candidate entity. The Adam optimizer can adaptive learning rates, it as an optimizer for this model. Dropout can randomly discard some neurons in the neural network, making the model not so complicated and preventing the model from overfitting.

In order to use Google open source gensim and Word2vec training word vector in CCWR embedding, the Word2vec parameters are shown in Table 5.

Table 5. Word2vec parameter

Model	Example
Model	Skip-gram
Window_size	5
Word vector dimensions	128
Sampling	Negative sampling

4.4   Experimental Results and Analysis

The major evaluation criteria in this model is the F1 value, the data comes from the ccks2019ERL task. The model use BILSTM-CNN model combines with SPSL and CCWR, and the baseline is BILSTM+CNN+SPSL model. The experimental results are shown in Tables 6 and 7.

Table 6. F1 values for multiple NEL models

Model	P	R	F1
SVM	0.6428	0.5430	0.5887
ME	0.6107	0.5227	0.5633
LSTM	0.7122	0.5341	0.6104
BILSTM	0.6933	0.5854	0.6348
BILSTM-CRF	0.7302	0.6245	0.6732
BiLSTM-CNN+BIO	0.7260	0.7035	0.7146
BILSTM-CNN+SPSL	0.7399	0.7528	0.7463

The SVM model in this paper refers to Chieu et al. [2], the model of ME refers to Chieu et al. [3], the BILSTM-CRF refers to Lample et al. [9], and the BiLSTM-CNN reference Chiu [14].

Table 6 shows that the F1 value of LSTM model is 0.0217 and 0.0471 more than the traditional SVM model and the ME model. Compared with the traditional method (SVM, ME), the neural network method can avoid the weight error propagation, and it can learn complex functions. Compared with the traditional method (SVM, ME), LSTM-CNN does not need complex artificial features, only need to use SPSL to label entity, and then automatically dig context information representation when input to LSTM-CNN. Table 6 shows that the *F1* value of BILSTM is 0.0244 higher than the *F1* value of LSTM, because BILSTM can obtain the semantic information of the Chinese text from the past and the semantic information from back to front. The reason why the CNN model is 0.0414 higher than the CRF model is: CRF solve ERL task will be strict with feature selection, cannot accommodate any context information, and has a great dependence on the corpus. The convolutional layer in the CNN model can be used for feature extraction, which can correctly identify the entity boundary and extract the feature vector at a deeper level, so it improves the performance of ER. The pooling layer in the CNN model compresses features, simplifies network computational complexity, and extracts key features, so it improves the performance of EL. At the same time, it solves the problem of uppercase and lowercase characters, border blurring and text language diversification.

Table 7. Embedding experiment under baseline

Model	P	R	F1
Baseline + char embedding	0.7455	0.6856	0.7143
Baseline + word embedding	0.7247	0.6603	0.6910
Baseline + CCWR embedding	0.7399	0.7528	0.7463

Baseline model is LSTM-CNN+SPSL. Table 7 shows: The F1 value of CCWR embedding is 0.0320 and 0.0553 higher than char embedding and word embedding. This is because the CNN+CCWR is more sensitive to characters and more concerned with position representation. LSTM+CCWR is a better representation of semantic information. It combines the characteristics of word embedding and character embedding to deeply understand semantic information.

5 Conclusion

In this paper, the CCWR embedding method is proposed, which enriches context semantics information and strengthens the ability of feature representation. Compared with the method of character embedding, it explicitly uses the information of the order of word and word. Compared with the method of word embedding, ERL does not have word segmentation errors, therefore CCWR improve the accuracy of ER. The SPSL fine-grained sequence annotation makes the entity feature vector hierarchical and decomposes into entity start position vector and the end position vector.

In the ER, the model pays more attention to the feature of the starting position of the entity or the feature of the end position of the entity. In terms of entity disambiguation, compared with the traditional cosine similarity method, the cosine similarity method cannot make full use of the context and the semantic information of two words.

In the EL, SPSL fully expresses the entity features so that entity disambiguation is not affected by noise. Tables 6 and 7 show that the joint SPSL and CCWR embedding models for Chinese short text entity recognition and linking have improved by 0.0731 compared to LSTM-CRF.

In summary, this paper proposes that SPSL and CCWR enrich the text context information and strengthen the ability of entity feature representation.

Acknowledgment. This work was financially supported by the Natural Science Foundation Youth Project of Hunan Province (No. 2019JJ50520).

References

1. Zhang, Y., Yang, J.: Chinese NER using lattice LSTM. arXiv preprint arXiv:1805.02023 (2018)
2. Morwal, S., Jahan, N., Chopra, D.: Named entity recognition using hidden Markov model (HMM). Int. J. Nat. Lang. Comput. (IJNLC) **1**(4), 15–23 (2012)
3. Chieu, H.L., Ng, H.T.: Named entity recognition: a maximum entropy approach using global information. In: Proceedings of the 19th International Conference on Computational Linguistics, vol. 1, pp. 1–7. Association for Computational Linguistics (2002)
4. Ju, Z., Wang, J., Zhu, F.: Named entity recognition from biomedical text using SVM. In: 2011 5th International Conference on Bioinformatics and Biomedical Engineering, pp. 1–4. IEEE (2011)
5. Huang, Z., Xu, W., Yu, K.: Bidirectional LSTM-CRF models for sequence tagging. arXiv preprint arXiv:1508.01991 (2015)
6. Turian, J., Ratinov, L., Bengio, Y.: Word representations: a simple and general method for semi-supervised learning. In: Proceedings of the 48th Annual Meeting of the Association for Computational Linguistics, pp. 384–394. Association for Computational Linguistics (2010)
7. Collobert, R., Weston, J., Bottou, L., et al.: Natural language processing (almost) from scratch. J. Mach. Learn. Res. **12**, 2493–2537 (2011)

8. Kim, Y.: Convolutional neural networks for sentence classification. arXiv preprint arXiv: 1408.5882 (2014)

9. Lample, G., Ballesteros, M., Subramanian, S., et al.: Neural architectures for named entity recognition. arXiv preprint arXiv:1603.01360 (2016)

10. Peters, M.E., Ammar, W., Bhagavatula, C., et al.: Semi-supervised sequence tagging with bidirectional language models. arXiv preprint arXiv:1705.00108 (2017)

11. Bagga, A., Baldwin, B.: Entity-based cross-document co-referencing using the vector space model. In: Proceedings of the 36th Annual Meeting of the Association for Computational Linguistics and 17th International Conference on Computational Linguistics, vol. 1, pp. 79–85. Association for Computational Linguistics (1998)

12. Fleischman, M., Hovy, E.: Fine grained classification of named entities. In: Proceedings of the 19th International Conference on Computational Linguistics, vol. 1, pp. 1–7. Association for Computational Linguistics (2002)

13. Miller, G.A., Charles, W.G.: Contextual correlates of semantic similarity. Lang. Cogn. Process. 6(1), 1–28 (1991)

14. Chiu, J.P.C., Nichols, E.: Named entity recognition with bidirectional LSTM-CNNs. Trans. Assoc. Comput. Linguist. 4, 357–370 (2016)

15. Zeng, Y., Yang, H., Feng, Y., Wang, Z., Zhao, D.: A convolution BiLSTM neural network model for Chinese event extraction. In: Lin, C.-Y., Xue, N., Zhao, D., Huang, X., Feng, Y. (eds.) ICCPOL/NLPCC -2016. LNCS (LNAI), vol. 10102, pp. 275–287. Springer, Cham (2016). https://doi.org/10.1007/978-3-319-50496-4_23

16. Parada, C., Dredze, M., Jelinek, F.: OOV sensitive named-entity recognition in speech. In: Twelfth Annual Conference of the International Speech Communication Association (2011)

17. Kim, Y., Jernite, Y., Sontag, D., et al.: Character-aware neural language models. In: Thirtieth AAAI Conference on Artificial Intelligence (2016)

18. Levy, O., Goldberg, Y.: Neural word embedding as implicit matrix factorization. In: Advances in Neural Information Processing Systems, pp. 2177–2185 (2014)

19. Chen, X., Xu, L., Liu, Z., et al.: Joint learning of character and word embeddings. In: Twenty-Fourth International Joint Conference on Artificial Intelligence (2015)

20. Cao, K., Rei, M.: A joint model for word embedding and word morphology. arXiv preprint arXiv:1606.02601 (2016)

21. Zheng, X., Chen, H., Xu, T.: Deep learning for Chinese word segmentation and POS tagging. In: Proceedings of the 2013 Conference on Empirical Methods in Natural Language Processing, pp. 647–657 (2013)

A Reading Assistant System for Blind People Based on Hand Gesture Recognition

Qiang Lu[1,2], Guangtao Zhai[1,2](✉), Xiongkuo Min[1,2], and Yucheng Zhu[1,2]

[1] Shanghai Institute for Advanced Communication and Data Science,
Shanghai, China
[2] Shanghai Jiao Tong University, Shanghai, China
{erislu,zhaiguangtao,minxiongkuo,zyc420}@sjtu.edu.cn

Abstract. A reading assistant system for blind people based on hand gesture recognition is proposed in this paper. This system consists of seven modules: camera input module, page adjustment module, page information retrieval module, hand pose estimation module, hand gesture recognition module, media controller and audio output device. In the page adjustment module, Hough line detection and local OCR (Optical Character Recognition) are used to rectify text orientation. In the hand gesture recognition module, we propose three practical methods: geometry model, heatmap model and keypoint model. Geometry model recognizes different gestures by geometrical characteristics of hand. Heatmap model which is based on image classification algorithm uses CNN (Convolutional Neural Network) to classify various hand gestures. To simplify the networks in heatmap model, we extract 21 keypoints from a hand heatmap and make them a dataset of points coordinates for training classifier. These three methods can get good results of gesture recognition. By recognizing gestures, our designed system can realize perfect reading assistant function.

Keywords: Reading assistant · OCR · Hand gesture recognition · Convolutional pose machine · Heatmap · Keypoint · Geometry

1 Introduction

For now, reading is still the most important way for people to gain useful information. However, there are millions of visually impaired people in the world and some of them are blind [10]. It is a great loss for these people to be unable to read the text. And few practical electrical devices are developed in the market to help them read in other ways. This paper proposes a new reading assistant system for blind people to help them read freely in phonetic form.

Supported in part by Shanghai Municipal Commission of Science and Technology (17JC1402900), and in part by the National Natural Science Foundation of China under Grants 61831015.

G. Zhai et al. (Eds.): IFTC 2019, CCIS 1181, pp. 200–211, 2020.
https://doi.org/10.1007/978-981-15-3341-9_17

Several reading assistant systems have been constructed for blind people. Brabyn et al. [2] have designed a remote reading system for sightless persons. Portable camera-based reading system has been proposed to help visually impaired people read labels on hand-held objects [11,16]. Jirasuwankul [9] proposes a projective transform technique to perform anti-skew of text. Sabab et al. [12] designed a smart device with a multimodal system that can convert any document to the interpreted form to a blind. Felix et al. [6] developed an Android mobile app that focuses on voice assistant, image recognition for the blind. In the filed of hand gesture recognition, Islam et al. [8] use deep convolutional neural network for extracting efficient features of hand images and Multiclass Support Vector Machine is used for recognizing gesture. Sun et al. [14] establish Gaussian mixture model according to the skin colors. Moreover, it also segments hand gestures by combining with AdaBoost classifier based on Haar features. Chaman et al. [5] propose a gesture recognition method by combining skin color detection with finger detection.

Although some systems are designed to help blind people read, there are still many limitations in practical usage [7]. Complex background will decrease accuracy and degrade user experience in many scenarios. Besides, the biggest deficiency is lack of interaction between human and machine. So we design the gesture recognition module in our system to realize human machine interaction. At runtime the webcam will capture the image of reading material for visually impaired people, and corresponding audio resource will be prepared well backstage. Meanwhile this system is monitoring the video from the webcam. If a specific gesture appears, this system will response immediately, such as playing a specified recording or stopping playing.

The rest of the paper is organized as follow. Section 2 presents the overall framework of our system firstly. The page adjustment method and three gesture recognition methods are introduced in the latter part of Sect. 2. In Sect. 3 experimental results including practical effects, gesture recognition accuracy, runtime performance and some other indices are shown in detail. Section 4 concludes this article with future research direction.

2   Methodology

This paper proposes a prototype system of reading assistant for blind people. As we can see from Fig. 1, this system consists of seven functional modules. The camera input module is a common webcam which captures images and transmits them to subsequent components. Page adjustment module is used to adjust text orientation.

Page information retrieval module mainly finds corresponding audio resource from the Internet according to the captured image. As for traditional printed books, Optical Character Recognition (OCR) can be used to extract text from white pages. Then Text To Speech (TTS) will synthesize voice on the basis of the extracted text and convey it to media controller. If this book read by blind person has been scanned completely and audio database is made already,

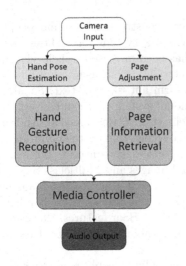

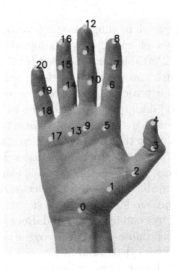

Fig. 1. Overall framework of this reading assistant system

Fig. 2. Output keypoints [13]

page information retrieval module will extract feature points of current page and retrieve them in database. Once feature points are matched successfully, the right audio data will be sent to media controller. In other special situations, books with a QR code in each page could be printed specially for the blind. When contents of one page are too complex to recognize, page information retrieval module can still gain audio resources by scanning this QR code.

We have selected Convolutional Pose Machine (CPM) as the method of hand pose estimation which has high accuracy and efficiency in most situations. So we apply it as a pre-processing procedure of gesture recognition. In convolutional pose machine, deep convolutional network is incorporated into pose machine framework to learn image features for better recognizing performance [15]. The belief map of each component is used to express the spatial constraint between the components. Belief map and feature map are passed along as data across the network [3]. To avoid vanishing gradients during deep network, each stage has supervisory training. Convolutional pose machine is an end-to-end architecture used to predict each part of an object [4]. The output are keypoints of all parts which can be seen from Fig. 2. Heatmaps from multiple convolutional layers can also represent the hand skeleton well.

Hand gesture recognition module is the crucial part of the whole system. In view of actual using situation of blind people, we propose three practical methods: geometry model, hand heatmap model and hand keypoint model. Media controller takes outputs of page information retrieval module and hand gesture recognition module. Signal from page information retrieval module is the audio data while signal from hand gesture recognition module is controlling signal. Media controller changes the playing status according to controlling signal. Audio output device is a loudspeaker used to convert these signals to sounds.

2.1   Page Adjustment

Blind persons don't know how to put books with right direction below the camera due to loss of vision. In many cases the images captured by cameras will contain scripts with different skew angles. To increase the robustness to the skew of text, we should do the skew detection and correction. So this module is used to adjust text orientation when the system detects text.

As for some slant text, it is necessary to rotate it to a desired position where text lines are horizontal. Firstly we should take the Fourier transform of the text image. In frequency domain, high frequency components represent the detail and texture information of the image and low frequency components represent the contour information of the image [1]. The Fourier transform of an image is usually represented by an amplitude image which can be calculated by Eq. (1) below.

$$Mag = \sqrt{Re(DFT)^2 + Im(DFT)^2}. \tag{1}$$

The range of amplitude is very large, so log function is used to narrow it down. Every pixel value of Fourier transform image should be normalized to $[0, 1]$ and then multiply by 255 for display. Binarization processing is required for subsequent line detection. The next step is Hough transformation for detecting lines. Usually three lines can be detected in the image: a horizontal line, a vertical one and a slant one with a certain angle. We should gain the degree value and spin original image opposite with this value. In some cases, more than one slant lines appear in the image. The degree value of each angle is supposed to be calculated, and the average should be selected as the rotating degree.

After finishing adjusting slant text, whether the text is upside down should be determined. At present a common method is to compare the recognition rate of original and inverted images. In this way we should implement OCR twice which is poor in real-time performance. What we has adopted is local OCR which means selecting a square region from a text image and making it a test sample. By scanning this sample and inverted one, whether the text image is upside down can be confirmed based on the recognition rates of these two scanning processes. Higher recognition rate means the text is in the right direction. So subsequent global OCR could be conducted with this right direction.

2.2   Geometry Model

Variation of relative position among hand joints contribute to different gestures. A hand comprises five fingers and each of them has four joints. In daily life, every finger is generally in two states: bend and straight. Different fingers in different states lead to different gestures. In order to determine which state a finger is in, we have introduced two mathematical concepts: angle and distance. 21 keypoints shown in Fig. 2 are adopted as representation of a hand.

For example, distance between Point 0 and Point 1 can be calculated.

$$d(P_0, P_1) = \sqrt{(x_1 - x_0)^2 + (y_1 - y_0)^2}. \tag{2}$$

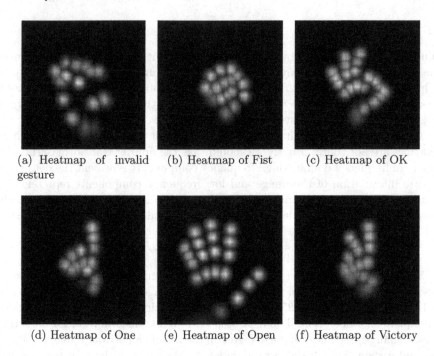

(a) Heatmap of invalid gesture   (b) Heatmap of Fist   (c) Heatmap of OK

(d) Heatmap of One   (e) Heatmap of Open   (f) Heatmap of Victory

Fig. 3. Heatmaps

Where, P_0 denotes Point 0 with (x_0, y_0) as its coordinate and P_1 denotes Point 1 with (x_1, y_1) as its coordinate. If distance between Point 8 and Point 0 is shorter than that between Point 5 and Point 0, we can say index finger is bend generally. Whether Point 5, Point 6, Point 7 and Point 8 are collinear can also indicate the state of index finger. With Point 6 as the vertex, line 5–6 and line 6–7 as each side, the angle of 5-6-7 can show collinear state of these three points. Cosine law is applied here to calculate the cosine of $\angle 6$.

$$\cos \angle 6 = \frac{d(P_5, P_6)^2 + d(P_6, P_7)^2 - d(P_5, P_7)^2}{2 \times d(P_5, P_6) \times d(P_6, P_7)}. \tag{3}$$

If $\cos \angle 6 < -0.9$, it can be consumed that Point 5, Point 6 and Point 7 are collinear. In a similiar way, if Point 5, Point 6 and Point 7 are collinear has been confirmed, there is no doubt that index finger is straight.

By this way, state of every finger can be determined. Through the establishment of congruent relationship between states of fingers and hand gestures, our system is able to recognize each gesture. However, conditions of recognizing gesture precisely through geometry model are very rigor. As far as blind people are concerned, gestures they try to pose may be not so standard in geometry, because they can't see whether their gestures are standard enough. Hence, we propose the next two models.

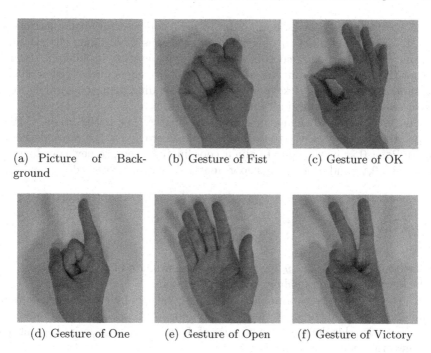

(a) Picture of Back-ground (b) Gesture of Fist (c) Gesture of OK

(d) Gesture of One (e) Gesture of Open (f) Gesture of Victory

Fig. 4. Gestures

2.3 Hand Heatmap Model

In recent years, convolutional neural network (CNN) is usually used to classify different types of images. Hand gesture region could be splitted out and stored locally as a database to train a classifier for gesture classification. For a wide range of images, CNN can extract potential features for the classification model. But for the blind, when interacting with this system, they don't know if their hands are placed in a right place. Maybe the background is extremely complex which will lead to false detection. Therefore, to eliminate influences of background, we have chosen hand heatmaps from hand pose estimation module as the input of convolutional neural network. Heatmaps of five gestures are shown in Fig. 3. We design a convolutional neural network with two convolution layers whose kernel size is 5×5 and two full-connected layers. Training images are resized to 100×100 and the model can achieve convergence quickly on condition of setting suitable learning rate.

2.4 Hand Keypoint Model

From hand heatmaps we can see, white dots in these pictures represent hand joints and black region is unimportant background. In other words, the black region is redundant for CNN to extract image features. What really matters are positions of hand keypoints. So in order to reduce unnecessary computation, we use 21 keypoints of hand to replace input heatmaps for training the classifier. Certainly hand keypoints coordinates of the same gesture in different pictures are not the same due to different scales. But what remains unchanged is relative location relation among 21 keypoints. So before sending them to neural network, it is necessary to transform all these coordinates.

The first step is to change origin of coordinates. The initial origin of these keypoints coordinates is the top left corner of camera scene. For convenience, we choose Point 0 as new origin of palmar coordinate system, so new coordinate of Point 0 is $(0,0)$ and other points coordinates should be processed.

$$
\begin{aligned}
P_i(x_i, y_i) &\rightarrow P'_i(x'_i, y'_i), \\
x'_i = x_i - x_0, &\ y'_i = y_i - y_0, \\
i &= 0, 1, 2 \ldots 20.
\end{aligned}
\tag{4}
$$

In the above formula, (x_i, y_i) is the original coordinate of Point i while (x'_i, y'_i) is the new coordinate in palmar coordinate system.

The second step is to normalize coordinate values of all keypoints. In views of different scales, distance between two keypoints of adjacent joints may vary a lot. To revise this deviation, maximum of distance between each keypoint and the new origin should be calculated.

$$
\begin{aligned}
d_M &= \max dis(P'_i, P'_0) \\
&= \max \sqrt{x'^2_i + y'^2_i}, \\
i &= 0, 1, 2 \ldots 20.
\end{aligned}
\tag{5}
$$

Where, d_M denotes the maximum of distance. Then present coordinates divided by this maximum are the normalized ones.

$$
\begin{aligned}
P'_i(x'_i, y'_i) &\rightarrow P''_i(x''_i, y''_i), \\
x''_i = x'_i/d_M, &\ y''_i = y'_i/d_M, \\
i &= 0, 1, 2 \ldots 20.
\end{aligned}
\tag{6}
$$

The last optional step is to guarantee that most keypoints are in the same quadrant which will simplify the subsequent calculation further. Point 9 is chosen as the reference because it is close to the palm of the hand. If x-coordinate of Point 9 is minus, change signs of x-coordinates of all points once and do the same things to y-coordinates according to the sign of y-coordinate of Point 9.

These modified keypoints coordinates, 42 numbers will be converted to a tensor and be transmitted to simple CNN with two convolution layers, two pooling layers whose pool size is 2×2 and two full-connected layers. The trained model can achieve high train efficiency and good accuracy.

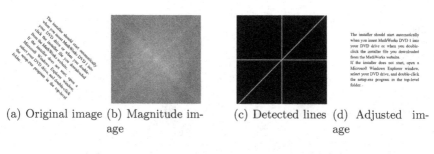

(a) Original image (b) Magnitude image (c) Detected lines (d) Adjusted image

Fig. 5. Page adjustment

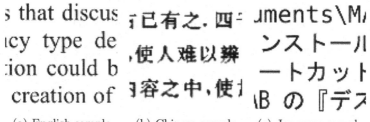

(a) English sample (b) Chinese sample (c) Japanese sample

Fig. 6. Local OCR

3 Experimental Results

3.1 Effects of Page Adjustment

Visual effects of page adjustment are shown in Fig. 5. Figure 5(a) is original slant text image and Fig. 5(d) is the adjusted one. Figure 5(b) is the magnitude image of Fourier transform and Fig. 5(c) shows three lines detected in magnitude image after binarization. We can see that slant text has been adjusted to horizontal successfully.

We have selected 300 text samples for testing local OCR and 3 examples are displayed in Fig. 6. As for English, Chinese and Japanese, local OCR has achieved accuracy of 97%, 95% and 94%. Local OCR is able to recognize more words in right direction of characters than in opposite direction so local OCR is a efficient way to determine book direction.

3.2 Effects of Hand Gesture Recognition

To help the blind interact with our system, five gestures are selected as operating commands to control the media player. These gestures are shown in Fig. 4.

Heatmaps of these gestures have been exhibited in Fig. 3. For each gesture, we have collected about 1500 pictures of heatmaps with labels. In consideration of possible misoperation of blind people, we also gather more than 1000 heatmaps

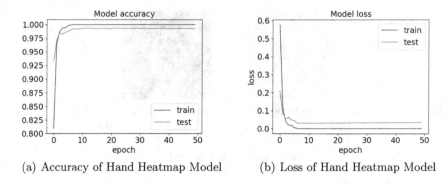

(a) Accuracy of Hand Heatmap Model   (b) Loss of Hand Heatmap Model

Fig. 7. Performance of hand heatmap model

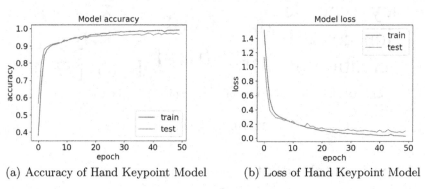

(a) Accuracy of Hand Keypoint Model   (b) Loss of Hand Keypoint Model

Fig. 8. Performance of hand keypoint model

of various invalid gestures. In this way the dataset of hand gesture heatmaps including more than 8000 pictures is built completely. Then this dataset is put into a convolutional neural network for training a gesture classifier. Training graphs of accuracy and loss are shown in Fig. 7. At last test accuracy has reached 99.42%.

At the same time of collecting heatmaps, we also have saved processed keypoints coordinates locally and made them a dataset of hand gesture keypoints. Each picture of gestures corresponds to 42 float-point coordinates and then this dataset is used to train a classifier. Test accuracy of keypoints model achieves 97.77% and Fig. 8 demonstrates variations of accuracy and loss in the process of training.

From Figs. 7 and 8 we can see, heatmap model could get a pretty high accuracy after a few epochs. Training time of keypoint model is much shorter which can be seen from Table 1.

Table 1. Training process of heatmap model and keypoint model

Epochs	Test accuracy (%)		Training time (s)	
	Heatmap	Keypoint	Heatmap	Keypoint
20	99.36	95.80	72.82	9.34
35	99.40	96.98	157.34	15.36
50	99.42	97.77	179.38	19.60

Table 2. Running performance of three models

Model	Average frame rate (fps)	Accuracy of recognition (%)
Geometry	27.50	84.60
Heatmap	22.00	92.45
Keypoint	24.45	94.32

As can be seen from Table 2, every method has its own unique strength and unavoidable drawback. Running frame rate of geometry model is the highest but its accuracy is low. In contrast, heatmap model can realize satisfactory precision with a relatively low fps (frame per second). Keypoint model has both high accuracy and frame rate. However its training process and predicting process are relatively complex, which is because keypoints coordinates from CPM should be normalized to be used.

3.3 Overall Working Effects

Our system is a reading assistant system used to help visually impaired people read. We have assigned five gestures as basic buttons of media playing: play, pause, stop, last song and next one. In this system, if one gesture is detected in consecutive 8 frames, the corresponding command will be executed. In consideration of runtime frame rate and recognition accuracy, responding time of one command is about one third second. Actual running effects are like the Figs. 9 and 10.

For the blind, what they need to do are turning on this system, opening a book and posing the gestures. The system will firstly locate text regions, recognize every character and then generate sounds according to these text. Meanwhile, if a specific gesture appears under the camera, it will be recognized rightly and corresponding command will be carried out. As a result, the text they want to know will be read aloud.

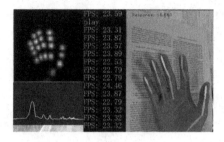
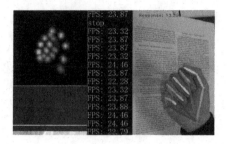

Fig. 9. Effect of play   **Fig. 10.** Effect of stop

4   Conclusion

For the purpose of helping blind people read in phonetic form, this reading assistant system is designed. Unlike any other reading assistant platform, this system has realized interaction between human and machine on the basis of hand gesture. Hough line detection and local OCR is adopted to adjust text orientation and three gesture recognition methods are proposed. From the experimental results, geometry model is the simplest one which require a small amount of computing resource. Heatmap model based on image classification of CNN is the most common one. Keypoint model can simplify training network and save computing power with effective detection.

Certainly, to improve robustness and efficiency of our system, affine transform is essential when the position of camera is not optimum and data enhancement is necessary in training hand heatmap model. New or modified model is needed in recognizing precisely gestures of special angles for blind people. All these are the future tasks for us to finish.

References

1. Abche, A.B., Yaacoub, F., Maalouf, A., Karam, E.: Image registration based on neural network and Fourier transform. In: 2006 International Conference of the IEEE Engineering in Medicine and Biology Society, pp. 4803–4806. IEEE (2006)
2. Brabyn, J., Crandall, W., Gerrey, W.: Remote reading systems for the blind: a potential application of virtual presence. In: 1992 14th Annual International Conference of the IEEE Engineering in Medicine and Biology Society, vol. 4, pp. 1538–1539. IEEE (1992)
3. Cao, Z., Hidalgo, G., Simon, T., Wei, S.E., Sheikh, Y.: OpenPose: real-time multi-person 2D pose estimation using part affinity fields. arXiv preprint arXiv:1812.08008 (2018)
4. Cao, Z., Simon, T., Wei, S.E., Sheikh, Y.: Realtime multi-person 2D pose estimation using part affinity fields. In: Proceedings of the IEEE Conference on Computer Vision and Pattern Recognition, pp. 7291–7299 (2017)
5. Chaman, S., D'souza, D., D'mello, B., Bhavsar, K., D'souza, T.: Real-time hand gesture communication system in Hindi for speech and hearing impaired. In: 2018 Second International Conference on Intelligent Computing and Control Systems (ICICCS), pp. 1954–1958. IEEE (2018)

6. Felix, S.M., Kumar, S., Veeramuthu, A.: A smart personal AI assistant for visually impaired people. In: 2018 2nd International Conference on Trends in Electronics and Informatics (ICOEI), pp. 1245–1250. IEEE (2018)
7. Hu, M., Chen, Y., Zhai, G., Gao, Z., Fan, L.: An overview of assistive devices for blind and visually impaired people. Int. J. Robot. Autom. **34**(5), 580–598 (2019)
8. Islam, M.R., Mitu, U.K., Bhuiyan, R.A., Shin, J.: Hand gesture feature extraction using deep convolutional neural network for recognizing American sign language. In: 2018 4th International Conference on Frontiers of Signal Processing (ICFSP), pp. 115–119. IEEE (2018)
9. Jirasuwankul, N.: Effect of text orientation to OCR error and anti-skew of text using projective transform technique. In: 2011 IEEE/ASME International Conference on Advanced Intelligent Mechatronics (AIM), pp. 856–861. IEEE (2011)
10. O'day, B.L., Killeen, M., Iezzoni, L.I.: Improving health care experiences of persons who are blind or have low vision: suggestions from focus groups. Am. J. Med. Qual. **19**(5), 193–200 (2004)
11. Rajput, R., Borse, R.: Alternative product label reading and speech conversion: an aid for blind person. In: 2017 International Conference on Computing, Communication, Control and Automation (ICCUBEA), pp. 1–6. IEEE (2017)
12. Sabab, S.A., Ashmafee, M.H.: Blind reader: an intelligent assistant for blind. In: 2016 19th International Conference on Computer and Information Technology (ICCIT), pp. 229–234. IEEE (2016)
13. Simon, T., Joo, H., Matthews, I., Sheikh, Y.: Hand keypoint detection in single images using multiview bootstrapping. In: CVPR (2017)
14. Sun, J.H., Ji, T.T., Zhang, S.B., Yang, J.K., Ji, G.R.: Research on the hand gesture recognition based on deep learning. In: 2018 12th International Symposium on Antennas, Propagation and EM Theory (ISAPE), pp. 1–4. IEEE (2018)
15. Wei, S.E., Ramakrishna, V., Kanade, T., Sheikh, Y.: Convolutional pose machines. In: Proceedings of the IEEE Conference on Computer Vision and Pattern Recognition, pp. 4724–4732 (2016)
16. Yi, C., Tian, Y., Arditi, A.: Portable camera-based assistive text and product label reading from hand-held objects for blind persons. IEEE/ASME Trans. Mechatron. **19**(3), 808–817 (2014)

Intrusion Detection Based on Fusing Deep Neural Networks and Transfer Learning

Yingying Xu[1], Zhi Liu[1(✉)], Yanmiao Li[2], Yushuo Zheng[3], Haixia Hou[2],
Mingcheng Gao[2], Yongsheng Song[4], and Yang Xin[2]

[1] School of Information Science and Engineering, Shandong University,
Qingdao 266237, China
liuzhi@sdu.edu.cn
[2] Center of Information Security,
Beijing University of Posts and Telecommunications, Beijing 100876, China
[3] High School Attached to Shandong Normal University, Jinan 250100, China
[4] Kedun Technology Co., Ltd., Yantai 265200, China

Abstract. Intrusion detection is the key research direction of network
security. With the rapid growth of network data and the enrichment
of intrusion methods, traditional detection methods can no longer meet
the security requirements of the current network environment. In recent
years, the rapid development of deep learning technology and its great
success in the field of imagery have provided a new solution for network
intrusion detection. By visualizing the network data, this paper proposes an intrusion detection method based on deep learning and transfer
learning, which transforms the intrusion detection problem into image
recognition problem. Specifically, the stream data visualization method
is used to present the network data in the form of a grayscale image, and
then a deep learning method is introduced to detect the network intrusion
according to the texture features in the grayscale image. Finally, transfer
learning is introduced to improve the iterative efficiency and adaptability
of the model. The experimental results show that the proposed method
is more efficient and robust than the mainstream machine learning and
deep learning methods, and has better generalization performance, which
can detect new intrusion methods more effectively.

Keywords: Deep learning · Intrusion detection · Convolutional neural
network · Transfer learning

1 Introduction

Network intrusion is one of the biggest threat to the cyberspace, and it refers
to any behaviors that attempt to compromise the confidentiality, integrity, and
availability of a host or network [1]. However, the current cyber security situation is not optimistic. The traditional intrusion detection models are often in a
passive situation for new intrusion methods, and cannot effectively solve various unknown intrusion detection problems. Therefore, it is imperative to explore

© Springer Nature Singapore Pte Ltd. 2020
G. Zhai et al. (Eds.): IFTC 2019, CCIS 1181, pp. 212–223, 2020.
https://doi.org/10.1007/978-981-15-3341-9_18

and research more accurate and efficient intelligent network intrusion detection solutions [2].

The intrusion detection system (IDS) is the "anti-theft alarm" in the field of computer network security. Its aim is to protect the system by combining alarms issued when the network security is threatened and inspection entities capable of responding to the alarm and taking corresponding actions. IDS can detect, ensure, and identify unauthorized use, replication, change, and destruction of information systems. There are currently three main types of intrusion detection: misuse based (sometimes called signature-based) [3], anomaly based [4], and mixed [5].

This paper combines the deep learning and transfer learning to apply deep convolutional neural networks to intrusion detection systems. The structure of this paper is organized as follows: The work related to existing intrusion detection methods are reviewed in Sect. 2. In Sect. 3, the proposed method are introduced, and followed by the experimental evaluations and comparison in Sect. 4. The conclusions are presented in Sect. 5.

2   Related Work

In recent years, with the rapid development of Internet and Internet of Things (IoT), more and more researches have been applied to network intrusion detection. Common models include deep confidence network (DBN), convolution neural network (CNN), and recurrent neural network (RNN) [6–8].

Staudemeyer et al. [9] proposed a long short-term memory (LSTM) intrusion detection model and proved that the LSTM classifier has advantages over other classifiers when detecting denial of service (DoS) attacks and probing attacks. In order to solve the problem of high false alarm rate, Kim et al. [10] proposed a language modeling based method to improve the performance of LSTM-based host intrusion detection system. Agarap et al. [11] introduced the support vector machine (SVM) in the final output layer of the gated recurrent unit (GRU) model to replace Softmax, and used the model for binary classification of intrusion detection. Kolosnjaji et al. [12] proposed a neural network consisting of convolution and feedforward neural structures, which provides a hierarchical feature extraction method for vectorizing the features of n-gram instructions and convolution.

Due to the problems of large amount of information, huge data volume, and long training time, local optimization is prone to occur in network data processing. Therefore, Zhao et al. [13] developed an intrusion detection method based on DBN and probabilistic neural network (PNN). By comparing different DBN structures, adjusting the number of layers, and hidden elements in the network model, Gao et al. [14] obtained a four-layer DBN model and obtained a better intrusion detection results. Tan et al. [15] designed a DBN-based Ad hoc network intrusion detection model and performed simulation experiments on the NS2 platform. Wang et al. [16] proposed an end-to-end network data classification method based on one-dimensional CNN. The method integrates feature extraction, feature selection, and classifier into a unified end-to-end framework and automatically learns the nonlinear relationship between the original input and the expected output [17]. Yin et al. [18] proposed the RNN-IDS and used the NSL-KDD data set to evaluate the performance in binary classification and multiple classification, as well as the influence of the number of neurons and different learning rates on the performance of the model.

3   Proposed Method

Convolutional neural network is an artificial neural network and has become a research hotspot in the field of image recognition and speech analysis [19]. The weight-sharing network structure makes it more similar to biological neural networks, thus reducing the complexity of the network model and decreasing the number of weights. Convolutional networks are specifically designed to recognize two-dimensional shapes that are highly invariant to translation, scaling, tilting, or other forms of deformation [20]. Therefore, images can be used directly as input to the model, avoiding complex feature extraction and data reconstruction in traditional machine learning algorithms. This advantage is more obvious when the network input is a multi-dimensional image.

CNN has three main ways to reduce network training parameters: local perception, weight sharing, and pooling [21]. The most powerful part of CNN is its ability to learn different levels of features from a large amount of data. Therefore, this paper proposes a CNN-based intrusion detection model, which introduces CNN into the field of intrusion detection by converting network data into grayscale images. Figure 1 is the system block diagram of the proposed CNN-IDS, which mainly includes three parts: data preprocessing, model training and intrusion detection.

(1) Data preprocessing

The purpose of data preprocessing is to convert raw network traffic data into grayscale images, which mainly include numeralization and normalization. The purpose of numeralization is to convert character data into numeric values for subsequent processing. This paper uses the traditional one-hot encoding method to digitize character data and labels.

Although the processed data can be trained, the numerical values in the sample are quite different, which will affect the convergence speed and training effect of the model. Therefore, normalization is required. The commonly used Min-Max method is used in the proposed system.

(2) Model training

Figure 2 shows the classic LeNet network [22] and the network architecture adopted in this paper will be continuously adjusted based on this. The specific model parameters will be determined through subsequent experiments, and the model with good prediction effect on the training set will be selected as the final detection model.

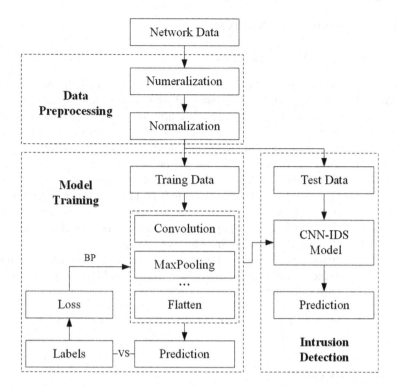

Fig. 1. The system block diagram of the proposed CNN-IDS.

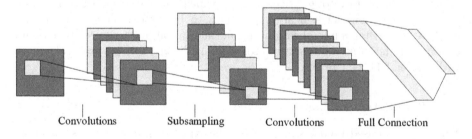

Fig. 2. An example of the LeNet model structure [22].

(3) Intrusion detection

The trained CNN-IDS model is used to perform intrusion detection on the test data, and the detection performance of the models under different parameters are compared. Finally, the best performance model is selected as the final detection model.

4   Experiment and Analysis

4.1   Dataset and Evaluate Metics

The data used in this paper is the KDD CUP 99 data set [23]. As the one of the most widely used intrusion detection datasets, it contains 4.9 million attack records, and each containing 41 network traffic characteristics and one label. The network traffic characteristics include basic features, content features, and traffic statistics features. The labels are mainly divided into five categories: DoS, normal, probe, remote to local (R2L), and user to root (U2R).

The KDD CUP 99 data set also includes a data set with 10% training subsets and test subsets, and the test subsets including 17 attack types that do not appear in the 10% training set, so they can be used to test the generalization capabilities of the model. In this paper, the model is first trained using a 10% training set, and then fine-tuned using the test subset.

In order to evaluate the effect of the model, some commonly used evaluation metrics are used, including accuracy, precision, recall, and f1-scores. The f1-score is used as the main indicator:

$$
\begin{aligned}
F1\text{-}score &= 2 \times Precision \times Recall/(Precision + Recall) \\
&= 2 \times TP/(2 \times TP + FN + FP).
\end{aligned}
\tag{1}
$$

4.2   Data Processing

Due to the CNNs have better image processing capabilities, this paper treats network traffic data into grayscale images, which translates intrusion detection problems into image classification problems to better exploit the advantages of CNN [24]. After numeralization and normalization, each record in the data contains 119 features. Adding two zeros padding directly forms a grayscale image with a size of 11×11 pixels.

4.3   Parameter Settings

Model training used 150000 samples from the experimental data set as a training set, 10000 samples as a validation set, and 10000 samples as a test set. An important indicator of whether a model is valid is the degree of convergence of training set losses. At the same time, the accuracy of the test set is also compared to determine whether it is over-fitting. In order to determine the appropriate model structure, this paper adopts a simple to complex strategy, starting from the simplest three-layer network, and continuously increasing the depth of the model. The results of ten experiments are shown in the Table 1, where the numbers in the model structure represent the output dimensions of each layer, and the last row is the test accuracy.

It can be seen from the Table 1, 6 layer structure and 7 layers structure with similar convergence efficiency and accuracy of the model selection, layer number of smaller models more often competitive advantage, because the deployment

Table 1. Iterative loss corresponding to different model structures.

	32-1152-2	32-64-256-2	32-64-256-128-2	32-64-256-128-64-2	32-64-256-128-64-32-2
1	0.3544	1.4109	3.9069	6.2213	8.0365
2	0.3513	1.0841	2.3240	3.6576	4.8646
3	0.3298	0.9542	1.8129	2.6463	3.0567
4	0.3103	0.8286	1.3818	1.9001	1.8237
5	0.3008	0.7035	0.9705	1.2824	1.0342
6	0.2903	0.6154	0.7031	0.7791	0.5420
7	0.2860	0.5525	0.5335	0.3466	0.3056
8	0.2911	0.5080	0.4707	0.1952	0.1946
9	0.2750	0.4864	0.4436	0.1391	0.1387
10	0.3042	0.4797	0.3959	0.1055	0.1188
Accuracy	90.3%	92.5%	98.5%	99.01%	99.1%

and training time will be reduced. Therefore, the 6 - layer structure is adopted as the detection model in this paper.

As the most important super parameter, it needs to be determined first. In general, the method of violence test is adopted, that is, multiple learning rate values are selected for testing, and the optimal solution is determined through the loss curve. In this paper, [0.00001, 0.00005, 0.0001, 0.0003] was selected as the comparison interval to select the appropriate learning rate. As can be seen from the Fig. 3, among the four learning rates, 0.00001 is too small and does not converge after 30 iterations; 0.0003 is too large and the curve is uneven. The difference between 0.00005 and 0.001 is not significant, but the former curve is smoother. Therefore, the learning rate of the CNN model in this paper is 0.00005.

The purpose of the optimizer is to optimize the gradient update process. In the deep learning model, the most commonly used optimizers are SGD, RMSprop, Adagrad, Adadelta, Adam, and Nadam. At the learning rate of 0.00005, each optimizer behaves as shown in Fig. 4. It can be seen that the Adadelta, SGD, and Adagrad optimizers perform poorly and the optimization results are not ideal, while RMSprop, Adam, and Nadam have better optimization effects. In this paper, Adam is used as the model optimizer.

The purpose of regularization is to reduce the overfitting degree of the model and improve the generalization ability of the model. The regularization of CNN model mainly adopts L2 and Dropout methods. The following three situations are discussed here: (1) Only L2 regularization is used, (2) Dropout only, (3) Combine L2 and Dropout. The specific model performance is shown in the Fig. 5. It can be seen from the figure that simply using L2 regularization can get a better fitting effect after 25 iterations, and only using Dropout method there is an overfitting phenomenon. After combining the two methods, you can not only get better training results, but also achieve faster data fitting.

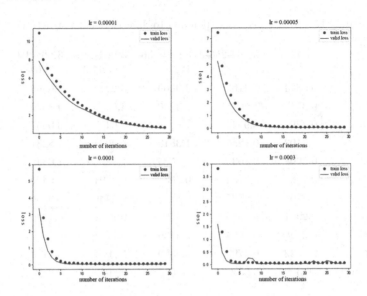

Fig. 3. Learning rate-loss map.

4.4   Comparison

Through the above experimental work, the six-layer structure of the model 32-64-256-128-64-2 are determined, 0.0005 are selected as the learning rate, adam are used as the optimization algorithm, and the L2 and Dropout regularization [25] methods are used. In order to reflect the effectiveness and superiority of the model, it needs to be compared with the state of the art machine learning and deep learning algorithms. This paper mainly selects SVM, decision tree (DT), kNN, LSTM, and DBN models for comparison [26]. The training set used is all KDD Cup 99 10% data set, experimental indicators using accuracy, precision, recall and f1-score. The comparison results are shown in Table 2.

Table 2. Comparison of results from different methods.

Method	Accuracy	Precision	Recall	F1-score
SVM	82.4%	74.0%	80.3%	0.77
KNN	85.2%	89.6%	75.6%	0.82
DT	91.9%	98.9%	86.0%	0.92
DBN	93.5%	92.3%	93.7%	0.93
LSTM	96.9%	98.8%	90.4%	0.95
Ours	97.9%	99.9%	97.7%	0.98

From the Table 2, we can see that different algorithms have good intrusion detection effects. Relatively speaking, the performance of traditional machine

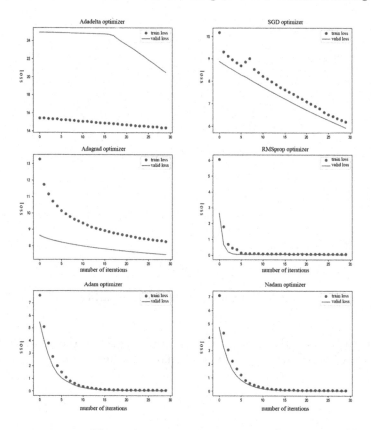

Fig. 4. Optimizer effect comparison.

learning algorithms is inferior to deep learning algorithms. The CNN-IDS proposed in this paper performs best in the models, reflecting the effectiveness and superiority of CNN in the problem of intrusion detection.

4.5 Transfer Learning and Fine-Tuning

In practical applications, the data that the model needs to process is much larger and more complex than the that in the experiment. In order to adapt the intrusion detection model to the new application environment, an effective solution is to introduce transfer learning [27].

The "corrected" data set in the KDD CUP 99 has 17 intrusion types not found in the training set, so it can be used to evaluate the generalization ability of the algorithm model to detect and identify new sample data. The corresponding f1-score were calculated using the sample numbers of six different sizes: 50000, 100000, 150000, 200000, 250000 and 300000. The results obtained are shown in the Fig. 6. It can be seen that the average f1-score obtained is 0.94, which

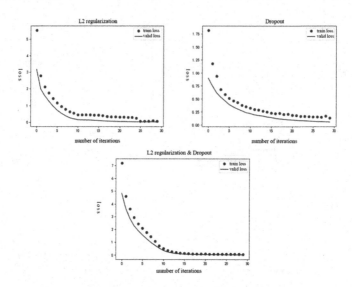

Fig. 5. Regularization method comparison.

is significantly lower than the 0.98 in the training set of the previous model. Therefore, it is necessary to fine tune the model for new data.

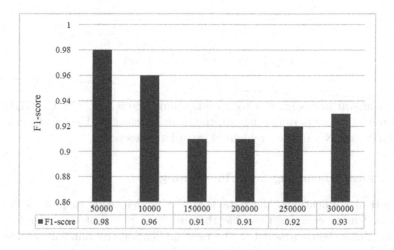

Fig. 6. The f1-score of different sample numbers.

Since the sample types of the two data sets used in this experiment are basically the same and the features are similar, the model used in the fine-tuning process is the same as the original model. The most suitable training method is obtained by testing different fine-tuning layers. Starting from the classification layer, experiments are performed one at a time and the parameters of the remaining layers are locked. In Fig. 6, the f1-score drops the fastest when the number of samples is from 50000 to 150000, so we uses the 100000 samples between them as the training data set. The test data is the first 300000 samples of the "corrected" data set. The experimental results are shown in the Table 3. When the number of training layers is increased to 3, the improvement of test effect is already limited.

Table 3. Fine-tuning experimental results.

Train layers	Weights	Rate	F1-score	Train time
1	65	0.10%	0.91	100 s
2	8321	13.80%	0.95	140 s
3	41217	68.70%	0.96	200 s

Considering that intrusion detection belongs to the field of network security, the harm of missed detection is far greater than the harm of error detection. Therefore, although there are many parameters that require training, three layers of fine tuning are still used in the experiment. Figure 7 shows the results of retesting after the transfer learning. The experimental results prove the validity and reliability of the proposed method. The fine-tuned model has been greatly improved and can better cope with the threat of new intrusion.

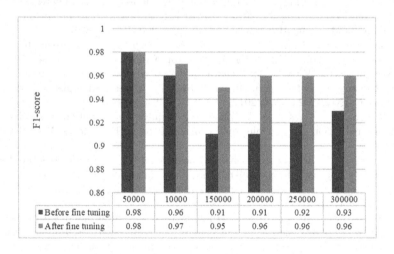

	50000	10000	150000	200000	250000	300000
Before fine tuning	0.98	0.96	0.91	0.91	0.92	0.93
After fine tuning	0.98	0.97	0.95	0.96	0.96	0.96

Fig. 7. Comparison of results before and after fine tuning.

5 Conclusion

In this paper, the network traffic data visualization method is used to transform the intrusion detection problem into image recognition problem, and then the deep learning technology is introduced into the network security field. The working principle of deep learning model is introduced in depth, and an intrusion detection model based on convolutional neural network is proposed. Based on the proposed model, the structure and hyperparameter values of the model are determined by comparing multiple tests using the KDD CUP 99 10% data set. The trained model is compared with the existing advanced intrusion detection methods to prove the validity of the experimental model. At the same time, transfer learning is introduced and the model is fine-tuned with new data, so that the intrusion detection model can obtain better detection capability at a faster speed to adapt to the rapid changes of the actual network.

Acknowledgment. This work was supported in part by the National Key R&D Program (No. 2018YFC0831006 and 2017YFB1400102), the Key Research and Development Plan of Shandong Province (No. 2017CXGC1503 and 2018GSF118228).

References

1. Buczak, A.L., Guven, E.: A survey of data mining and machine learning methods for cyber security intrusion detection. IEEE Commun. Surv. Tutor. **18**(2), 1153–1176 (2016)
2. Torres, P., Catania, C., Garcia, S., Garino, C.G.: An analysis of recurrent neural networks for botnet detection behavior. In: 2016 IEEE Biennial Congress of Argentina (ARGENCON), pp. 1–6. IEEE, Buenos Aires, June 2016
3. Alrawashdeh, K., Purdy, C.: Toward an online anomaly intrusion detection system based on deep learning. In: 2016 15th IEEE International Conference on Machine Learning and Applications (ICMLA), pp. 195–200. IEEE (2016)
4. Shone, N., Ngoc, T.N., Phai, V.D., Shi, Q.: A deep learning approach to network intrusion detection. IEEE Trans. Emerg. Top. Comput. Intell. **2**(1), 41–50 (2018)
5. Saxe, J., Berlin, K.: eXpose: a character-level convolutional neural network with embeddings for detecting malicious URLs, file paths and registry keys. arXiv preprint arXiv:1702.08568 (2017)
6. Xiao, Y., Xing, C., Zhang, T., Zhao, Z.: An intrusion detection model based on feature reduction and convolutional neural networks. IEEE Access **7**, 42210–42219 (2019)
7. Labonne, M., Olivereau, A., Polvé, B., Zeghlache, D.: A cascade-structured meta-specialists approach for neural network-based intrusion detection. In: 2019 16th IEEE Annual Consumer Communications & Networking Conference (CCNC), pp. 1–6. IEEE (2019)
8. Kasongo, S.M., Sun, Y.: A deep learning method with filter based feature engineering for wireless intrusion detection system. IEEE Access **7**, 38597–38607 (2019)
9. Staudemeyer, R.C.: Applying long short-term memory recurrent neural networks to intrusion detection. S. Afr. Comput. J. **56**(1), 136–154 (2015)
10. Kim, G., Yi, H., Lee, J., Paek, Y., Yoon, S.: LSTM-based system-call language modeling and robust ensemble method for designing host-based intrusion detection systems. arXiv preprint arXiv:1611.01726 (2016)
11. Agarap, A.F.M.: A neural network architecture combining gated recurrent unit (GRU) and support vector machine (SVM) for intrusion detection in network traffic data. In: Proceedings of the 2018 10th International Conference on Machine Learning and Computing, pp. 26–30. ACM (2018)

12. Kolosnjaji, B., Zarras, A., Webster, G., Eckert, C.: Deep learning for classification of malware system call sequences. In: Kang, B.H., Bai, Q. (eds.) AI 2016. LNCS (LNAI), vol. 9992, pp. 137–149. Springer, Cham (2016). https://doi.org/10.1007/978-3-319-50127-7_11

13. Zhao, G., Zhang, C., Zheng, L.: Intrusion detection using deep belief network and probabilistic neural network. In: 2017 IEEE International Conference on Computational Science and Engineering (CSE) and IEEE International Conference on Embedded and Ubiquitous Computing (EUC), vol. 1, pp. 639–642. IEEE (2017)

14. Gao, N., Gao, L., Gao, Q., Wang, H.: An intrusion detection model based on deep belief networks. In: 2014 Second International Conference on Advanced Cloud and Big Data, pp. 247–252. IEEE (2014)

15. Tan, Q.S., Huang, W., Li, Q.: An intrusion detection method based on DBN in ad hoc networks. In: Wireless Communication and Sensor Network: Proceedings of the International Conference on Wireless Communication and Sensor Network (WCSN 2015), pp. 477–485. World Scientific (2016)

16. Wang, W., Zhu, M., Wang, J., Zeng, X., Yang, Z.: End-to-end encrypted traffic classification with one-dimensional convolution neural networks. In: 2017 IEEE International Conference on Intelligence and Security Informatics (ISI), pp. 43–48. IEEE (2017)

17. Dubey, S., Dubey, J.: KBB: a hybrid method for intrusion detection. In: 2015 International Conference on Computer, Communication and Control (IC4), pp. 1–6. IEEE (2015)

18. Yin, C., Zhu, Y., Fei, J., He, X.: A deep learning approach for intrusion detection using recurrent neural networks. IEEE Access 5, 21954–21961 (2017)

19. Al-Qatf, M., Lasheng, Y., Al-Habib, M., Al-Sabahi, K.: Deep learning approach combining sparse autoencoder with SVM for network intrusion detection. IEEE Access 6, 52843–52856 (2018)

20. Li, Z., Qin, Z., Huang, K., Yang, X., Ye, S.: Intrusion detection using convolutional neural networks for representation learning. In: Liu, D., Xie, S., Li, Y., Zhao, D., El-Alfy, E.-S.M. (eds.) ICONIP 2017. LNCS, vol. 10638, pp. 858–866. Springer, Cham (2017). https://doi.org/10.1007/978-3-319-70139-4_87

21. Wang, Z.: Deep learning-based intrusion detection with adversaries. IEEE Access 6, 38367–38384 (2018)

22. LeCun, Y., Bottou, L., Bengio, Y., Haffner, P., et al.: Gradient-based learning applied to document recognition. Proc. IEEE 86(11), 2278–2324 (1998)

23. Siddique, K., Akhtar, Z., Khan, F.A., Kim, Y.: KDD Cup 99 data sets: a perspective on the role of data sets in network intrusion detection research. Computer 52(2), 41–51 (2019)

24. Wang, W., et al.: HAST-IDS: learning hierarchical spatial-temporal features using deep neural networks to improve intrusion detection. IEEE Access 6, 1792–1806 (2018)

25. Srivastava, N., Hinton, G., Krizhevsky, A., Sutskever, I., Salakhutdinov, R.: Dropout: a simple way to prevent neural networks from overfitting. J. Mach. Learn. Res. 15(1), 1929–1958 (2014)

26. Xin, Y., et al.: Machine learning and deep learning methods for cybersecurity. IEEE Access 6, 35365–35381 (2018)

27. Weiss, K.R., Khoshgoftaar, T.M.: Analysis of transfer learning performance measures. In: 2017 IEEE International Conference on Information Reuse and Integration (IRI), pp. 338–345. IEEE (2017)

The Competition of Garbage Classification Visual Recognition

Yiyi Lu, Hanlong Guo, Jiaming Ma, and Zhengrong Ge[✉]

Internet Department, China Telecom Shanghai Branch, Shanghai, China
gezhengrong.sh@chinatelecom.cn

Abstract. This paper introduces the details of "Tianyi Cup" Intelligent Environmental Protection challenge. It includes the assumption of different topics, the selection and preparation of garbage data sets, the segmentation and preprocessing of garbage data sets, as well as the results of competition run on the final data set. And put forward some reasonable suggestions to promote the level of garbage classification.

Keywords: Image classification · Garbage classification · Train set · Test set

1 Introduction

The sustainable development of economy, society and environment are commonly problems facing the international community. Circular economy, low-carbon development, green development and building a socialist harmonious society have always been the development concept advocated by China. China has a large population base, and the output of domestic waste produced every day cannot be underestimated. With the rapid development of society, garbage classification has become a very important aspect of urban management. The State Council issued a plan to promote garbage classification on March 30 2017, and set a goal for the recycling rate in cities where household garbage is sorted to reach 35 percent by 2020 [1].

In order to realize ecological development, it is necessary to reduce the pollution of domestic waste to the environment from the source. It is an important measure to implement the classification policy of Municipal Solid Waste (MSW) in community to realize the reduction, resource and harmless treatment of MSW. In recent years, the relevant departments of the state have repeatedly emphasized the implementation of classified treatment of living waste and taken certain measures, but found that the effect is not obvious from the actual life. Shanghai started to carry out the mandatory garbage classification from July 1, 2019 in accordance with the Shanghai Household Waste Management Regulation [2]. In order to achieve the purpose of classified collection, classified transportation and classified disposal of household garbage, residents and garbage disposal staffs need to assist in garbage classification from the front-end of life (residential garbage dumping) to the back-end of life (garbage collection and disposal stations) [3]. In the process of garbage classification, image recognition can be applied to greatly improve the efficiency and accuracy of the work.

G. Zhai et al. (Eds.): IFTC 2019, CCIS 1181, pp. 224–234, 2020.
https://doi.org/10.1007/978-981-15-3341-9_19

2 Building of Garbage Datasets

Shanghai, as one of the first pilot city for classified treatment of MSW, put forward that the policy of classified treatment of MSW should be implemented in the urban center first in the early 21st century. In order to ensure the correctness of the policy, in 2000, the first pilot communities in the center of Shanghai began the pilot work, and more than 3700 residential areas were forced to implement the classified management measures of domestic waste. However, due to the long-term habits and concepts of the citizens, as well as the restrictions of the terminal treatment technology and mechanical treatment capacity of the midway transportation, the effect of classified collection is not significant.

The current Household Waste Management Regulation requires residents and garbage disposal staffs to sort garbage by hand. Residents and garbage disposal staffs can be confused about how to determine the correct way to dispose of a large variety of garbage. The motivation for this competition was to find an automatic method for sorting garbage. This has the potential to make garbage classification more efficient and help reduce waste. This will not only have positive environmental effects but also beneficial economic effects.

2.1 Garbage Categories Setups

According to the current garbage classification standards, garbage can be roughly divided into four different types: (1) green color (compostable) - waste from food, fruit, vegetable or leaf; (2) red color (hazardous) - waste from electronic, medicine, can of pesticide or battery; (3) yellow color (recyclable) - paper, plastic and metal; and (4) blue color (general) - foam food containers, plastic bag and others stuff that unworthy to recycle.

Following the household waste management regulation, the garbage dataset should include such classes: plastic, paper, metal, glass and other categories, and each garbage category is divided into 10 different categories for reference. In [4], Yang etc. created a dataset that took images of a single piece of garbage with six classes consisting of glass, paper, metal, plastic, cardboard, and trash with around 400–500 images for each class.

In order to further deepen the understanding of garbage classification and further promote the awareness of garbage classification, we built garbage datasets with four categories, such as recyclable garbage, hazardous garbage, wet garbage and dry gar-bage. Hazardous garbage, wet garbage and dry garbage belong to non-recyclable garbage. For each category, there include several type items as follows.

Recyclable Garbage (Fig. 1):
1. Plastics: beverage bottle, shampoo bottle, edible oil drums, yogurt boxes, plastic bowls (basins), plastic playways, etc.
1. Paper: cardboard box, newspaper, envelope, printing paper, advertisement leaflet, etc.
2. Glass: wine bottle, windowpanes, medicine bottle, soy sauce bottle, cooking bottle, etc.

3. Metals (copper, iron, aluminum, etc.): pop cans, metal components, milk powder cans, etc.
4. Clothing: clothes, bed sheets, quilts, shoes, towels, plush toys, etc.
5. Electronics: TV, washing machine, air conditioner, refrigerator, computer, camera, mobile phone, charger, children's electric toy, remote controller, CD, digital music player, U disk, etc.

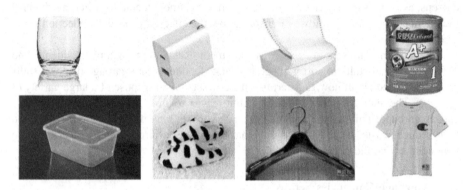

Fig. 1. Samples of recyclable items.

Hazardous Garbage (Fig. 2):

Batteries (mercury-containing, nickel-hydrogen and nickel-cadmium batteries, etc.) rechargeable batteries, button cell, storage batteries, fluorescent tubes, mercury-containing fluorescent tubes, energy-saving lamps, other mercury thermometers, medicines, paint cans, pesticide cans, X-ray films and other photosensitive films.

Fig. 2. Sample of hazardous garbage.

Wet Garbage (Fig. 3):

1. Grain and its products: rice, flour, beans and other grains and their processed foods.
2. Fruits and vegetables: melons, green leafy vegetables, root vegetables, mushroom and other vegetables, as well as the pulp and peel of various fruits.
3. Meat: eggs, chickens, ducks, pigs, beef, mutton and processed meat and eggs.

4. Aquatic products: fish, shrimp, shellfish (without shells) and their processed foods.
5. Contents of canned food.
6. Seasonings: sugar, salt, monosodium glutamate, starch, chili sauce and other sauces.
7. Snacks: cakes, candy, nuts, cheese.
8. Dry goods: air drying, drying food, dried mushrooms, red dates, dried longans, etc.
9. Brewing beverage: instant beverage powder, tea bag, tea residue and Chinese medicine residue.
10. Potted plants: flowers, branches and leaves.

Fig. 3. Sample of wet garbage.

Dry Garbage (Fig. 4):
1. In addition to the above three categories of garbage.
2. Garbage with indistinguishable categories.
3. Napkin, diaper, sanitary napkin paper, thin plastic bag, paper with more serious pollution, lime soil, large bones, shells, ceramic fragments.

Fig. 4. Sample of dry garbage.

2.2   The Building of Garbage Datasets

2.2.1   Item Category Detection Dataset

Suggested that an intelligent garbage bin is set in the community, it requires to detect whether a single item placed in the garbage bin belongs to recyclable garbage or not. Deep learning can be applied to learn to classify garbage with the help of garbage dataset [5].

To train and learn a single item category classification model, it requires a lot of pictures of related category. The dataset should meet the following requirements:

1. The dataset should include more than 30,000 pictures.
2. 10,000 pictures for recyclable items, and 20,000 pictures for non-recyclable items.

For the items in the picture, they need to be marked out with four-category labels (recyclable, harmful waste, wet garbage, and dry garbage) according to the category label of the four categories in Sect. 2.1.

2.2.2   Multiple-Item Categories Monitoring Dataset

The garbage disposal station is suggested to provide with an intelligent garbage sorting conveyor belt (conveying equipment with similar functions) to detect the category of each item on the conveyor belt or a specific category of items on the conveyor belt. The dataset for such scenario should meet the following requirements:

1. There are more than 100 million pictures, and more than 2,000 pictures for each category.
2. The background of each picture is basically the same and simple.
3. There is more than one item on each picture and the brightness is the same (preferably the light source is the same).
4. Mark out the items in the picture with four categories labels.

The pictures in the dataset are stored in jpeg files. The pictures come from the Internet and cannot be used for any commercial purposes.

3   Garbage Classification Competition

3.1   Preprocessing of the Data for Competition

The task of the competition is to distinguish between recyclable and non-recyclable garbage (hazardous garbage, dry garbage and wet garbage). In order to avoid the possibility of submitting manually tagged results by contestants, the test set is divided into "open" and "closed" parts (6:4). The closed test set is given to the contestants 24–48 h before the end of the competition. The contestants submit their result on the closed test set within a very short period as the final result.

The final folder name and the approximate number of pictures included in the data set for this competition are shown in the Table 1.

Table 1. First round competition dataset

	Test_Public	Test_Private	Train Set
Recyclable sample	2506 pieces	1671 pieces	About 6265 pieces
Non-recyclable sample	2932 pieces	1954 pieces	About 7329 pieces

Test_Public Dataset

1. In the "recyclable" folder of positive samples, a specified number of pictures are randomly selected from each subfolder and put into the folder named "test_public_positive" (no subfolder is required).
2. In the "non-recyclable" folder of negative samples, a specified number of pictures are randomly selected from each sub-folder and put into the folder named "test_public_negative" (no sub-folder is required).
3. Create a new answer_public.csv file as the public answer file. The first column is "pic_id", which is the file name of all pictures in the "test_public_positive" and "test_public_negative" folders. The second column is "label", which is the first "0" or "1" of the picture file name.
4. Then, change the first place of the first column "pic_id" to "2", thus completing the production of the public answer set.
5. Put all the pictures in the "test_puplic_positive" and "test_public_negative" folders into one folder "test_public" and change the first digit of the file name of all the pictures to "2"

Test_Private Dataset

1. In the "recyclable" folder of positive samples, a specified number of pictures are randomly selected from the remaining pictures in each subfolder and put into the folder named "test_private_positive" (no subfolder is required).
2. In the negative sample [non-recyclable] folder, a specified number of pictures are randomly selected from the remaining pictures in each subfolder and put into the folder named "test_private_negative" (no subfolder is required).
3. Create a new answer_private.csv file as the private answer file. The first column is "pic_id", which is the file name of all pictures in the "test_private_positive" and "test_private_negative" folders. The second column is "label", which is the first "0" or "1" of the picture file name.
4. Then, change the first place of the first column "pic_id" to "3", thus completing the production of the private answer set.
5. Put all the pictures in the "test_private_positive" and "test_private_negative" folders into one folder "test_private", and change the first digit of the file name of all the pictures to "3". At last, fill in the "green space" in the "ideal data description" below for the size and quantity of the final folder.

Make a Training Set

The remaining pictures of each subfolder in the positive sample "recyclable" and negative sample "non-recyclable" folders are put into a folder named "train set"

Participants need to deal with less of possible picture quality problems and less of picture duplication problems.

Training Set index
The training set is about 1.8G in size and contains all the pictures in the folder "trainset", totaling 13,371 pieces of pictures.

1. 6259 positive samples are pictures with the file name beginning with "10".
2. 7112 negative samples are pictures with the file name beginning with "00".

Test Set Index
Test sets are divided into test public sets and test private sets:

1. The test public set is about 417 M in size and contains all the pictures in the folder "test_public", totaling more than 5,300 pieces of pictures.
2. The size of the test set private is about 271 M, which is all the pictures in the folder "test_private", totaling more than 3500 pieces of pictures.

Data Processing Indicators for the Second Round
The data processing principle for the second round was quite the same as the first round. In order to increase the difficulty of the competition, we added more noise in the test set. The method to add noise was to add shields and twist the items randomly. The number of pictures for each category of recyclable garbage and non-recyclable garbage are shown in Figs. 5 and 6.

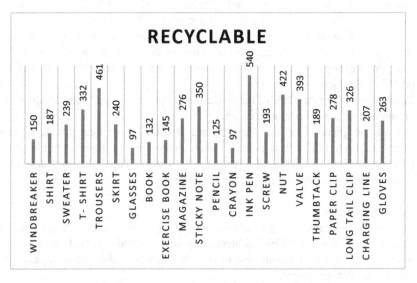

Fig. 5. Number of recyclable garbage.

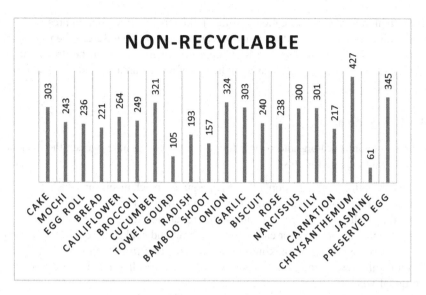

Fig. 6. Number of non-recyclable garbage.

3.2 Test Selection Index

The area under Receiver Operating Characteristic Curve (ROC) curve, namely AUC, was used for classification evaluation. ROC analysis is a binary classification model, namely a model with only two types of output results. The boundary between classes must be defined by a threshold. The case prediction of the binary classification model has four outcomes:

1. True Positive (TP): The forecast is positive, but in fact it is also positive.
2. Fake Positive (FP): The forecast is positive, but it is actually negative.
3. True Negative (TN): The forecast is negative, but in fact it is also negative.
4. Fake Negative (FN): The forecast is negative, but it is actually positive.

True Positive Rate (TPR): In all samples that are actually positive, the ratio that is correctly judged to be positive.

$$TPR = TP/(TP + FN)$$

Fake Positive Rate (FPR): In all samples that are actually negative, the ratio that is wrongly judged to be positive.

$$FPR = FP/(FP + TN)$$

Given a binary classification model and its threshold, a (X = FPR, Y = TPR) coordinate point can be calculated from the (positive/negative) real and predicted values of all samples. Diagonal lines from (0, 0) to (1,1) divide ROC space into upper left/lower right regions. Points above this line represent a good classification result (better than random classification), while points below this line represent a poor classification result (worse than random classification). Therefore, when comparing different classification models, the ROC curve of each model can be drawn, and the area under the curve can be used as an indicator of the advantages and disadvantages of the models.

3.3   Rule Setting

In order to avoid cheating and over fitting as much as possible, we set the evaluation rules and rankings as follows:

1. The evaluation ranking adopts the private/public mechanism, in which the private ranking corresponds to a certain proportion of the data in the submitted result file, and the public ranking corresponds to the remaining data.
2. Provide the opportunity for each team to submit and evaluate the ranking five times a day, update the public ranking in real time, ranking from high to low. If the team submits the results multiple times a day, the new result version will overwrite the original version.
3. Due to the influence of the generalization performance of the use model, the submission with the highest score in the public list may not have the highest score in private, so it is necessary for the contestants to select two result files that consider both the generalization performance and the model score from their effective submission to enter the private list evaluation.
4. The private ranking will be announced after the end of the competition. The final effective results and effective ranking of the competition will be based on the private ranking as the final objective ranking.

We will review the [work notebook] submitted by the top 30 contestants in the final objective ranking, including the complete code and process documents. The review requirements are as follows:

- Submitted [work notebook] as required.
- No cheating.
- Can be successfully reproduced in k-lab.
- Subjective evaluation results as comprehensive ranking.

3.4   Test Result

After the data testing process is completed, we make a feedback statistic on the results of data testing.

In the Test set preparation phase, the recognized form of garbage in the real scene is simulated through certain processing methods. After testing, the data can better support the competition and solve the problems.

Data sets behave as follows in commonly used public models:

Model	Epoch	AUC
ResNet50	00020	0.93239211
Inception	00020	0.91119868
X-ception	00020	0.92133434

Notice: The above models show the results of debugging by different testers.

The model data feedback of dataset fully shows the liquidity and variability of dataset itself. With the progress of people's living standard and the development of people's life style, the standard of garbage classification is in a dynamic change. Therefore, it is very important for the whole competition system to keep the garbage classification data set in an open state. At the same time, the openness of the data set also conforms to the scientific principle, which ensures the scientific value of this competition.

3.5   Performance of the Participants

The number of the participant was 194 with 170 teams. There were 1297 submissions in total. The final performance of the competition is as following (Fig. 7):

	rill	0.99030670	27	19-09-06 12:04
	东道主	0.98998849	40	19-09-06 12:04
	初学者	0.98858940	13	19-09-06 12:04

Fig. 7. Top 3 teams.

The Rank 1 team used ensemble model to solve the problem. The ensemble model consists of "Resnext101 32x8d", "Resnext101 32x16d", "EfficientNet b3" and "EfficientBet b4" with different weights. The performance of these models is as followings:

	Resnext101 32x8d	Resnext101 32x16d	EfficientNet b3	EfficientNet b4
Single model	0.9858	0.9866	0.9837	0.9845
Ensemble	0.9903(Rank1)			

The Rank 2 team used 448*448 large scale ensemble model to solve the problem. The ensemble model consists of "Resnet152" and "Inceptionv3" with different weights. The performance of these models are as followings:

	Resnet152	Inceptionv3
Single model	0.98556	0.98514
Ensemble	0.989988 (Rank2)	

The Rank 3 team used SEnet154 model with domain adaptation to solve the problem. The performance of the model is 0.988589.

4 Conclusion

The model data feedback of dataset fully shows the liquidity and variability of dataset itself. With the progress of people's living standard and the development of people's life style, the standard of garbage classification is in a dynamic change. Therefore, it is very important for the whole competition system to keep the garbage classification data set in an open state. At the same time, the openness of the data set also conforms to the scientific principle, which ensures the scientific value of this competition. The dataset will be open to the public after the competition and can be requested via email.

References

1. http://english.www.gov.cn/policies/latest_releases/2017/03/30/content_281475612021190.htm
2. Web news. https://www.china-admissions.com/blog/china-waste-management-shanghai/
3. Ning, Y., Cao, Y.: Analysis of Shanghai living garbage classification and processing. In: 5th International Conference on Education, Management and Information Technology, pp. 56–61 (2019)
4. Thung, G., Yang, M.: Classification of trash for recyclability status (2016)
5. Mittal, G., Yagnik, K.B., Garg, M., Krishnan, N.C.: Spotgarbage: smartphone app to detect garbage using deep learning. In: Proceedings of the 2016 ACM International Joint Conference on Pervasive and Ubiquitous Computing, New York, pp. 940–945 (2016)

An Academic Achievement Prediction Model Enhanced by Stacking Network

Shaofeng Zhang$^{(\boxtimes)}$, Meng Liu, and Jingtao Zhang

University of Electronic Science and Technology of China, Chengdu, China
{2017221005016,2017221004027,jtzhang}@std.uestc.edu.cn

Abstract. This article focuses on the use of data mining and machine learning in AI education to achieve better prediction accuracy of students' academic achievement. So far, there are already many well-built gradient boosting machines for small data sets prediction, such as light-GBM, XGBoost, etc. Based on this, we presented and experimented a new method in a regression prediction. Our Stacking Network combines the traditional ensemble models with the idea of deep neural network. Compared with the original Stacking method, Stacking Network can infinitely increase the number of layers, making the effect of Stacking Network much higher than that of traditional Stacking. Simultaneously, compared with deep neural network, this Stacking Network inherits the advantages of the Boosting machines. We have applied this approach to achieve higher accuracy and better speed than the conventional Deep neural network. And also, we achieved a highest rank on the Middle School Grade Dataset provided by Shanghai Telecom Corporation.

Keywords: Machine learning · Data mining · Stacking network

1 Introduction

With the development of artificial intelligence (AI), emerging applications as "AI + education" and "intelligent classroom" are gradually coming into people's sight. Increasing number of schools integrate AI assistants into the classroom. At present, China is entering the era of "intelligent education". In a traditional classroom, due to time and resources constraints, teachers and parents can not balance between students' learning status and academic progress, and pay less attention to the large number of data that illustrates students' actual problems and situations. Intelligence education combines the traditional education industry with the latest artificial intelligence algorithm, deeply excavates the students' performance data of historical answers at various points, and predicts the students' academic performance in the examination.

In recent years, there are some performance predictions based on Neural Networks [6–8], yet many are limited by high dimensions and few data sets, which makes tuning parameters of neural networks more difficult and easier to fall into a local minimum.

© Springer Nature Singapore Pte Ltd. 2020
G. Zhai et al. (Eds.): IFTC 2019, CCIS 1181, pp. 235–245, 2020.
https://doi.org/10.1007/978-981-15-3341-9_20

The conventional prediction models based on boosting machines often perform well on small data sets with short training time, but the accuracy is not restricted by overfitting. Therefore, we propose a combination of the deep neural network and the stacking to improve the prediction accuracy.

In this research, we predicted the exam results and trend in the last semester of junior middle school students, by using the information of relevant exams and test points throughout the middle school. Datasets were provided by Shanghai Telecom Corporation.

Our work combines the fundamental idea of gradient boosting machines and the uses of the Idea of Nonlinear Transform from Deep Neural Network. The presented work optimized the required time for training, and increased accuracy. Section 2 presents data sets cleaning and baseline models constructed by Light-GBM [1], XGBoost [2], Catboost [10]. Ensembled algorithms including Blending, Stacking were used to decrease the Mean Absolute Error (MAE), followed by the use of Stacking network. Section 3 explains the different outcomes from the above approaches and rationalized the reasoning. We have concluded with future scope in Section 4.

2   Related Work

In the following, we first review recent work on boosting learning algorithm. We then review the two main ensemble trick.

2.1   Boosting Learning Algorithm

Most recent boosting learning algorithm XGBoost [2] utilize its second derivative exceed the Gradient Boosting Decision Tree [34]. Catboost [10] rely on its excellent handling of discrete features, which make it become one of the most useful data science competition algorithm. Others adopted this idea, LightGBM [1] proposed an algorithm based on histogram calculation make it more efficiently and quickly.

2.2   Ensemble Trick

Since the 1980s, many scholars have studied ensemble learning. Before that, there have been two excellent ensemble methods, Stacking [3] and Blending [5]. These approaches have shown ensemble trick has huge improvement compared with single model.

3   Model Structure

3.1   Structure-Stacking Network

Bagging. Bagging is a technique for reducing generalization errors by combining several models. The main idea is to train several different models separately, and then let all models vote on the output of the test samples.

$$
\begin{aligned}
avg(((g_t(x) - f_t(x)))^2 &= avg(g_t^2 - 2g_tf + f^2) \\
&= avg(g_t^2) - 2Gf + f^2 \\
&= avg(g_t^2) - G^2 + (G - f)^2 \\
&= avg(g_t^2) - 2G^2 + G^2 + (G - f)^2 \\
&= avg(g_t^2 - 2g_tG + G^2) + (G - f)^2 \\
&= avg((g_t - G)^2) + ((G - f))^2.
\end{aligned}
\tag{1}
$$

As can be seen from the formula (1), the difference of Bagging depends on the effect of a single model and the difference between models.

Traditional Stacking. Stacking is a hierarchical model ensemble method, which uses the prediction results learned by the basic model to retrain the sub-learner (usually linear model) and assign weight to the results of each model.

Fig. 1. Traditional stacking method.

Stacking Network. Compared to stacking, conventional Deep Neural Network require too many parameters to estimate, cause resulting in worse accuracy especially in small datasets. On the other hand, Stacking only re-predicts the prediction results of the basic models, and can only increase the breadth of Stacking by adding the basic models, but can not increase the depth of the overall structure. Here, we propose a Stacking Network, which can preserve both the depth of the deep network and the accuracy of the integration model. Given a data set, using the 5-fold Stacking method, each fold uses different model training and predicts the validation data.

Given a data set, using the 5-fold Stacking method, each fold uses different model training and predicts the validation data.

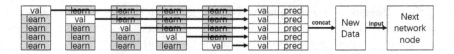

Fig. 2. Stacking network node structure.

Figure 2 is a classic network node in our staking network. For a single node, it will convey primitive features (X) to the next node as well as labels (Y) created from K-fold prediction.

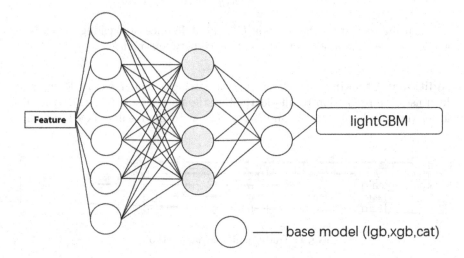

Fig. 3. Stacking network structure.

Figure 3 is our Stacking Network structure. Each node is a single model, such as LightGBM, XGBoost and CatBoost. The structure between the nodes in the previous layer and the nodes in the next layer is shown in Fig. 2.

3.2   Single Model-LightGBM

The basic idea of histogram algorithm is to discretize the continuous floating-point eigenvalues into K integers and construct a histogram with a width of K. When traversing the data, the discretized values are used as the cumulative statistics of the index in the histogram. After traversing the data once, the histogram accumulates the required statistics. Then, according to the discrete values of the histogram, it traverses to find the optimal segmentation points.

Because LightGBM has high training speed and accuracy, we use LightGBM as a single node in the middle layer of Stacking Network.

Algorithm 1: Histogram-based Algorithm	**Algorithm 2:** Gradient-based One-Side Sampling

Input: I: training data, d: max depth
Input: m: feature dimension
$nodeSet \leftarrow \{0\}$ ▷ tree nodes in current level
$rowSet \leftarrow \{\{0,1,2,...\}\}$ ▷ data indices in tree nodes
for $i = 1$ **to** d **do**
  for $node$ **in** $nodeSet$ **do**
    $usedRows \leftarrow rowSet[node]$
    for $k = 1$ **to** m **do**
      $H \leftarrow$ new Histogram()
      ▷ Build histogram
      for j **in** $usedRows$ **do**
        bin $\leftarrow I.f[k][j].$bin
        $H[$bin$].y \leftarrow H[$bin$].y + I.y[j]$
        $H[$bin$].n \leftarrow H[$bin$].n + 1$
      Find the best split on histogram H.
    ...
  Update $rowSet$ and $nodeSet$ according to the best split points.
  ...
	Input: I: training data, d: iterations
	Input: a: sampling ratio of large gradient data
	Input: b: sampling ratio of small gradient data
	Input: $loss$: loss function, L: weak learner
	models $\leftarrow \{\}$, fact $\leftarrow \frac{1-a}{b}$
	topN $\leftarrow$ a $\times$ len(I) , randN $\leftarrow$ b $\times$ len(I)
	for $i = 1$ **to** d **do**
	  preds $\leftarrow$ models.predict(I)
	  g $\leftarrow loss(I,$ preds), w $\leftarrow \{1,1,...\}$
	  sorted $\leftarrow$ GetSortedIndices(abs(g))
	  topSet $\leftarrow$ sorted[1:topN]
	  randSet $\leftarrow$ RandomPick(sorted[topN:len(I)], randN)
	  usedSet $\leftarrow$ topSet + randSet
	  w[randSet] $\times$ = fact ▷ Assign weight $fact$ to the small gradient data.
	  newModel $\leftarrow$ L(I[usedSet], $-$ g[usedSet], w[usedSet])
	  models.append(newModel)

Fig. 4. LightGBM algorithm [1].

3.3   Single Model-XGBoost

It is an efficient and scalable implementation of gradient boosting framework by (Friedman, 2001) (Friedman et al., 2000). Compare with Gradient Boosting Decision Tree [34], It has two main advantages. First shrinkage learning rate, when XGBoost completes an iteration, it will multiply the weight of the leaf node by the coefficient, mainly to weaken the influence of each tree, so that there is more learning space behind. Second, the traditional GBDT only uses first-order derivative information when optimizing, and XGBoost performs second-order Taylor expansion on the cost function to obtain first-order and second-order derivatives.

Algorithm 1: Exact Greedy Algorithm for Split Finding

Input: I, instance set of current node
Input: d, feature dimension
$gain \leftarrow 0$
$G \leftarrow \sum_{i \in I} g_i$, $H \leftarrow \sum_{i \in I} h_i$
for $k = 1$ **to** m **do**
　　$G_L \leftarrow 0$, $H_L \leftarrow 0$
　　for j **in** *sorted(I, by $\mathbf{x}_{jk}$)* **do**
　　　　$G_L \leftarrow G_L + g_j$, $H_L \leftarrow H_L + h_j$
　　　　$G_R \leftarrow G - G_L$, $H_R \leftarrow H - H_L$
　　　　$score \leftarrow \max(score, \frac{G_L^2}{H_L+\lambda} + \frac{G_R^2}{H_R+\lambda} - \frac{G^2}{H+\lambda})$
　　end
end
Output: Split with max score

Fig. 5. XGBoost algorithm-1 [2].

Algorithm 2: Approximate Algorithm for Split Finding

---

for $k = 1$ **to** m **do**
　　Propose $S_k = \{s_{k1}, s_{k2}, \cdots s_{kl}\}$ by percentiles on feature k.
　　Proposal can be done per tree (global), or per split(local).
end
for $k = 1$ **to** m **do**
　　$G_{kv} \leftarrow= \sum_{j \in \{j | s_{k,v} \geq \mathbf{x}_{jk} > s_{k,v-1}\}} g_j$
　　$H_{kv} \leftarrow= \sum_{j \in \{j | s_{k,v} \geq \mathbf{x}_{jk} > s_{k,v-1}\}} h_j$
end
Follow same step as in previous section to find max
score only among proposed splits.

Fig. 6. XGBoost algorithm-2 [2].

Algorithm 3: Sparsity-aware Split Finding

Input: I, instance set of current node
Input: $I_k = \{i \in I | x_{ik} \neq \text{missing}\}$
Input: d, feature dimension
Also applies to the approximate setting, only collect statistics of non-missing entries into buckets
$gain \leftarrow 0$
$G \leftarrow \sum_{i \in I} g_i, H \leftarrow \sum_{i \in I} h_i$
for $k = 1$ **to** m **do**
$\quad$ // enumerate missing value goto right
$\quad G_L \leftarrow 0, \ H_L \leftarrow 0$
$\quad$ **for** j in sorted(I_k, ascent order by x_{jk}) **do**
$\quad\quad G_L \leftarrow G_L + g_j, \ H_L \leftarrow H_L + h_j$
$\quad\quad G_R \leftarrow G - G_L, \ H_R \leftarrow H - H_L$
$\quad\quad$ score $\leftarrow \max(score, \frac{G_L^2}{H_L+\lambda} + \frac{G_R^2}{H_R+\lambda} - \frac{G^2}{H+\lambda})$
$\quad$ **end**
$\quad$ // enumerate missing value goto left
$\quad G_R \leftarrow 0, \ H_R \leftarrow 0$
$\quad$ **for** j in sorted(I_k, descent order by x_{jk}) **do**
$\quad\quad G_R \leftarrow G_R + g_j, \ H_R \leftarrow H_R + h_j$
$\quad\quad G_L \leftarrow G - G_R, \ H_L \leftarrow H - H_R$
$\quad\quad$ score $\leftarrow \max(score, \frac{G_L^2}{H_L+\lambda} + \frac{G_R^2}{H_R+\lambda} - \frac{G^2}{H+\lambda})$
$\quad$ **end**
end
Output: Split and default directions with max gain

Fig. 7. XGBoost algorithm-3 [2].

3.4 Single Model-CatBoost

Catboost is a Boosting algorithm with high precision. It handles multiple feature offsets based on XGBoost and utilizes the residual fitting idea of GBDT.

Algorithm 1: Updating the models and calculating model values for gradient estimation

---

input : $\{(\mathbf{X}_k, Y_k)\}_{k=1}^n$ ordered according to σ, the number of trees I;

$M_i \leftarrow 0$ for $i = 1..n$;

for $iter \leftarrow 1$ **to** I **do**

    for $i \leftarrow 1$ **to** n **do**

        for $j \leftarrow 1$ **to** $i - 1$ **do**

            $g_j \leftarrow \frac{d}{da}Loss(y_j, a)|_{a=M_i(\mathbf{X}_j)}$;

        $M \leftarrow LearnOneTree((\mathbf{X}_j, g_j)$ for $j = 1..i - 1)$;

        $M_i \leftarrow M_i + M$;

return $M_1 \ldots M_n$; $M_1(\mathbf{X}_1), M_2(\mathbf{X}_2) M_n(\mathbf{X}_n)$

Fig. 8. Catboost algorithm [10].

4   Experiment and Analysis

In this section, we evaluate proposed ensemble trick. We select the data set provided by Shanghai Telecom Company. The data set contains the exam scores of 500 students in a middle school in junior high school. The training set contains the exam scores of the student (without the last exam in the last year), the distribution of test sites for each test and the basic information of the student. The test label we need to predict is the last exam score in the last year.

In our experiment, we choose CatBoost as our first layers, XGBoost as our second layers. The last two layers' models are LightGBM. CatBoost can effectively use discrete features, and the prediction error is slightly lower than Light-GBM and XGBoost, so as the first layer, it can effectively reduce information loss. Each tree in LightGBM algorithm can use the gradient of the previous tree to sample, which greatly reduces the speed of training and prediction. Therefore, we use LightGBM algorithm as the second two layers of stacking network to improve the speed of prediction of our model.

Table 1. Final results from different model.

Merge function	Metric function	
	MAE	10*log10(MSE)
lightGBM	3.4585	7.4621
XGBoost	3.4265	7.4348
CatBoost	3.3742	7.4068
Blending	3.2695	7.3821
Stacking	3.1486	7.3210
Stacking network	3.0856	7.2746

As shown in Table 1, the performance of single gradient boosting machines are not as good as that of the Blending and Stacking models, while the MAE of Stacking Network is much better than that of Blending and Stacking. When Stacking network re-integrates the integrated model, it not only increases the width of the model, but also increases the depth of the model, which is analogous to the Deep Neural Network. It is noteworthy that the output of the deep neural network node to the next node is a single value, while the output of Stacking Network to the next node model is the data predicted by the model.

5   Conclusion and Future Work

In this paper, we proposed a novel approach for predicting students' academic achievement (exam results) based on Stacking Network. Stacking Network is a method of multi-level integration built upon integration models. It combines traditional gradient boosting machines with deep neural network to make up for the deficiency that traditional Stacking can only be integrated twice at most. Stacking Network can not only be used in this competition, but also provides an overall holistic idea to overcome the accuracy shortage of gradient boosting machines as well as curse of dimensionality of the Deep Neural Network. In our case, experiments have demonstrated that Stacking Network is much more effective than traditional Stacking and Blending on small data sets. The Stacking Network multilayer integration training is relatively slow, but the prediction time depends on the time required for the last layer model prediction. In this case, if time allows, such method will increase the training accuracy.

As a future work, It is important to reduce training time by optimizing the network structure of Stacking Network and find a way to reduce the MAE caused by the initial several layers of Stacking Network.

References

1. Ke, G., Meng, Q., Finley, T., et al.: Lightgbm: a highly efficient gradient boosting decision tree. In: Advances in Neural Information Processing Systems, pp. 3146–3154 (2017)
2. Chen, T., Guestrin, C.: XGBoost: a scalable tree boosting system. In: Proceedings of the 22nd ACM SIGKDD International Conference on Knowledge Discovery and Data Mining, pp. 785–794. ACM (2016)
3. Lemley, M.A., Shapiro, C.: Patent holdup and royalty stacking. Tex. L. Rev. **2006**, 85 (1991)
4. Breiman, L.: Bagging predictors. Mach. Learn. **24**(2), 123–140 (1996)
5. Fauconnier, G., Turner, M.: The Way We Think: Conceptual Blending and the Mind's Hidden Complexities. Basic Books, New York (2008)
6. Rowley, H.A., Baluja, S., Kanade, T.: Neural network-based face detection. IEEE Trans. Pattern Anal. Mach. Intell. **20**(1), 23–38 (1998)
7. Specht, D.F.: A general regression neural network. IEEE Trans. Neural Netw. **2**(6), 568–576 (1991)
8. Krogh, A., Vedelsby, J.: Neural network ensembles, cross validation, and active learning. In: Advances in Neural Information Processing Systems, pp. 231–238 (1995)

9. Li, J., Chang, H., Yang, J.: Sparse deep stacking network for image classification. In: Twenty-Ninth AAAI Conference on Artificial Intelligence (2015)
10. Prokhorenkova, L., Gusev, G., Vorobev, A., et al.: CatBoost: unbiased boosting with categorical features. In: Advances in Neural Information Processing Systems, pp. 6638–6648 (2018)
11. Odom, M.D., Sharda, R.: A neural network model for bankruptcy prediction. In: 1990 IJCNN International Joint Conference on Neural Networks, pp. 163–168. IEEE (1990)
12. Rose, S.: Mortality risk score prediction in an elderly population using machine learning. Am. J. Epidemiol. **177**(5), 443–452 (2013)
13. Grady, J., Oakley, T., Coulson, S.: Blending and metaphor. Amst. Stud. Theory Hist. Linguist. Sci. Ser. **4**, 101–124 (1999)
14. Freund, Y., Iyer, R., Schapire, R.E., et al.: An efficient boosting algorithm for combining preferences. J. Mach. Learn. Res. **4**(Nov), 933–969 (2003)
15. Schapire, R.E.: A brief introduction to boosting. In: IJCAI, vol. 99, pp. 1401–1406 (1999)
16. Solomatine, D.P., Shrestha, D.L.: AdaBoost. RT: a boosting algorithm for regression problems. In: 2004 IEEE International Joint Conference on Neural Networks (IEEE Cat. No. 04CH37541), vol. 2, pp. 1163–1168. IEEE (2004)
17. Kudo, T., Matsumoto, Y.: A boosting algorithm for classification of semi-structured text. In: Proceedings of the Conference on Empirical Methods in Natural Language Processing, pp. 301–308 (2004)
18. Yosinski, J., Clune, J., Bengio, Y., et al.: How transferable are features in deep neural networks? In: Advances in Neural Information Processing Systems, pp. 3320–3328 (2014)
19. Esteva, A., Kuprel, B., Novoa, R.A., et al.: Dermatologist-level classification of skin cancer with deep neural networks. Nature **542**(7639), 115 (2017)
20. Glorot, X., Bengio, Y.: Understanding the difficulty of training deep feedforward neural networks. In: Proceedings of the Thirteenth International Conference on Artificial Intelligence and Statistics, pp. 249–256 (2010)
21. Hecht-Nielsen, R.: Theory of the backpropagation neural network. In: Neural Networks for Perception, pp. 65–93. Academic Press (1992)
22. Maas, A.L., Hannun, A.Y., Ng, A.Y.: Rectifier nonlinearities improve neural network acoustic models. In: Proceedings of ICML, vol. 30, no. 1, p. 3 (2013)
23. Psaltis, D., Sideris, A., Yamamura, A.A.: A multilayered neural network controller. IEEE Control Syst. Mag. **8**(2), 17–21 (1988)
24. Kalchbrenner, N., Grefenstette, E., Blunsom, P.: A convolutional neural network for modelling sentences. arXiv preprint arXiv:1404.2188 (2014)
25. Saposnik, G., Cote, R., Mamdani, M., et al.: JURaSSiC: accuracy of clinician vs risk score prediction of ischemic stroke outcomes. Neurology **81**(5), 448–455 (2013)
26. Holland, P.W., Hoskens, M.: Classical test theory as a first-order item response theory: application to true-score prediction from a possibly nonparallel test. Psychometrika **68**(1), 123–149 (2003)
27. Liu, Y., An, A., Huang, X.: Boosting prediction accuracy on imbalanced datasets with SVM ensembles. In: Ng, W.-K., Kitsuregawa, M., Li, J., Chang, K. (eds.) PAKDD 2006. LNCS (LNAI), vol. 3918, pp. 107–118. Springer, Heidelberg (2006). https://doi.org/10.1007/11731139_15
28. Chawla, N.V., Lazarevic, A., Hall, L.O., Bowyer, K.W.: SMOTEBoost: improving prediction of the minority class in boosting. In: Lavrač, N., Gamberger, D., Todorovski, L., Blockeel, H. (eds.) PKDD 2003. LNCS (LNAI), vol. 2838, pp. 107–119. Springer, Heidelberg (2003). https://doi.org/10.1007/978-3-540-39804-2_12

29. Bühlmann, P., Hothorn, T.: Boosting algorithms: regularization, prediction and model fitting. Stat. Sci. **22**(4), 477–505 (2007)
30. Bagnell, J.A., Chestnutt, J., Bradley, D.M., et al.: Boosting structured prediction for imitation learning. In: Advances in Neural Information Processing Systems, pp. 1153–1160 (2007)
31. Du, X., Sun, S., Hu, C., et al.: DeepPPI: boosting prediction of protein-protein interactions with deep neural networks. J. Chem. Inf. Model. **57**(6), 1499–1510 (2017)
32. Lu, N., Lin, H., Lu, J., et al.: A customer churn prediction model in telecom industry using boosting. IEEE Trans. Industr. Inf. **10**(2), 1659–1665 (2012)
33. Bühlmann, P., Hothorn, T.: Twin boosting: improved feature selection and prediction. Stat. Comput. **20**(2), 119–138 (2010)
34. Friedman, J.H.: Stochastic gradient boosting. Comput. Stat. Data Anal. **38**(4), 367–378 (2002)

Quality Assessment

Blind Panoramic Image Quality Assessment Based on Project-Weighted Local Binary Pattern

Yumeng Xia[2], Yongfang Wang[1,2(✉)], and Peng Ye[2]

[1] Shanghai Institute for Advanced Communication and Data Science,
Shanghai University, Shanghai 200444, China
yfw@shu.edu.cn
[2] School of Communication and Information Engineering,
Shanghai University, Shanghai 200444, China

Abstract. The majority of existing objective panoramic image quality assessment algorithms are based on peak signal to noise ratio (PSNR) or structural similarity (SSIM). However, they are not highly consistent with human perception. In this paper, a new blind panoramic image quality assessment metric is proposed based on project-weighted gradient local binary pattern histogram (PWGLBP), which explores the structure degradation in sphere by combining with the nonlinear transformation relationship between the projected plane and sphere. Finally, support vector regression (SVR) is adopted to learn a quality predictor from feature space to quality score space. The experimental results demonstrate the superiority of our proposed metric compared with state-of-the-art objective PIQA methods.

Keywords: Panoramic image · Blind quality assessment · Local binary pattern · Nonlinear transformation · Support vector regression

1 Introduction

As virtual reality (VR) technologies are developing rapidly, more and more VR applications require high quality panoramic images and videos. However, owing to the process of capture, compression and transmission in real communication system, the distortions of panoramic images and videos are bound to occur [1]. It is crucial to design accurate quality assessment algorithms for panoramic images and videos in the future development of VR.

Panoramic image/video records views in all possible directions at the same time around one central position which can provide free viewing for viewer with the help of VR Helmet-Mounted Display (HMD) [2]. The panoramic images/videos are sphere, but they are always projected to a 2-D format when storing, image processing, compressing and coding. As shown in Fig. 1, there are actually lots of projection formats such as equirectangular projection (ERP), cube map projection (CMP) [3], icosahedron projection (ISP) [4], segmented sphere projection (SSP) [5], etc. Hence, panoramic image quality assessment (PIQA) is discriminate with traditional 2-D IQA because the

© Springer Nature Singapore Pte Ltd. 2020
G. Zhai et al. (Eds.): IFTC 2019, CCIS 1181, pp. 249–258, 2020.
https://doi.org/10.1007/978-981-15-3341-9_21

measurement of distortions in 2-D plane cannot linearly reflects the human perception in sphere space.

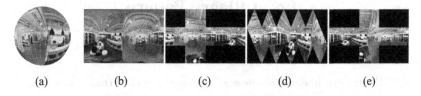

(a)   (b)   (c)   (d)   (e)

Fig. 1. Projection formats. (a) panoramic image. (b) ERP. (c) CMP. (d) ISP. (e) SSP.

During the past years, several panoramic IQA algorithms are proposed by considering the nonlinear mapping relationship between the observation (sphere) and representation (2-D plane) space, such as Spherical-peak signal to noise ratio (S-PSNR) [6], Craster Parabolic Projection-PSNR (CPP-PSNR) [7], weighted spherical-PSNR (WS-PSNR) [8] and spherical-structural similarity (S-SSIM) [9]. S-PSNR measures distortion based on sampled point set on spherical surface, where sample values are interpolated by the corresponding neighboring pixels on projected plane. In CPP-PSNR, pixels in the original and distorted images are re-mapped to a Craster Parabolic Projection (CPP), followed by PSNR calculate. Unlike S-PSNR and CPP-PSNR, WS-PSNR method reflects the error in sphere accurately by multiplying weight and the error of each pixel on projection planes without any mapping the pixels to a new domain. S-SSIM index combines the relationship between the sphere and projected 2-D plane with structural similarity indicators of 2-D plane to avoid the interference brought by projection. Although these works achieve good performance in predicting panoramic images, they are known for their limitations. Firstly, they are derived from traditional 2-D PSNR [10] and SSIM [11], which are not highly consistent with human perception. Secondly, as traditional full-reference (FR) IQA methods, PSNR and SSIM require original reference images completely available. Unfortunately, undistorted reference images are usually unavailable in the receiving end of the communication system. To overcome the shortcomings of the existed PIQA algorithms, we propose a new blind PIQA model. In order to reflect the accurate quality degradation in observation space, distortion measured in representation space is weighted by nonlinear relationship between the 2-D plane and spherical surface.

The remainder of this paper is organized as follows. In Sect. 2, we introduce our blind PIQA model. We compare the performance of the proposed model with state-of-the-art IQA metrics in Sect. 3. Finally, we draw conclusions in Sect. 4.

2 Proposed Method

In this paper, we propose a blind panoramic image quality assessment model. The framework of our model is shown in Fig. 2.

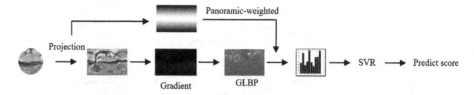

Fig. 2. Framework of proposed blind PIQA model.

First, local binary pattern (LBP) [13] is used to describe the structure information of image by encoding the image microstructures to form LBP code in the image gradient domain (GLBP) [14]. The weight relationship between the observation and representation space are then considered. Furthermore, the weight map is merged with the GLBP code, the weighted sum of all pixel points with the same GLBP value is counted to form project-weighted histogram (PWGLBP). Human visual system (HVS) can catch different information from an image on different scales [15]. Hence, beside the original image scale, the coarser scale is formed by downsampling by a factor of 2 in each dimension. Finally, support vector regression (SVR) with radial basis function (RBF) kernel is adopted to mapping the feature to quality score.

2.1 Transformation Relationship Between the Spherical Surface and Projected Plane

Figure 3 shows an example of projecting uniformly distributed samples from ERP format (representation space) to spherical surface (observation space). When uniformly distributed samples are mapped from ERP to spherical surface, the area of pixel is stretched [16]. Therefore, the relationship between representation and observation space (i.e., weight $\omega(i,j)$) could be denoted as the stretched ratio of area from projected planes to spherical surface.

In equirectangular projection shown in Fig. 3, assuming (x, y) and (θ, φ) as the ERP and spherical location respectively in continuous spatial domain, the projection relationship is:

$$\begin{cases} x = f(\theta, \varphi) = \theta \\ y = g(\theta, \varphi) = \varphi \end{cases} \tag{1}$$

where $\theta \in (-\pi, \pi)$, $\varphi \in \left(-\frac{\pi}{2}, \frac{\pi}{2}\right)$. The area stretching ratio (SR) of ERP plane can be defined as [8]:

$$SR(x, y) = \cos(\varphi) = \cos(y) \tag{2}$$

Due to the digital image is a sampled representation of a continuous image, equal (2) should be discretized for digital image. The pixel (i, j) in digital image can be

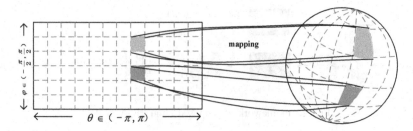

Fig. 3. The mapping from ERP to spherical space.

Fig. 4. Example of ERP weight distribution.

regarded as a block, which corresponds to a micro area bounded by $\{(w,h)|i-0.5 \leq w \leq i+0.5, j-0.5 \leq h \leq j+0.5\}$. Thus, the SR (i.e., weight $\omega(i,j)$) of digital image can be deduced as:

$$\omega(i,j) = SR(x(i,j), y(i,j)) \qquad (3)$$

Assuming the $M \times N$ denotes the resolution of ERP image and $M = 2N$. The relationship between two locations can be defined as [8]:

$$\begin{cases} x = \frac{2(i+0.5-M/2)\pi}{M} \\ y = \frac{(j+0.5-N/2)\pi}{N} \end{cases} \qquad (4)$$

Consequently, the weights are calculated as:

$$\omega(i,j) = cos\frac{(j+0.5-N/2)\pi}{N} \qquad (5)$$

Figure 4 shows the weight distributions for ERP format, where higher intensity level indicates larger weights.

2.2 Project-Weighted Histogram Based on GLBP (PWGLBP)

Structure information is a vital factor to be considered for panoramic IQA since HVS is sensitive to structure degradation. Existing studies show that second order derivatives of images can effectively catch the variation of the tiny structure which has influence on visual perceptual quality of panoramic images [19].

Specifically, the first-order derivative information is captured by calculating the gradient map of ERP format image with the utilization of Prewitt filters. In order to extract the second-order derivative information, the rotation invariant uniform LBP [20] is employed on gradient map to encode each pixel to form GLBP code. The rotation invariant uniform GLBP pattern at each pixel is deduced as [14]:

$$
GLBP_{P,R} = \begin{cases} \sum_{i=0}^{P-1} s(g_i - g_c), & u(GLBP_{P,R}) \leq 2 \\ P+1, & others \end{cases} \tag{6}
$$

$$
s(g_i - g_c) = \begin{cases} 1, & g_i - g_c \geq 0 \\ 0, & g_i - g_c < 0 \end{cases} \tag{7}
$$

$$
u(GLBP_{P,R}) = \|s(g_{P-1} - g_c) - s(g_0 - g_c)\| + \sum_{i=0}^{P-1} \|s(g_i - g_c) - s(g_{i-1} - g_c)\| \tag{8}
$$

where P means the number of neighbor and R is the radius of neighborhood. g_c and g_i represent the gradient magnitudes of the center pixel and its neighbors, respectively. And u is a uniform index which is used to calculate the number of bitwise transitions. Although the value of u reduces the types of the GLBP patterns to $P+2$, experiments show that rotation invariant uniform patterns are enough to represent the most structure features of images [20]. As shown in Fig. 5, the introduction of different kinds of distortion may change the original GLBP patterns distinctly.

After obtaining the GLBP code of ERP format images, in order to reflect the structure degradation in observation space accurately, we accumulated the projected weights $\omega(i,j)$ of pixels with the same GLBP pattern, which can be defined as the project-weighted GLBP histogram (PWGLBP):

$$
h_{p-w}(k) = \sum_{i=1}^{N} \omega_i \varphi(GLBP_{P,R}(i), k) \tag{9}
$$

ω_i is mapping weights, N is the number of pixels, k is GLBP value, where function $\varphi(\cdot)$ is defined as:

$$
\varphi(x, y) = \begin{cases} 1, & x = y \\ 0, & otherwise \end{cases} \tag{10}
$$

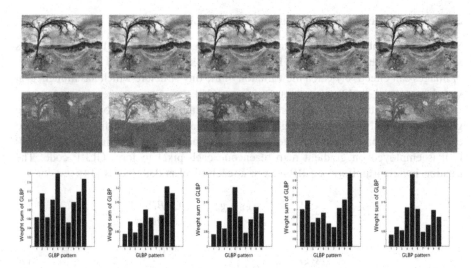

Fig. 5. GLBP maps of different distortion type. The first row is original image, JPEG, JPEG2000, Gaussian noise, Gaussian blur. The second row is corresponding GLBP feature map of image in the first row. The third row shows the panoramic-weighted histograms of first row.

P is set to 8 and R is set to 1, thus there would be 10 bins for each $h_{p-w}(k)$. Considering the perception property of HVS, $h_{p-w}(k)$ which obtained from different 5 image scales, constitute the structural features of distorted panoramic images. Figure 5 shows the histograms obtained from different kinds of distortion types, we can see that distortions change the original histogram distinctly, thus proposed PWGLBP can effectively describe the structure degradation caused by various distortions.

It is worth mentioning that our image quality assessment model can extend to video quality assessment. Since panoramic videos are usually filmed in a gentle way to avoid uncomfortable feelings of viewers, the average quality of each frame can be regarded as the quality of whole video [18]. Besides, in this paper, images in subjective database are equirectangular projection format, but if the transformation relationship of other projections is available, our model can also be extended to other projection formats.

3   Experimental Results

We use a public subjective panoramic image quality assessment database [12] to verify the performance of our algorithm. The precedent metrics for panoramic and traditional image are also implemented on the database for comparison.

3.1   Database and Evaluation Methodology

In the PIQA database [12], sixteen raw reference images are captured by professional photographers and available under Creative Commons copyright. Then four types of distortions with five distortion levels for each type are introduced to generate 320 distorted images, including JPEG compression, JPEG2000 compression, Gaussian blur and white Gaussian noise.

The indicators of measuring performance of the methods are the Spearman rank order correlation coefficient (SRCC), Kendall rank-order correlation coefficient (KRCC), Pearson linear correlation coefficient (PLCC) and root mean square error (RMSE). SRCC, KRCC and PLCC evaluation indicators describe the consistency between subjective and objective scores, and greater values show superior correlation with human perception. And smaller RMSE value corresponds the smaller deviations between subjective and objective scores.

3.2 Performance Evaluation

The LIBSVM package [17] is applied in the proposed methods. In details, ε-SVR based on a redial basis function (RBF) kernel is utilized (c = 1, g = 1). In the experiments, 80% of the distorted images used for training and remaining 20% used for testing. This random training-testing split is repeated 1000 times, and the median performance is recorded. We compare our PWGLBP method with state-of-the-art full-reference PIQA models: CPP-PSNR, WS-PSNR, S-PSNR, S-SSIM and traditional 2-D models PSNR and SSIM. Besides, performance of histogram obtained directly from GLBP with gradient magnitude as weight map is also listed for comparison. Moreover, CPP-PSNR, WS-PSNR and S-PSNR are adopted to assess the quality of the panoramic image by using Y component only. The prediction performance is listed in Table 1, where the top metric is highlighted in boldface.

Table 1. Performance comparison of PIQA metrics on panoramic image quality database.

Metrics	SRCC	KRCC	PLCC	RMSE
PSNR [10]	0.4979	0.3382	0.5080	1.8211
SSIM [11]	0.3483	0.2492	0.4373	1.9014
S-PSNR [6]	0.5319	0.3588	0.5319	1.7904
CPP-PSNR [7]	0.5182	0.3502	0.5186	1.8078
WS-PSNR [8]	0.5032	0.3414	0.5044	1.8256
GLBP	0.7928	0.6042	0.8117	1.2251
PWGLBP	**0.8226**	**0.6304**	**0.8449**	**1.1394**

In the light of the experimental results, the proposed PWGLBP model has higher consistency with human perception than other metrics. From Table 1, we can find that the CPP-PSNR, WS-PSNR, S-PSNR, considering the panoramic characteristics out-perform the PSNR. Owing to using project-weighted GLBP histogram to describe the degradation of structure information in images, proposed model is superior to GLBP histogram without panoramic characteristic.

The scatter plots of the objective scores of testing set versus mean opinion score (MOS) are shown in Fig. 6. Since the random training-testing split is repeated 1000 times in our experiment, a set of testing sets is randomly selected for the drawing of

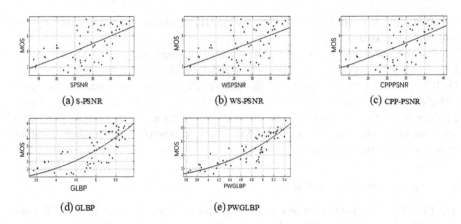

Fig. 6. Scatter plots of predicted scores and MOS. The horizontal axis denotes predicted quality scores and vertical axis is the MOS values. First row: S-PSNR, WS-PSNR and CPP-PSNR; second row: GLBP and PWGLBP.

Table 2. Performance of different distortion types on panoramic image quality database.

Metrics	SRCC	KRCC	PLCC	RMSE
JPEG	0.8571	0.6952	0.8820	1.0803
JPEG2000	0.8929	0.7333	0.9089	0.8974
GN	0.9536	0.8421	0.9686	0.4538
GB	0.9500	0.8386	0.9604	0.5359

Fig. 6. Compared to full-reference PIQA models, our model achieves better performance. These results are consistent with the performance comparison in Table 1. In addition, it is worth noting that even GLBP without panoramic weighted has better performance than state-of-the-art PIQA models, which means GLBP is more suitable for describing distortion than PSNR and SSIM in panoramic images. And with the utilization of relationship between spherical surface and 2-D plane, PWGLBP performs better than GLBP. Inspired by this, features which are more consistent with human perception could be used in the design of PIQA model.

In this experiment, we analyze the performance of the proposed model on four different types of distortion. The evaluation indicators are listed in Table 2. We can see that the proposed model is effective across different types of distortion. Because compression and blur would directly degrade the primary structure and edge information [21], our model can capture that change with the help of panoramic-weighted LBP.

4 Conclusion

In this paper, we proposed a new blind PIQA metric based on PWGLBP. Firstly, we exploit the nonlinear relationship between the representation and observation space. Then, combined with the nonlinear relationship, project-weighted histogram obtained from the GLBP feature map of images is proposed to describe the structural information degradation in observation space. Experimental results on the public PIQA database demonstrate that proposed method achieves better performance than other existing algorithms.

Acknowledgment. This work was supported by Natural Science Foundation of China under Grant No. 61671283, 61301113.

References

1. Battisti, F., Carli, M., Le Callet, P., Paudyal, P.: Toward the assessment of quality of experience for asymmetric encoding in immersive media. IEEE Trans. Broadcast. **64**(2), 392–406 (2018)
2. Yang, S., Zhao, J., Jiang, T.: An objective assessment method based on multi-level factors for panoramic videos. In: IEEE Visual Communications and Image Processing, pp. 1–4 (2017)
3. Greene, N.: Environment mapping and other applications of world projections. IEEE Comput. Appl. **6**(11), 21–29 (1986)
4. Choi, K.P., Zakharchenko, V.: Test sequence formats for virtual reality video coding. In: JVET-C0050, 3rd JVET Meeting, Geneva, CH, June 2016
5. Li, J., Wen, Z., Li, S.: Novel tile segmentation scheme for omnidirectional video. In: Proceedings of the 2016 IEEE International Conference on Image Processing, pp. 370–374 (2016)
6. Yu, M., Lakshman, H., Girod, B.: A framework to evaluate omnidirectional video coding schemes. In: Proceedings of the 2015 IEEE International Symposium Mixed Augmented Reality, pp. 31–36 (2015)
7. Zakharchenko, V., Choi, K.P., Park, J.H.: Omnidirectional video quality metrics and evaluation process. In: Data Compression Conference, p. 472 (2017)
8. Sun, Y., Lu, A., Yu, L.: Weighted-to-spherically uniform quality evaluation for omnidirectional video. IEEE Signal Process. Lett. **24**(9), 1408–1412 (2017)
9. Chen, S., Zhang, Y., Li, Y., Chen, Z., Wang, Z.: Spherical structural similarity index for objective omnidirectional video quality assessment. In: IEEE International Conference on Multimedia and Expo (ICME), San Diego, CA, pp. 1–6 (2018)
10. Hore, A., Ziou, D.: Image quality metrics: PSNR vs. SSIM. In: 20th International Conference on Pattern Recognition, Istanbul, pp. 2366–2369 (2010)
11. Wang, Z., Bovik, A.C., Sheikh, H.R.: Image quality assessment: from error visibility to structural similarity. IEEE Trans. Image Process. **13**(4), 600–612 (2004)
12. Duan, H., Zhai, G., Min, X., Zhu, Y., Fang, Y., Yang, X.: Perceptual quality assessment of omnidirectional images. In: IEEE International Symposium on Circuits and Systems (ISCAS), Florence, pp. 1–5 (2018)
13. Liu, T., Liu, K.: No-reference image quality assessment by wide-perceptual-domain scorer ensemble method. IEEE Trans. Image Process. **27**(3), 1138–1151 (2018)

258   Y. Xia et al.

14. Li, Q., Lin, W., Fang, Y.: No-reference quality assessment for multiply-distorted images in gradient domain. IEEE Signal Process. Lett. **23**(4), 541–545 (2016)
15. Gu, K., Li, L., Lu, H., Min, X., Lin, W.: A fast reliable image quality predictor by fusing micro- and macro-structures. IEEE Trans. Industr. Electron. **64**(5), 3903–3912 (2017)
16. Zou, W., Yang, F., Wan, S.: Perceptual video quality metric for compression artifacts: from two-dimensional to omnidirectional. IET Image Proc. **12**(3), 374–381 (2018)
17. Chang, C.C., Lin, C.J.: LIBSVM: a library for support vector machines. ACM Trans. Intell. Syst. Technol. **2**(3), 1–27 (2011)
18. Xu, M., Li, C., Liu, Y., Deng, X., Lu, J.: A subjective visual quality assessment method of panoramic videos. In: IEEE International Conference on Multimedia and Expo (ICME), Hong Kong, pp. 517–522 (2017)
19. Zhang, B., Gao, Y., Wang, Y.: Local derivative pattern versus local binary pattern: face recognition with high-order local binary pattern descriptor. IEEE Trans. Image Process. **19**(2), 533–544 (2010)
20. Ojala, T., Pietikainen, M., Maenpaa, T.: Multiresolution gray-scale and rotation invariant texture classification with local binary pattern. IEEE Trans. Pattern Anal. Mach. Intell. **24**(7), 971–987 (2002)
21. Wu, J., Lin, W., Shi, G., Liu, A.: Perceptual quality metric with internal generative mechanism. IEEE Trans. Image Process. **22**(1), 43–54 (2013)

Blind 3D Image Quality Assessment Based on Multi-scale Feature Learning

Yongfang Wang[1,2(✉)], Shuai Yuan[2], Yumeng Xia[2], and Ping An[1,2]

[1] Shanghai Institute for Advanced Communication and Data Science,
Shanghai University, Shanghai 200444, China
yfw@shu.edu.cn
[2] School of Communication and Information Engineering, Shanghai University,
Shanghai 200444, China

Abstract. 3D image quality assessment (3D-IQA) plays an important role in 3D multimedia applications. In recent years, convolutional neural networks (CNN) have been widely used in various images processing tasks and achieve excellent performance. In this paper, we propose a blind 3D-IQA metric based on multi-scale feature learning by using multi-column convolutional neural networks (3D-IQA-MCNN). To address the problem of limited 3D-IQA dataset size, we take patches from the left image and right image as input and use the full-reference (FR) IQA metric to approximate a reference ground-truth for training the 3D-IQA-MCNN. Then we put the patches from left image and right image into the pre-trained 3D-IQA-MCNN and obtain two quality feature vectors based on multi-scale. Finally, by regressing the quality feature vectors onto the subjective mean opinion score (MOS), the visual quality of 3D images is predicted. Experimental results show that the proposed method achieves high consistency with human subjective assessment and outperforms several state-of-the-art 3D-IQA methods.

Keywords: 3D image quality assessment · Convolutional neural networks · Multi-scale features · No-reference · Feature extraction

1 Introduction

With the rapid development of 3D image processing technology, the demand for digital 3D images has increased obviously [1]. The distortions of 3D image would be introduced in the process of capture, compression and transmission. As a result, 3D image quality assessment (3D-IQA) is an important and challenging research problem in the 3D multimedia applications.

The 2D image quality assessment (2D-IQA) has been widely studied, and many image quality metrics were proposed, such as peak signal-to-noise ratio (PSNR) and structural similarity (SSIM) [2]. Compared with 2D-IQA tasks, it is very hard to deal with 3D-IQA for lacking of understanding of 3D visual perception, so it is not desirable to simply extend the 2D quality metrics to the 3D cases for 3D-IQA. 3D perceptual quality factors, such as depth perception [6] and visual comfort [5] should be considered for 3D-IQA.

© Springer Nature Singapore Pte Ltd. 2020
G. Zhai et al. (Eds.): IFTC 2019, CCIS 1181, pp. 259–267, 2020.
https://doi.org/10.1007/978-981-15-3341-9_22

3D-IQA approaches are categorized into three classes considering whether the reference images are used, which are full-reference (FR) 3D-IQA approaches, reduced-reference (RR) 3D-IQA approaches, and no-reference (NR) or blind 3D-IQA approaches. Although FR 3D-IQA offers an effective way to evaluate quality difference, in many cases, the reference image is not available. Recently, researchers have focused on the NR 3D-IQA methods. Gu et al. [8] proposed an NR 3D-IQA model by using no-linear addition, saliency-based parallax compensation, and binocular dominance weighting. Gu et al. [9] also proposed an NR 3D-IQA approach by extending a 2D sharpness metric to 3D-IQA with binocular rivalry. Su et al. [10] proposed an NR 3D-IQA method based on natural scene statistics (NSS), which used energy of wavelet coefficients to construct a convergent cyclopean map and extracted the bivariate, univariate, and correlation NSS as features and used support vector regression (SVR) for 3D image quality prediction. Shao et al. [7] proposed an NR 3D-IQA metric, which uses binocular guided quality lookup and visual codebook and applied a blind quality pooling to predict 3D image quality score. Shao et al. [11] also learned the mechanism of primary visual cortex in the human brain and constructed dictionaries which reflect the action potentials of local and global receptive fields. The 3D image quality was predicted through pooling the coded coefficients.

In recent years, Convolutional Neural Networks (CNN) have demonstrated notable progress in various computer vision tasks. Oh et al. [11] proposed a 3D-IQA approach by utilizing deep neural networks. Deep neural network-based method is data-driven, which do not use hand-crafted features of prior domain knowledge about the human visual system (HVS). However, the CNN-based methods have some limitations. First, the shallow network structure limits the receptive field. In view of the small receptive filed, the model exploits the contextual information over small image region. Second, masses of existing methods depend on the single-scale context information to predict 3D image quality score instead of multi-scale information.

In order to solve the aforementioned issues, we proposed a new blind 3D-IQA metric, which considers multi-scale feature learning by using multi-column CNN. Our contribution consists of three parts. First, we used multi-column CNN to extract multi-scale features from 3D image, which has better performances and stability. Secondly, we used a deeper structure, which enlarged the receptive field and exploited contextual over large image region. Finally, in the training stage, we divided each 3D image pair (left and right image) into patches and employed the SSIM [2] values of corresponding patches as labels. In the testing stage, we put patches from 3D image pair into the pre-trained model and obtained two quality feature vectors with multi-scale. Then we regressed the quality feature vectors onto the subjective mean opinion score (MOS) to predict the visual quality of 3D images and achieved state-of-the-art performance.

The rest of the paper is organized as follows. In Sect. 2, we describe our proposed model. In Sect. 3, we present experiment results. Finally, we draw conclusions in Sect. 4.

2   Proposed Blind 3D-IQA Based on Multi-scale Feature Learning

Motivated by the success of multi-column deep neural networks [15] in image classification and multi-column CNN [16] in image super-resolution, we present blind 3D image quality assessment by using multi-column CNN (3D-IQA-MCNN), which is used to learn multi-scale features in NR 3D-IQA. The network architecture of our model is shown in Fig. 1. The multi-column block consists of three-column convolution layers with different kernel sizes.

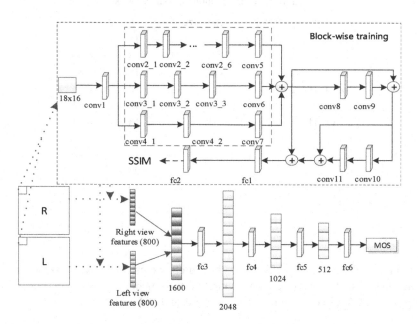

Fig. 1. Network architecture of the proposed 3D-IQA-MCNN. Red box denotes multi-column block. In the multi-column block, we cascade six convolution layers with kernel size 3 × 3, three convolution layers with kernel size 5 × 5, two convolution layers with kernel size 7 × 7. (Color figure online)

For an input 3D image, we split its left image and right image to 18 × 16 small patches without overlaps at first. Then, we train the proposed model on these patches and use the SSIM values of corresponding patches as labels similar to [11]. At the end, we put the patches from left image and right image into the pre-trained 3D-IQA-MCNN and get two quality feature vectors. By regressing the quality feature vectors onto the MOS, the visual quality of 3D images is predicted.

2.1   Multi-column Network Architecture

The network configuration is listed in Table 1. At the input layer of the network, we use a convolution layer with 64 filters of kernel size 5 × 5 to extract coarse features.

Then we use a multi-column block to extract the multi-scale features from the patches. The convolutional layers can be described as

$$\mathcal{F}(x, W_l) = \sigma(W_l x + b_l) \tag{1}$$

where W_l and b_l denote learnable weights and biases of the convolution layer. x denote the input of the convolution layer. The activation function σ is Leaky Rectified Linear Unit (Leaky ReLU) [17].

Table 1. Configurations of our network

Layer name	Padding	Kernel size	Stride	Output size
conv1	2	5×5	1	$18 \times 16 \times 64$
conv2_1–2_6	1	3×3	1	$18 \times 16 \times 64$
conv3_1–3_3	2	5×5	1	$18 \times 16 \times 64$
conv4_1–4_2	3	7×7	1	$18 \times 16 \times 64$
conv5–7	0	1×1	1	$18 \times 16 \times 64$
conv8–11	1	3×3	1	$18 \times 16 \times 64$
fc1	–	–	–	1024
fc2	–	–	–	1
fc3	–	–	–	2048
fc4	–	–	–	1024
fc5	–	–	–	512
fc6	–	–	–	1

In proposed 3D-IQA-MCNN, for each column, we use the different kernel sizes to extract features. The detailed architecture of this model is shown in Fig. 1. The receptive field $\mathcal{R}$ of convolution layer can be calculated as

$$\mathcal{R} = \kappa + (\kappa - 1) \times (n - 1) \tag{2}$$

where κ denotes size of convolving kernel and n denotes the number of convolution layers in each column. According to (2), six convolution layers with kernel size 3×3, three convolution layers with kernel size 5×5, and two convolution layers with kernel size 7×7 have the same receptive field.

In order to learn more robust features, features extracted from the different column are fused at the same receptive field. In order to produce more non-linear combinations, we add a convolution layer with kernel size 1×1 at the last layer of each column. Then we use element-wise summation to fuse the feature maps from different columns.

2.2 Quality Prediction

At first, we train the 3D-IQA-MCNN on the patches and used the SSIM values as labels similar to [11]. The SSIM provides a good local quality score for training owing to using the luminance, contrast, and structural similarity.

The image patch score is predicted by minimizing the following MSE loss function, which can be expressed as

$$\mathcal{L}(\Theta) = \frac{1}{2N} \sum\nolimits_{i=1}^{N} \|F(X_i; \Theta) - F_i\|^2 \tag{3}$$

where Θ denotes a set of learnable parameters in the model and N denotes the number of training samples in each batch. X_i and F_i denote the input image patch and its SSIM value respectively. $F(X_i; \Theta)$ stands for the predict score of X_i with network weight Θ.

Then we put the patches from left image and right image into the pre-trained 3D-IQA-MCNN and get two quality feature vectors. Finally, we use four fully connected (FC) layers to map the quality feature vectors to the subjective MOS. The FC layers configuration is also listed in Table 1. The image score is predicted by minimizing the following MSE loss, which can be expressed as

$$\mathcal{L}(\Theta) = \frac{1}{2N} \sum\nolimits_{i=1}^{N} \|\mathcal{P}(X_i; \Theta) - \mathcal{M}_i\|^2 \tag{4}$$

where X_i and $\mathcal{M}_i$ denote the input image patch and its subjective MOS respectively. Θ denotes the learnable parameters in the FC layers and N denotes the number of training samples in each batch. $\mathcal{P}(X_i; \Theta)$ stands for the predict score of X_i with network weight Θ.

3   Experimental Results

In this section, we demonstrate experiments to validate the effectiveness of the proposed model. We use deep learning framework PyTorch [18] to train the 3D-IQA-MCNN. The LIVE 3D database Phase I [19] and LIVE 3D database Phase II [12] is used to train and evaluate our model. Five different types of distortions at various levels, JPEG 2000 compression (JP2K), JPEG compression (JPEG), white noise (WN), Gaussian blurring (BLUR), and fast fading (FF) are symmetrically applied to reference stereo pairs at various levels in the LIVE Phase I, while the same distorted types are both symmetrically and asymmetrically applied to reference stereo pairs at various levels in the LIVE Phase II. Besides, both two databases provide differential MOS (DMOS) of the stereo pairs, higher DMOS indicates lower quality.

At the first training stage, we first extract 16×18 patches from the left images and right images without overlapping. Note that all images in the databases are in 640×360 format, and we obtain 800 patches from a single image. Since different image patches have different quality values, we compute SSIM value for them and take SSIM value as label. Then we train the model via the mini-batch stochastic gradient descent (SGD) with backpropagation [20]. We set batch size of SGD to 64, momentum parameter to 0.9, and weight decay to 10^{-4}. We use the method described in [21] to initialize the weight of the convolutional filters. We train our model with different learning rates. Learning rate is changed to 0.1 of the previous one at the interval of 20 epochs. It begins with 0.1 and stops at 0.00001. After reaches 0.00001, it becomes a fixed value. Due to a large learning rate used during the training session, we adopt the gradient clipping [22] to restrain gradient explosion. Specifically, the gradients are clipped to 0.4. Finally, our model is obtained by training after 200 epochs.

Table 2. PLCC comparison on the LIVE 3D database phase I

PLCC	JPEG	JP2K	WN	BLUR	FF	ALL
Chen [23]	0.877	0.616	0.914	0.925	0.734	**0.883**
Gorley [24]	0.758	0.427	0.749	0.806	0.593	0.502
You [3]	0.831	0.173	0.892	0.697	0.711	0.803
Benoit [4]	0.834	0.146	**0.928**	0.861	0.671	0.858
Lin [14]	0.799	0.196	0.925	0.811	0.700	0.784
Sazzad [13]	0.774	0.565	0.803	0.628	0.694	0.624
Chen [12]	**0.885**	**0.655**	0.920	**0.934**	**0.775**	0.881
Proposed	**0.913**	**0.749**	**0.957**	**0.955**	**0.823**	**0.932**

Table 3. SROCC comparison on the LIVE 3D database phase I

PLCC	JPEG	JP2K	WN	BLUR	FF	ALL
Chen [23]	**0.904**	**0.592**	**0.943**	0.925	0.761	**0.928**
Gorley [24]	0.819	0.427	0.776	0.776	**0.797**	0.298
You [3]	0.863	0.133	0.916	0.699	0.654	0.805
Benoit [4]	0.838	0.135	0.933	0.807	0.627	0.852
Lin [14]	0.839	0.207	0.928	**0.935**	0.658	0.856
Sazzad [13]	0.721	0.526	0.807	0.597	0.705	0.624
Chen [12]	0.818	0.577	0.920	0.899	0.605	0.890
Proposed	**0.881**	**0.751**	**0.964**	**0.944**	**0.784**	**0.923**

Table 4. PLCC comparison on the LIVE 3D database phase II

PLCC	JPEG	JP2K	WN	BLUR	FF	ALL
Chen [23]	0.806	**0.803**	**0.910**	**0.871**	0.797	0.823
Gorley [24]	0.599	0.576	**0.893**	0.742	0.760	0.197
You [3]	0.845	0.657	0.885	0.540	0.802	0.731
Benoit [4]	0.767	0.574	0.875	0.825	0.777	0.769
Lin [14]	0.744	0.583	0.909	0.671	0.699	0.642
Sazzad [13]	0.645	0.531	0.657	0.721	0.727	0.669
Chen [12]	**0.859**	0.802	0.855	0.864	**0.815**	**0.854**
Proposed	**0.835**	**0.834**	0.847	**0.976**	**0.843**	**0.856**

Table 5. SROCC comparison on the LIVE 3D database phase II

PLCC	JPEG	JP2K	WN	BLUR	FF	ALL
Chen [23]	0.837	**0.830**	**0.956**	**0.907**	**0.891**	0.836
Gorley [24]	0.746	0.657	**0.918**	0.828	0.846	0.376
You [3]	0.841	0.674	0.873	0.520	0.787	0.719
Benoit [4]	0.757	0.577	0.876	0.836	0.815	0.770
Lin [14]	0.718	0.613	0.907	0.711	0.701	0.638
Sazzad [13]	0.625	0.479	0.647	0.775	0.725	0.648
Chen [12]	**0.851**	0.779	0.883	0.848	0.805	0.864
Proposed	**0.856**	**0.820**	0.833	**0.908**	**0.875**	**0.869**

At the second training stage, we first put the patches from left image and right image into the pre-trained 3D-IQA-MCNN and obtain two quality feature vectors. The size of each feature vector is 1×800. Then we use four FC layers to map the quality feature vector to the subjective MOS, which are also optimized via mini-batch SGD. The learning rate is set to 0.001, which is a fixed value. Finally, the model is obtained by training after 50 epochs.

We deploy Pearson linear correlation coefficient (PLCC) and Spearman rank order correlation coefficient (SROCC) to evaluate the performance of 3D-IQA algorithms. PLCC and SROCC serve as a measure of prediction accuracy and monotonicity respectively. The higher values of PLCC and SROCC represent the better performance. For the simulation, 80% of the LIVE database is randomly selected for use in training. The remaining 20% is used for testing. In the experiments, we completely separated the training and testing sets from the database with respect to each reference image.

We compare our method with five FR 3D-IQA metrics and two state-of-the-art NR 3D-IQA metrics. They are Chen [23], Gorley [24], You [3], Benoit [4], Lin [14], Sazzad [13], and Chen [12]. The experimental results are listed in Tables 2, 3, 4 and 5.

Tables 2, 3, 4 and 5 show SROCC and PLCC comparison results, respectively, where the best two algorithms has been highlighted in boldface. From Tables 2, 3, 4 and 5, we can see that our method outperforms state-of-the-art methods in PLCC value and SROCC value. Note that the proposed method achieves the best correlation scores among the NR 3D-IQA algorithms.

The proposed 3D-IQA-MCNN also shows acceptable performance on the distorted 3D images with WN in LIVE Phase II. This is because the WN is symmetrically and asymmetrically applied to reference stereo pairs in the LIVE Phase II and the model parameters are weakly trained by the randomly distributed noise, as such distortion is highly uncorrelated with the overall spatial characteristics. However, the proposed approach improves the performance on most individual and overall distortion types in terms of both PLCC and SROCC.

4 Conclusion

In this work, we investigate a cascaded multi-column convolutional neural network for NR 3D-IQA. Our method extracts multi-scale features from the left and right image to predict the 3D image quality. By taking the large scale of image patches as training set and taking its SSIM value as the label, our model can achieve promising quality prediction results. Experiments on two LIVE 3D datasets demonstrate the superiority of the proposed NR 3D-IQA methods.

Acknowledgment. This work was supported by Natural Science Foundation of China under Grant No. 61671283, 61301113.

References

1. Chen, L., Zhao, J.: Robust contourlet-based watermarking for depth-image-based rendering 3D images. In: 2016 IEEE International Symposium on Broadband Multimedia Systems and Broadcasting (BMSB), Nara, pp. 1–4 (2016)
2. Wang, Z., Bovik, A.C., Sheikh, H.R., Simoncelli, E.P.: Image quality assessment: from error visibility to structural similarity. IEEE Trans. Image Process. 13(4), 600–612 (2004)
3. You, J., et al.: Perceptual quality assessment for stereoscopic images based on 2D image quality metrics and disparity analysis. In: Proceedings of the International Workshop Video Processing Quality Metrics Consumer Electronics, pp. 1–6 (2010)
4. Benoit, A., et al.: Quality assessment of stereoscopic images. EURASIP J. Image Video Process. 2008, 1–13 (2009)
5. Tam, W.J., Speranza, F., Yano, S., Shimono, K., Ono, H.: Stereoscopic 3D-TV: visual comfort. IEEE Trans. Broadcast. 57(2), 335–346 (2011)
6. Lebreton, P., Raake, A., Barkowsky, M., Le Callet, P.: Evaluating depth perception of 3D stereoscopic videos. IEEE J. Sel. Top. Signal Process. 6(6), 710–720 (2012)
7. Shao, F., Lin, W., Wang, S., Jiang, G., Yu, M.: Blind image quality assessment for stereoscopic images using binocular guided quality lookup and visual codebook. IEEE Trans. Broadcast. 61(2), 154–165 (2015)
8. Gu, K., et al.: No-reference stereoscopic IQA approach: from nonlinear effect to parallax compensation. J. Elect. Comput. Eng 2012(pt.3), 436031.1–436031.12 (2012)
9. Gu, K., Zhai, G., Lin, W., Yang, X., Zhang, W.: No-reference image sharpness assessment in autoregressive parameter space. IEEE Trans. Image Process. 24(10), 3218–3231 (2015)
10. Su, C., Cormack, L.K., Bovik, A.C.: Oriented correlation models of distorted natural images with application to natural stereopair quality evaluation. IEEE Trans. Image Process. 24(5), 1685–1699 (2015)
11. Oh, H., Ahn, S., Kim, J., Lee, S.: Blind deep S3D image quality evaluation via local to global feature aggregation. IEEE Trans. Image Process. 26(10), 4923–4936 (2017)
12. Chen, M., Cormack, L.K., Bovik, A.C.: No-reference quality assessment of natural stereopairs. IEEE Trans. Image Process. 22(9), 3379–3391 (2013)
13. Sazzad, Z.M., et al.: Objective no-reference stereoscopic image quality prediction based on 2D image features and relative disparity. Adv. Multimed. 2012(8), 1–16 (2012)
14. Lin, Y., Wu, J.: Quality assessment of stereoscopic 3D image compression by binocular integration behaviors. IEEE Trans. Image Process. 23(4), 1527–1542 (2014)
15. Ciregan, D., Meier, U., Schmidhuber, J.: Multi-column deep neural networks for image classification. In: 2012 IEEE Conference on Computer Vision and Pattern Recognition, Providence, RI, pp. 3642–3649 (2012)
16. Shuai, Y., Wang, Y., Peng, Y., Xia, Y.: Accurate image super-resolution using cascaded multi-column convolutional neural networks. In: 2018 IEEE International Conference on Multimedia and Expo (ICME 2018), pp. 1–6, 23–27 July (2018)
17. Mass, A.L., et al.: Rectifier nonlinearities improve neural network acoustic models. In: ICMLW, vol. 30, no. 1 (2013)
18. https://pytorch.org/
19. Moorthy, A.K., et al.: Subjective evaluation of stereoscopic image quality. Signal Process. Image Commun. 28(8), 870–883 (2013)
20. Lecun, Y., Bottou, L., Bengio, Y., Haffner, P.: Gradient-based learning applied to document recognition. Proc. IEEE 86(11), 2278–2324 (1998)

21. He, K., Zhang, X., Ren, S., Sun, J.: Delving deep into rectifiers: surpassing human-level performance on imagenet classification. In: 2015 IEEE International Conference on Computer Vision (ICCV), Santiago, pp. 1026–1034 (2015)
22. Pascanu, R., et al.: On the difficulty of training recurrent neural networks. In: ICML, pp. 1310–1318 (2013)
23. Chen, M.-J., et al.: Full-reference quality assessment of stereopairs accounting for rivalry. Signal Process. Image Commun. **28**(9), 1143–1155 (2013)
24. Gorley, P., et al.: Stereoscopic image quality metrics and compression. In: Proceedings of the SPIE, vol. 6803 (2008)

Research on Influence of Content Diversity on Full-Reference Image Quality Assessment

Huiqing Zhang[1,2], Shuo Li[1,2(✉)], Zhifang Xia[1,3], Weiling Chen[4], and Lijuan Tang[5]

[1] Faculty of Information Technology, Beijing University of Technology, Beijing 100124, China
shuoli1025@163.com
[2] Engineering Research Center of Digital Community, Ministry of Education, Beijing 100124, China
[3] The State Information Center of China, Beijing 100045, China
[4] Fujian Key Lab for Intelligent Processing and Wireless Transmission of Media Information, Fuzhou University, Fuzhou 350000, China
[5] Jiangsu Vocational College of Business, Jiangsu 226000, China

Abstract. With the development of image quality assessment (IQA), the full-reference (FR) IQA algorithms are becoming more and more mature. This paper analyzes the performance of five FR natural scene (NS) IQA algorithms and three FR screen content (SC) IQA algorithms, and makes analysis on the eight IQA algorithms in terms of prediction performance and design principle. Experiments show, (1) the performance of perceptual similarity based IQA method proposed by Gu *et al.* is better; (2) the multi-scale technology plays an important role in improving the performance of algorithms. Finally, summarized and expounded the development direction of image quality evaluation.

Keywords: Full-reference image quality assessment · Natural scene images · Screen content images

1 Introduction

Digital images are inevitably affected by various distortions in the process of acquisition and processing, which lead to image degradation and affects people's subjective visual experience. Therefore, Image Quality Assessment (IQA) has important and wide applications in many fields such as image acquisition, image watermarking, image compression, image transmission, image recovery, image enhancement and image replication.

This work is supported by the Major Science and Technology Program for Water Pollution Control and Treatment of China (2018ZX07111005), Natural Science Research of Jiangsu Higher Education Institutions under grant (18KJB52).

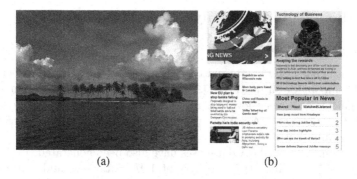

(a)   (b)

Fig. 1. An example of optical natural scene image and screen content image. (a) An optical natural scene image captured by a camera; (b) A screen content image captured by a screenshot software. (Color figure online)

With the rapid development of science and technology, the current image no longer refers to the Natural Scene (NS) images which are taken by cameras (Fig. 1(a)), but also includes Screen Content (SC) [1,2] images generated by computers and other electronic equipment (Fig. 1(b)). We can often encounter them in a variety of multimedia applications and services, such as online news and advertising, online education, electronic brochures, remote computing, cloud games and so on. The content of typical SC image includes natural images, text, graphics, logos, graphics and other image content. Therefore, SC images have some different characteristics from NS images. NS images are mostly thick lines with rich color changes and complex texture content. SC images are mostly thin lines with limited colors and simple shapes. Therefore, the methods used to evaluate the quality or distortion of NS images are not applicable to SC images necessarily. The corresponding algorithms should be designed according to the actual situation.

According to the amount of reference image information required in the design process of IQA algorithm, the objective IQA algorithm can be divided into three types [3]: Full-Reference (FR), Reduced-Reference (RR), and No-Reference (NR) IQA algorithm. Since the FR method can obtain all the information of the reference image, it is the most reliable method among IQA methods [4]. In addition, FR-IQA can also guide the coding [5–8] and optimization of NR-IQA [9–12,19]. This paper introduces three classical image databases firstly; then eight algorithms are analyzed from the performance results and algorithm principle-seach method has its own ideas and characteristics, five of which are about NS images and the other three about SC images. The performance of each algorithm is compared through experiments, Finally, we summarize the above. It provides reference for further understanding of FR-IQA algorithm and other two types of IQA algorithm.

2　Methodology and Analysis

2.1　Image Database

The ultimate goal of IQA is to match the subjective will of human beings, but the establishment of subjective IQA requires a lot of money and manpower, and the efficiency is very low. Therefore, a large number of researchers have explored the objective IQA model, which requires the establishment of artificial subjective evaluation IQA database to verify the performance of the objective IQA algorithm, so many IQA databases have also been generated [13–15]. Next, this paper will introduce three important IQA databases and compare their related data in Table 1.

In order to fully cope with the new distortions and mixed distortion types brought by emerging electronics products, data sets containing multiple distortion types are designed. Ponomarenko *et al.* design the TID2013 database [16], which is extended from the TID2008 database [17] and is currently the largest image database designed to test the performance of visual quality assessment algorithms. It contains 25 original images and 3000 distorted images, each reference image corresponds to 24 distortion types at 5 degradation levels. By conducting 985 subjective experiments on volunteers from five countries (Finland, France, Italy, Ukraine, and the United States), the Mean Objective Score (MOS) of the new database is collected as the true value of the distorted image quality score. The higher the MOS value, the better the image visual quality.

Table 1. Comparison of three typical database parameters.

Database	TID2013	SCID	HSNID
Number of reference images	25	40	20
Number of distorted images	3000	1800	600
The distortion types	24	9	6
Distortion levels	5	5	5
Quality indicator	MOS	MOS	MOS
Time of establishment	2013	2017	2018

The emergence of new electronic products not only brings new types of distortion and hybrid distortion, but also makes SC images enter our lives in large numbers. However, there are very few databases used to evaluate SC images specially. Therefore, Ni *et al.* developed the Screen Content Images Database (SCID) [18]. The SCID database contains 40 reference images and 1800 distorted images. These 40 reference images cover text, graphics, symbols, patterns, and natural images. Among them, the 1800 distorted images contain 9 distortion types at 5 degradation levels. Finally, the evaluation is performed by the double stimulation method, and MOS is given by the evaluators, where the larger the MOS value, the better the image quality.

So far, the subjective evaluation of the image database developed by various researchers has been implemented separately, so their image quality scores cannot be sufficiently consistent, which inevitably makes the experimental results unconvincing. To this end, Gu *et al.* establish the first Hybrid Screen content and Natural scene Image Database (HSNID) [19]. The HSNID consists of 600 distorted NS images and SC images generated with 20 reference images in 6 common distortion types at 5 degradation levels. The MOS value provided by the database using single stimulus method is obtained from 50 non-professional undergraduates or postgraduates. The MOS score ranged from the worst "1" to the best "5" with an interval of "1". They use a 5-point discrete scale for scoring. Finally, the MOS value is obtained by the average score of all people. The higher the MOS value, the better the image quality.

2.2 IQA Methodology

IQA methods can be divided into two categories [20]: (1) artificial subjective evaluation method, in which the image quality score is directly given by the observer. Since human beings are the end users of images, subjective evaluation is the ultimate standard for evaluating image quality, but it requires a lot of money and manpower, and cannot be realized in the systems that require real-time quality evaluating of images or video sequences. In addition, the subjective evaluation is also affected by the observer's own factors such as the observer's knowledge scope and personal hobbies, which makes the evaluation results unstable. (2) Objective evaluation method. This method is an evaluation algorithm developed by researchers to simulate subjective judgment. It simulates the human visual system, uses some mathematical tools to extract and operate the image, and finally evaluates the quality of the distorted images. It has the characteristics of simplicity, real-time, reusability and easy integration, and has become the research focus of image quality assessment. Nowadays, with the improvement of living standards, people's pursuit of visual effects has been constantly improved. Therefore, a variety of objective IQA algorithms have been generated.

NS IQA. Most traditional IQA metrics are primarily based on the assumption that the human visual system is well suited to extract structural information from the scene. The Structural Similarity Index (SSIM) [21] proposed by Wang *et al.* can be regarded as a milestone in the development of the IQA model. SSIM is developed based on the assumptions mentioned above, so a measure of structural similarity can provide a good approximation of perceived image quality. Wang *et al.* proposed a multi-scale extension of SSIM, MS-SSIM [22], in their later work, which also confirmed that MS-SSIM is better than single-scale results. Later, Wang and Li improved the original MS-SSIM into Information content Weighted SSIM Index (IW-SSIM) [23], and proposed a multi-scale information content weighting method based on NS image. Find the optimal convergence strategy for the design of the IQA algorithm. This method is inspired by visual information fidelity (VIF) [24]. From the perspective of information theory, this paper studies the IQA pooling problem. Based on the advanced statistical image model, the information content weighting method is established and combined with the scale IQA method is combined to perform image quality prediction. This new weighting method can significantly improve the performance of the IQA algorithm based on peak signal-to-noise ratio (PSNR) and mechanism similarity SSIM. In addition, the widely accepted VIF algorithm can be reinterpreted under the same information content weighting framework.

Wang et al. proposed that SSIM is widely accepted because of its good evaluation accuracy, pixel-level quality measurement, simple mathematical formula, easy analysis and optimization. Since it underestimates the influence of edge damage, the result is poor prediction ability for relative quality of blurred images and white noise images. It is well known that edges are critical to visual perception and play an important role in the recognition of image content. Therefore, Liu et al. proposed a new IQA method based on gradient similarity. Gradient Similarity Measurement (GSM) [25] measures the change in gradient similarity between contrast and structure in the case of considering the transmission of visual information and gradients that are conducive to scene understanding. Finally, the overall image quality score is obtained by adaptively integrating the effects of brightness changes and contrast structure changes. The proposed method makes up for the situation that the existing related schemes are not well considered in the edge information, and has the robustness and high precision, which makes it more practical to use and embed in various optimization processes of image processing tasks.

In previous studies, some researchers studied the relationship between Visual Saliency (VS) and IQA, and found that the proper combination of VS information can make the IQA algorithm more effective. Zhang et al. proposed a simple but very effective Visual Saliency-based Index (VSI) by analyzing the relationship between image VS changes and perceived quality degradation [26]. They believe that VS mapping can be used not only as a weight function in the pooling stage, but also as a feature map to characterize the quality of local image regions. The gradient magnitude changes are then subtly combined with the visually significant changes caused by the distortion to predict the image quality score. Compared to the previously proposed IQA objective algorithm, the VSI performs better and has lower computational complexity in terms of space and time.

The existing IQA algorithm (especially the FR-IQA method) employs a common two-stage structure. The goal of the first phase is to estimate local distortion. The second stage is to convert the local distortion map into a visual quality score, the pooling process. Despite the numerous pooling strategies, they ignore the impact of image content on distortion. Compared with the rapid development of local distortion measurement, few people are committed to effective pooling scheme research. Gu et al. designed a new pooling model by analyzing the influence of image content and distortion on image distortion distribution [27]. Taking into account the distribution of distortion position, distortion intensity, frequency variation and histogram variation, four models combining multi-scale, sort-based weighting, frequency variation inducing regulator and entropy multiplication and multiplication are combined to obtain the overall quality score. Compared to the traditional average pooling strategy, this strategy has improved performance by an average of 15%, and achieved the best overall performance in terms of numerical and statistical significance.

In order to further optimize the image quality evaluation method and enhance the visual effect, the image quality is quickly and reliably estimated in the quality monitoring of the real-time encoding and transcoding system. Gu *et al.* proposed an image quality evaluation method based on Perceptual SIMilarity (PSIM) [28], which first extracts the gradient amplitude of the input reference image and the distorted image, and then calculates the gradient amplitude at multiple scales. The similarity and color similarity of the value map, and finally the above visual degradation metrics are grouped by the perceptual-based pooling group to obtain an objective quality prediction score. Compared with Gradient Magnitude Similarity Deviation (GMSD) [29] and VSI [26], it has the following advantages: (1) using simple operators, PSIM has the best average performance and high computational efficiency; (2) a perceptual-based pooling scheme is proposed. The parameters used are coarsely adjusted and fine-tuned, and the differences in the microstructure, macrostructure and color information of the reference image and the distorted image are combined.

SC IQA. IQA technology has been widely used in the past decade. These methods performed well in NS image quality evaluation, but performed poorly in SC image quality evaluation. To this end, Gu *et al.* proposed a Saliency-guided Quality Measure of SC Images (SQMS) [30]. Unlike previous methods, SQMS focuses on the identification of salient regions. The method mainly relies on a simple convolution operator, first highlighting the degradation of the structure caused by different types of distortion, and then detecting the significant region of the distortion that is usually more noticeable, and finally gives the overall image quality prediction. Compared with other methods, the SQMS method has advantages such as superior performance, computational efficiency and parameter sensitivity.

In addition to saliency, gradient information also has some value in image evaluation. It is common practice to generate various features using gradient magnitudes, but the gradient directions are often ignored. Zhang *et al.* found that the gradient direction can significantly reflect the quality perception of SC images, and then propose a simple and effective SC image quality evaluation method based on Gradient Direction-based Index (GDI) [31]. Find out which direction in the gradient amplitude field can cause the largest change, and use the local information to extract the gradient direction and amplitude, and then use the deviation model for pooling, and finally get the image quality prediction score. The consistency of this method with the perception of the human visual system is superior to the most advanced quality assessment model at the time.

Structural changes are also an important consideration in IQA. In order to evaluate the quality of SC images more accurately, Gu *et al.* evaluated the quality of SC images by analyzing the structural changes caused by compression, transmission, etc., and proposed a Structural Variation based Quality Index (SVQI) [32]. They divide the structure into global structures and local structures, which correspond to the basic perception and detail perception of human beings. Factors related to basic perception (such as brightness, contrast, and complexity) are attributed to a global structure, and factors related to detail perception (such as edge information) are reduced to local structures. Finally, a measure of the global structure and local structural changes is systematically combined to predict the final quality score of the SC image. The SVQI model proposed by Gu *et al.* has more outstanding performance than the most advanced IQA model at that time.

2.3   Experimental Analysis

In order to compare the performance of the eight IQA algorithms mentioned in this paper, namely IW-SSIM, GSM, VSI, ADD-GSIM, PSIM, SQMS, GDI and SVQI. We use the Video Quality Experts Group (VQEG) [33] suggest the Pearson Linear Correlation Coefficient (PLCC) and Spearman Rank Order Correlation Coefficient (SROCC) two indicators. The closer the values of PLCC and SROCC are to 1, the better the performance of the corresponding IQA algorithm is. In addition, PLCC and SROCC respectively reflect the accuracy and monotony of IQA algorithm. It should be noted that before calculating PLCC and SROCC, the Logistic function suggested by VQEG should be used to carry out nonlinear mapping of objective mass fraction. Therefore, this paper uses the commonly used five-parameter function to express as follows:

$$f(q) = s_1(\frac{1}{2} - \frac{1}{1 + e^{s_2(q-s_3)}}) + s_4 q + s_5 \tag{1}$$

where, q and $f(q)$ are objective quality evaluation scores and their correlated mapping scores, and $s_i (i \in 1, 2, 3, 4, 5)$ is five free parameters fitted based on gauss-newton method.

Table 2. Performance comparison of the eight IQA methods on the three databases.

IQA	TID2013		SCID		HSNID		Average		W-Average	
	PLCC	SROCC	PLCC	SROCC	PLCC	SROCC	PLCC	SROCC	PLCC	SROCC
IW-SSIM	0.8319	0.7779	0.7880	0.7714	0.7485	0.7626	0.7895	0.7706	0.8078	0.7740
GSM	0.8464	0.7946	0.7065	0.6945	0.7611	0.7602	0.7713	0.7498	0.7902	0.7574
VSI	**0.8999**	**0.8965**	0.7698	0.7621	0.7906	0.7778	0.8201	0.8121	0.8442	**0.8383**
ADD-SSIM	0.8809	0.8287	0.7979	0.7876	0.8236	0.8229	**0.8341**	**0.8131**	0.8468	0.8144
PSIM	0.9080	**0.8926**	0.8658	0.8668	0.8327	0.8341	0.8688	0.8645	0.8852	0.8773
SQMS	0.7619	0.6789	0.8563	0.8320	0.8598	**0.8517**	0.8260	0.7875	0.8044	0.7495
GDI	0.5997	0.6632	0.7609	0.7491	0.8456	0.8581	0.7354	0.7568	0.6814	0.7140
SVQI	0.7335	0.5992	**0.8610**	**0.8386**	**0.8561**	0.8488	0.8169	0.7622	0.7899	0.7073

Table 2 lists the prediction results of eight IQA metrics tested on three databases, and bold the prediction scores obtained by the top two methods, among which red and black are the algorithms ranked first and second in performance respectively. In order to compare the performance of each IQA algorithm more comprehensively, we also listed two average performance evaluations [34], namely direct average and weighted average.

As shown in Table 2, the performance ranking of each IQA metric on TID2013 database is: PSIM > VSI > ADD-GSIM > SQMS > SVQI > GDI > GSM > IW-SSIM. It can be seen that only PSIM index has a prediction accuracy rate of over 90%, but it is slightly lower than VSI index in terms of monotonicity prediction. The performance ranking on SCID database is: PSIM > SVQI > SQMS > ADD-GSIM > VSI > GDI > IW-SSIM > GSM. It can be seen that PSIM index achieves the best score in both the accuracy prediction and the monotonicity prediction. It is important to note that only SQVI metric scores match PSIM. The ranking of performance on HSNID database is as follows: SQMS > GDI > SVQI > PSIM > ADD-GSIM > VSI > GSM > IW-SSIM. It can be seen that SQMS, GDI and SVQI have similar predictive abilities, but SQMS has the best performance in terms of accuracy prediction and GDI has the best performance in terms of monotonicity prediction.

Table 3. Comparison of computational cost (seconds/image) on the three testing image databases.

Metrics	IW-SSIM	GSM	VSI	ADD-GSIM	PSIM	SQMS	GDI	SVQI
TID2013	0.5358	0.0221	0.2277	0.1001	0.0679	0.0681	0.2028	0.8299
SCID	2.3195	0.0602	0.4749	0.4620	0.3378	0.3119	0.7193	4.2213
HSNID	1.1556	0.0484	0.3272	0.2303	0.1423	0.1534	0.2429	2.0247

The performance of PSIM method in TID2013 and SCID image databases is significantly better than that of other IQA methods tested. In terms of two average results, the performance of PSIM is significantly better than that of the ADD-GSIM model which ranks second in performance, and far better than that of the other six IQA metrics. IW-SSIM, GSM, VSI and ADD-GSIM combine large-scale gradient and significance to evaluate NS images, so the performance on TID2013 database is relatively better, while SVQI, SQMS and GDI use large-scale gradient to evaluate SC images, so the performance on SCID database is more superior. PSIM can achieve the best performance on both databases at the same time, because it combines the method of evaluating natural images and screen content images, and adopts multi-scale gradient to evaluate images. On HSNID database, PSIM did not achieve the best performance. The reason may be that PSIM did not consider that two different image types could be included in the same database at the same time. In future studies, we can consider the appropriate combination of large scale and small scale to evaluate the database of mixed image types. It is clear that PSIM has the best overall performance across the three databases, and the model scores very high on both the monotonicity and accuracy of its predictions.

Next, to measure the execution efficiency of each metric, we further measured the average running time of the eight IQA methods on three databases. The entire experimental environment is tested on a dell laptop with a CPU of 2.20 GHz and RAM of 4.00 GB × 2 using MATLAB R2016a software platform, and the results is recorded in Table 3. It is not hard to find that GSM has the shortest average running time, but its prediction scores are not satisfactory. By contrast, PSIM has superior performance and short running time. Obviously, PSIM works well because it uses simple operators and fuses the differences between the microstructures, macrostructures, and color information of reference and distorted images.

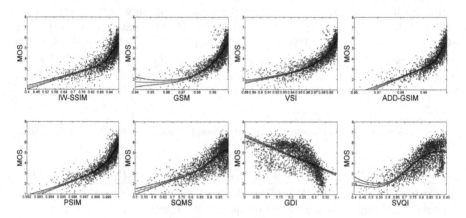

Fig. 2. Scatter plots of MOS versus IW-SSIM, GSM, VSI, ADD-GSIM, PSIM, SQMS, GDI, SVQI on the TID2013 database.

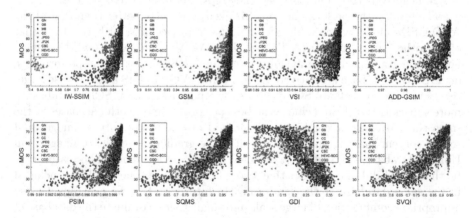

Fig. 3. Scatter plots of MOS versus IW-SSIM, GSM, VSI, ADD-GSIM, PSIM, SQMS, GDI, SVQI on the SCID database.

For a more intuitively and accurately show the relationship between the objective quality prediction score of IQA algorithm and MOS, we use two different types of scatter diagrams on TID2013 and SCID databases for illustration, which are scatter plot without distinguishing distortion types (as shown in Fig. 2) and scatter plot with distinguishing distortion types (as shown in Fig. 3). Figure 2 shows a scatter diagram drawn on the TID2013 database. Each scatter plot contains 3000 sample data points and two dashed lines in two colors: the red dotted line represents the Logistic function fitting curve, and the black dotted line represents the 95% confidence interval. It can be seen from Fig. 2 that PSIM is the best fitting data point for both vertical sample data points (in the same direction as the red dotted line) and horizontal sample data points (perpendicular to the red dotted line). Figure 3 shows a scatter map drawn on the SCID database. In each scatter diagram, points corresponding to different distortion types are distinguished by different colored graphs. As can be seen from Fig. 3, the points of each distortion type are evenly distributed on the PSIM map and tend to be more like a line, which shows that the PSIM method has the best adaptability to different distortion types.

3   Conclusion

In this paper, we first introduced three state-of-the-art databases and eight classic IQA methods in detail. They were designed to evaluate NS or SC image quality and achieved state-of-the-art performance at the time. After the analysis of the three experiments of calculating the PLCC and SROCC performance indexes, measuring the average running time of each algorithm and drawing scatter diagram for the eight algorithms of IW-SSIM, GSM, VSI, additive GSIM, PSIM, SQMS, GDI and SVQI on the databases of TID2013, SCID and HSNID, we found that PSIM algorithm combined with gradient information at multiple scales can achieve the best results in evaluating NSI image and SC image simultaneously. In previous studies, researchers selected various methods to evaluate images according to the different characteristics of NS images and SC images. Most of them used large-scale gradient combined with significance method to evaluate NC images, while most SC images used large-scale gradient method to evaluate images. Therefore, in future studies, the two methods can be appropriately weighted and combined with the actual situation to obtain better performance.

References

1. Lin, T., Zhang, P., Wang, S., Zhou, K., Chen, X.: Mixed chroma sampling-rate high efficiency video coding for full-chroma screen content. IEEE Trans. Circ. Syst. Video Technol. **23**(1), 173–185 (2013)
2. Gu, K., Zhai, G., Lin, W., Yang, X., Zhang, W.: Learning a blind quality evaluation engine of screen content images. Neurocomputing **196**(5), 140–149 (2016)
3. Lin, W., Kuo, C.C.J.: Perceptual visual quality metrics: a survey. J. Vis. Commun. Image Represent. **22**(4), 297–312 (2011)

4. Chu, J., Chen, Q., Yang, X.: Review on full reference image quality assessment algorithms. Appl. Res. Comput. **31**(1), 13–22 (2014)
5. Zhai, G., Cai, J., Lin, W., Yang, X., Zhang, W., Etoh, M.: Cross-dimensional perceptual quality assessment for low bit-rate videos. IEEE Trans. Multimed. **10**(7), 1316–1324 (2008)
6. Zhai, G., Cai, J., Lin, W., Yang, X., Zhang, W.: Three dimensional scalable video adaptation via user-end perceptual quality assessment. IEEE Trans. Broadcast. **54**(3), 719 (2008)
7. Wang, S., Gu, K., Zeng, K., Wang, Z., Lin, W.: Perceptual screen content image quality assessment and compression. In: IEEE International Conference on Image Processing, pp. 1434–1438, September 2015
8. Wang, S., Gu, K., Zeng, K., Wang, Z., Lin, W.: Objective quality assessment and perceptual compression of screen content images. IEEE Comput. Graphics Appl. **38**(1), 47–58 (2016)
9. Gu, K., Tao, D., Qiao, J., Lin, W.: Learning a no-reference quality assessment model of enhanced images with big data. IEEE Trans. Neural Netw. Learn. Syst. **29**(4), 1301–1313 (2018)
10. Gu, K., Zhou, J., Qiao, J., Zhai, G., Lin, W., Bovik, A.C.: No-reference quality assessment of screen content pictures. IEEE Trans. Image Process. **26**(8), 4005–4018 (2017)
11. Qian, J., Tang, L., Jakhetiya, V., Xia, Z., Gu, K., Lu, H.: Towards efficient blind quality evaluation of screen content images based on edge-preserving filter. Electron. Lett. **53**(9), 592–594 (2017)
12. Wang, S., et al.: Subjective and objective quality assessment of compressed screen content images. IEEE J. Emerg. Sel. Top. Circ. Syst. **6**(4), 532–543 (2016)
13. Sheikh, H.R., Wang, Z., Cormack, L., Bovik, A.C.: LIVE image quality assessment database release 2 (2005). http://live.ece.utexas.edu/research/quality
14. Le, P.C., Autrusseau, F.: Subjective quality assessment IRCCyN/IVC database (2005). http://www.irccyn.ec-nantes.fr/ivcdb/
15. Gu, K., Zhai, G., Lin, W., Yang, X., Zhang, W.: Hybrid no-reference quality metric for singly and multiply distorted images. IEEE Trans. Broadcast. **60**(3), 555–567 (2014)
16. Ponomarenko, N., et al.: Image database TID2013: peculiarities, results and perspectives. Sig. Process. Image Commun. **30**, 55–77 (2015)
17. Ponomarenko, N., et al.: TID2008-a database for evaluation of full-reference visual quality assessment metrics. Adv. Mod. Radioelectron. **10**, 30–45 (2009)
18. Ni, Z., Ma, L., Zeng, H., Chen, J., Cai, C., Ma, K.: ESIM: edge similarity for screen content image quality assessment. IEEE Trans. Image Process. **26**(10), 4818–4831 (2017)
19. Gu, K., Xu, X., Qiao, J., Jiang, Q., Lin, W., Thalmann, D.: Learning a unified blind image quality metric via on-line and off-line big training instances. IEEE Trans. Big Data (2019, to appear). https://doi.org/10.1109/TBDATA.2019.2895605
20. Liu, S., Wu, L., Gong, Y., Liu, X.: Overview of image quality assessment. Sciencepaper Online **6**(7), 501–506 (2004)
21. Wang, Z., Bovik, A.C., Sheikh, H.R., Simoncelli, E.P.: Image quality assessment: from error visibility to structural similarity. IEEE Trans. Image Process. **13**(4), 600–612 (2004)
22. Wang, Z., Simoncelli, E.P., Bovik, A.C.: Multiscale structural similarity for image quality assessment. In: The Thirty-Seventh Asilomar Conference on Signals, Systems and Computers, November 2003

23. Wang, Z., Li, Q.: Information content weighting for perceptual image quality assessment. IEEE Trans. Image Process. **20**(5), 1185–1198 (2011)
24. Sheik, H.R., Bovik, A.C.: Image information and visual quality. IEEE Trans. Image Process. **15**(2), 430–444 (2006)
25. Liu, A., Lin, W., Narwaria, M.: Image quality assessment based on gradient similarity. IEEE Trans. Image Process. **21**(4), 1500–1512 (2012)
26. Zhang, L., Shen, Y., Li, H.: VSI: a visual saliency-induced index for perceptual image quality assessment. IEEE Trans. Image Process. **23**(10), 4270–4281 (2014)
27. Gu, K., Wang, S., Zhai, G., Lin, W., Yang, X., Zhang, W.: Analysis of distortion distribution for pooling in image quality prediction. IEEE Trans. Broadcast. **62**(2), 446–456 (2016)
28. Gu, K., Li, L., Lu, H., Min, X., Lin, W.: A fast reliable image quality predictor by fusing micro- and macro-structures. IEEE Trans. Ind. Electron. **64**(5), 3903–3912 (2017)
29. Xue, W., Zhang, L., Mou, X., Bovik, A.C.: Gradient magnitude similarity deviation: a highly efficient perceptual image quality index. IEEE Trans. Image Process. **23**(2), 684–695 (2014)
30. Gu, K., et al.: Saliency-guided quality assessment of screen content images. IEEE Trans. Multimed. **18**(6), 1098–1110 (2016)
31. Ni, Z., Ma, L., Zeng, H., Cai, C., Ma, K.: Gradient direction for screen content image quality assessment. IEEE Signal Process. Lett. **23**(10), 1394–1398 (2016)
32. Gu, K., Qiao, J., Min, X., Yue, G., Lin, W., Thalmann, D.: Evaluating quality of screen content images via structural variation analysis. IEEE Trans. Vis. Comput. Graph. **24**(10), 2698–2701 (2018)
33. VQEG: Final report from the video quality experts group on the validation of objective models of video quality assessment, [EB/OL]. http://www.vqeg.org/
34. Gu, K., Zhai, G., Lin, W., Liu, M.: The analysis of image contrast: from quality assessment to automatic enhancement. IEEE Trans. Cybern. **64**(1), 284–297 (2016)

Screen Content Picture Quality Evaluation by Colorful Sparse Reference Information

Huiqing Zhang[1,2], Donghao Li[1,2(✉)], Shuo Li[1,2], Zhifang Xia[3], and Lijuan Tang[4]

[1] Faculty of Information Technology, Beijing University of Technology, Beijing 100124, China
lidonghao97@163.com
[2] Engineering Research Center of Digital Community, Ministry of Education, Beijing 100124, China
[3] The State Information Center of China, Beijing 100045, China
[4] Jiangsu Vocational College of Business, Jiangsu 226000, China

Abstract. With the rapid development of multimedia interactive applications, the processing volume of the screen content (SC) images is increasing day by day. The research on image quality assessment is the basis of many other applications. The focus of general image quality assessment (QA) research is natural scene (NS) images, now for the quality assessment research of SC images becomes very urgent and has received more and more attention. Accurate quality assessment of SC images helps improve the user experience. Based on these, this paper proposes an improved method using very sparse reference information for accurate quality assessment of SC images. Specifically, the proposed method extracts macroscopic, microscopic structure and color information respectively, and measures the differences in terms of macroscopic, microscopic features and color information between the original SC image and its distorted version, and finally calculates the overall quality score of the distorted SC image. The quality assessment model we built uses a dimension reduction histogram and only needs to transmit very sparse reference information. Experiments show that the proposed method has obvious superiority over the state-of-the-art relevant quality metrics in the visual quality assessment of SC images.

Keywords: Screen content image · Sparse reference · Image quality assessment

1 Introduction

With the rapid development of mobile Internet technology, some multi-client communication systems have also been developed. In these systems, different

This work is supported by the Major Science and Technology Program for Water Pollution Control and Treatment of China (2018ZX07111005), Natural Science Research of Jiangsu Higher Education Institutions under grant (18KJB52).

© Springer Nature Singapore Pte Ltd. 2020
G. Zhai et al. (Eds.): IFTC 2019, CCIS 1181, pp. 280–292, 2020.
https://doi.org/10.1007/978-981-15-3341-9_24

terminals can communicate with each other, thereby realizing the transmission and processing of video, images and other contents. In the field of transmission video, some traditional video coding algorithms [1–4] improve the visual quality of video. In the field of image processing, the visual quality of images has attracted more and more attention from many researchers, but in practical applications, due to limited resources of the system, various operation links such as compression [5], enhancement [6,7], restoration [8,9], transmission [10], etc., may introduce distortion, which affects the user experience. Image quality assessment is generally divided into two types of methods: subjective assessment and objective assessment. Subjective assessment means that the observer directly gives the quality score, because it is the assessment of the overall image quality made by the human audience, and the human audience is ultimate final user, so subjective assessment is decisive. However this method is time-consuming and laborious, and the assessment result is not stable, so it cannot be popularized in practical application. Based on these, reliable objective assessment has become the subject of many researchers. In order to get closer to human visual perception, many objective image QA models have been proposed based on low-level vision [11–13], brain theory [14,15], contrast enhancement [16,17], statistics [18] and so on.

It should be noted that these image quality assessment methods are proposed for NS images, but NS images are significantly different from SC images, and NS images have rich colors and complex textures, and the lines of NS images are thick. SC images have limited colors and simple shapes, and the lines of SC images are thin. In [19] and [20], it is shown that those QA models for NS images cannot well evaluate the quality of SC images. Therefore, the quality assessment method which is applicable to NS images is not necessarily applicable to SC images, so it is necessary to propose a quality assessment method for SC images. Recently, some objective models using a small amount of reference information have been proposed, such as RWQMS [21], RQMSC [22], PBM [23], which are feasible in evaluating SC image quality. However, compared with NS images, studies on SC image quality assessment are not very complete, either rich reference information is required (e.g., SPQA [19] and GDI [24]), or the prediction of monotonicity is not good enough (e.g., RWQMS [21], RQMSC [22], PBM [23]).

In order to solve the above problem, we propose a new method to accurately estimate SC image quality by using a small amount of original image information. Some studies have shown that adding gradient magnitude information to QA models is very suitable for extracting image structures, as mentioned in [11,12, 25–27]. In our method, the overall quality score is deduced by combining with the differences of the macroscopic, microscopic structure and color information between the original image and the corresponding distorted image. It should be noted that we use the method of histogram establishment to extract the feature of the images, which greatly reduces the amount of information that needs to be transmitted.

In this paper, we first briefly review some classical image quality assessment algorithms. Then we introduce our QA model and design method. Next, we compare our method with the classical image quality assessment methods and evaluate them with five quality metrics. Finally, it is concluded that the proposed method is superior to the state-of-the-art relevant methods.

2 Methodology

The design idea of the model proposed in this paper is to compare the original image with the distorted image, extracts their differences in macroscopic, microscopic structure and color information. By combining the information, we can evaluate the quality of SC images. Due to the differences between NS images and SC images, the effect of these methods is poor when applied to SC images. The macroscopic structure of the NS images is regarded as the outline of the image, while the main content of the SC images is concentrated in the text area. Based on the above, we propose a new method for SC images. The model proposed in this paper is shown in Fig. 1. Specifically, we extract the macroscopic structure and microscopic structure of the SC image firstly, as well as the color information, then use the histogram method to reduce the dimension of the reference information. Finally we combine the differences information to derive the overall quality score.

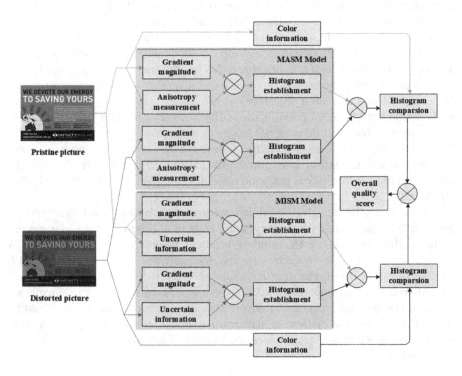

Fig. 1. Illustration of our QA model.

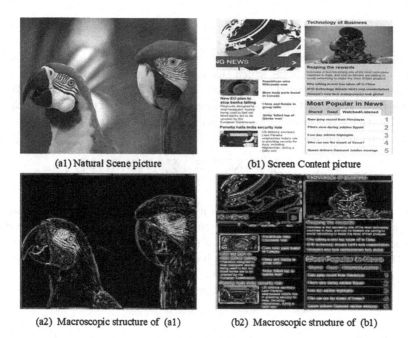

(a1) Natural Scene picture

(b1) Screen Content picture

(a2) Macroscopic structure of (a1)

(b2) Macroscopic structure of (b1)

Fig. 2. Comparisons the macroscopic structures of NS image and SC image. (a1) is the pristine NS image and (b1) is the pristine SC image. (a2) and (b2) are the macroscopic structures of (a1) and (b1).

2.1 Macroscopic Structure Measurement

In this paper, we fuse the gradient and anisotropy measurements to derive the macroscopic structure measurement (MASM) of images. The macroscopic structures of NS images and SC images can be extracted, as shown in Fig. 2.

(1) *Gradient magnitude.* To be specific, this work uses the Scharr operator [11,20,29] to calculate the gradient magnitude of a image. The gradient magnitude can be calculated by the following formulas:

$$G(\mathbf{X}) = \sqrt{G_x^2 + G_y^2}, \tag{1}$$

where

$$G_x = H_x \bigotimes \mathbf{X} = \frac{1}{16} \begin{bmatrix} +3 & 0 & -3 \\ +10 & 0 & -10 \\ +3 & 0 & -3 \end{bmatrix} \bigotimes \mathbf{X}, \tag{2}$$

$$G_y = H_y \bigotimes \mathbf{X} = \frac{1}{16} \begin{bmatrix} +3 & +10 & +3 \\ 0 & 0 & 0 \\ -3 & -10 & -3 \end{bmatrix} \bigotimes \mathbf{X}, \tag{3}$$

and $\mathbf{X}$ indicates a SC image signal; H_x and H_y represent the Scharr convolution masks; $\bigotimes$ denotes the convolution operator. Based on the above operation, the gradient magnitude can clearly display the structure information in the image.

(2) *Anisotropy measurement.* Generally speaking, the large anisotropy is the macroscopic structures which have intensity variation, and the small anisotropy stands for the structures with homogeneous scattering. We can use the structural tensor to calculate the anisotropy measurement. Specifically, the structural tensor is expressed as the matrix generated by the gradient magnitude of a SC image.

$$T(u) = \begin{pmatrix} \sum_v \langle \nabla_x \mathbf{X}_v, \nabla_x \mathbf{X}_v \rangle & \sum_v \langle \nabla_y \mathbf{X}_v, \nabla_x \mathbf{X}_v \rangle \\ \sum_v \langle \nabla_x \mathbf{X}_v, \nabla_y \mathbf{X}_v \rangle & \sum_v \langle \nabla_y \mathbf{X}_v, \nabla_y \mathbf{X}_v \rangle \end{pmatrix}, \tag{4}$$

where $v \in R(u)$ belongs to the neighborhood of a pixel u; ∇_x and ∇_x represent the partial differential operators which are in the horizontal and vertical directions; $\langle \cdot, \cdot \rangle$ is the inner product of a pair of vectors. $T(u)$ is a semi-positive definite symmetric 2×2 matrix, and it has two eigenvectors η_u and η'_u with two corresponding eigenvalues λ_u and $\lambda'_u (\lambda_u \geq \lambda'_u \geq 0)$.

Based on the above, the anisotropy measurement can be defined as the relative distance between λ_u and λ'_u,

$$A(u) = \frac{\lambda_u - \lambda'_u + \epsilon}{\lambda_u + \lambda'_u + \epsilon}, \tag{5}$$

where ϵ is a very small constant to avoid the case where the denominator is 0. It can be seen from Eq. 5 that the minimum value and the maximum value of A are 0 and 1. Based on the above calculation process, MASM model is defined as follows:

$$MASM(\mathbf{X}) = G \cdot A, \tag{6}$$

where G and A are generated by Eqs. 1–3 and Eqs. 4–5 respectively. In addition, MASM model can effectively suppress the minor texture of the image and retain the dominant structure of the image.

2.2   Microscopic Structure Measurement

In this section, we introduce the MISM model to detect the detailed texture in the image. MISM model consists of gradient magnitude and uncertain information.

(1) *Gradient magnitude.* The gradient magnitude has been introduced in the previous section and will not be repeated here.

(2) *Uncertain information.* We combine Gaussian and motion low-pass filters to measure the uncertain information. To be specific, a Gaussian filter can be expressed as

$$H_g(i,j) = \frac{1}{2\pi\theta^2} exp(-\frac{i^2 + j^2}{2\theta^2}), \tag{7}$$

where θ is the standard deviation that controls the smoothing strength; We can get the Gaussian filtered SC image $\mathbf{X}$ by Eq. 8:

$$\mathbf{X}_g = \mathbf{X} \bigotimes H_g. \tag{8}$$

As mentioned above, we derive uncertain information from the differences between $\mathbf{X}$ and $\mathbf{X}_g$. Here we use the standard version of gradient similarity:

$$SX_g = f(\mathbf{X}, \mathbf{X}_g) = \frac{(S(\mathbf{X}) - S(\mathbf{X}_g))^2}{S^2(\mathbf{X}) + S^2(\mathbf{X}_g)}. \tag{9}$$

The motion filter is shown as:

$$M(p,q) = \begin{cases} \frac{1}{t} & (p \cdot sin\theta + q \cdot cos\theta) = 0; \ p^2 + q^2 \leq \frac{t^2}{4} \\ 0 & otherwise \end{cases}, \tag{10}$$

where θ represents the specific direction of motion and t represents the amount of motion in pixels, and t is usually set to 9 in the motion filter, then we can get the motion filtered SC image $\mathbf{X}$ by Eq. 11:

$$\mathbf{X}_m = \mathbf{X} \bigotimes M. \tag{11}$$

Like gaussian filter, the uncertain information can be derived from motion filter:

$$SX_m = f(\mathbf{X}, \mathbf{X}_m) = \frac{(S(\mathbf{X}) - S(\mathbf{X}_m))^2}{S^2(\mathbf{X}) + S^2(\mathbf{X}_m)}. \tag{12}$$

Based on the above calculation process, we can define the uncertain information as follows:

$$U = \frac{1}{2}(SX_g + SX_m). \tag{13}$$

Finally, our MISM model is shown as:

$$MISM(\mathbf{X}) = G \cdot U, \tag{14}$$

where G and U are generated by Eqs. 1–3 and Eqs. 7–13 respectively.

2.3   Color Information Measure

In order to evaluate the quality of SC images more accurately, the color information is introduced. To be specific, we create a cell array named c with 3 elements, and the first element of array c is c_1:

$$c_1 = \frac{G_x^2}{G_x^2 + G_y^2 + eps}. \tag{15}$$

The second element of array c is c_2:

$$c_2 = \frac{G_y^2}{G_x^2 + G_y^2 + eps}. \tag{16}$$

The third element of array c is c_3:

$$c_3 = \begin{cases} 1 & if\ G(\mathbf{X}) = 0 \\ 0 & otherwise \end{cases}, \tag{17}$$

where $G(\mathbf{X})$, G_x and G_y are shown as Eqs. 1, 2 and 3, the value of *eps* is 2.2204×10^{-16}. Finally the color information can be expressed by

$$C(\mathbf{X}) = \frac{[G_{cx}, G_{cy}]}{G_{cx} + G_{cy}}, \tag{18}$$

where G_{cx} and G_{cy} can be expressed as follows:

$$G_{cx} = \sum c_1 + \frac{1}{2} \sum c_3 \qquad G_{cy} = \sum c_2 + \frac{1}{2} \sum c_3. \tag{19}$$

2.4 Overall Quality Measure

Based on the above analysis, in order to more evaluate the quality of SC images accurately, we combine macroscopic structure, microscopic structure and the color information to derive the overall quality score. For the sake of making the contrast effect between the original image and the distorted image better, we distinguish between important and unimportant components through non-linear mapping. Then the important components are associated with 1, the unimportant components are associated with 0. Finally, we choose the cumulative normal distribution function (CDF) as the non-linear mapping function.

$$F(x) = \frac{1}{\sqrt{2\pi}\sigma} \int_{-\infty}^{x} exp[-\frac{(t - \mu)^2}{2\sigma^2}]dt, \tag{20}$$

where μ is the adjustment threshold; x is the stimulus amplitude; σ is the parameter that controls the slope of the predicted probability change, and we set it to 0.05 according to previous experience.

We pass the MASM model and the MISM model into the CDF to get two mappings. In practice, we perform feature extraction on the original image $\mathbf{X}$ and the distorted image $\mathbf{X}'$ respectively. Before doing this, our method uses the histogram to represent the distribution, this can take consideration to the reference information transmission rate and quality assessment performance. Specifically, our range distribution of $\varphi([r_{min}, r_{max}])$ is divided into N equal- length gaps. The histogram can be obtained by setting:

$$h_i = |g_i|, \quad g_i = \{\varphi(\tau) \in S_i\}, \tag{21}$$

where $S_i = [r_{min} + (i - 1)\frac{\tilde{r}}{N}, r_{min} + i\frac{\tilde{r}}{N}]$, $\tilde{r} = r_{max} - r_{min}$. The histogram of lossless SC image can be expressed as follows:

$$H_{\mathbf{X}}(i) = h_i / \sum_{n=1}^{N} h_n. \tag{22}$$

Apply the above operation to the distorted image $\mathbf{X}'$, and we get the histogram of the distorted image: $H_{\mathbf{X}'}(i)$. By comparing the two histograms, the score of MASM model is as follows:

$$Q_{MASM}(\mathbf{X}, \mathbf{X}') = \frac{1}{N} \sum_{i=1}^{N} \left(\frac{min\{H_{\mathbf{X}}(i), H_{\mathbf{X}'}(i)\} + \epsilon}{max\{H_{\mathbf{X}}(i), H_{\mathbf{X}'}(i)\} + \epsilon} \right). \tag{23}$$

Then we will do the above operation with MISM mapping to derive $Q_{MISM}(\mathbf{X}, \mathbf{X}')$ which means the score of MISM model.

Next we divide the color information into two parts, as shown in Eqs. 24 and 25:

$$C^{\star}(\mathbf{X}) = C(\mathbf{X}) \quad if \ G > 0.05 \times max\{G(\mathbf{X})\}, \tag{24}$$

$$C^{*}(\mathbf{X}) = C(\mathbf{X}) \quad if \ G \leq 0.05 \times max\{G(\mathbf{X})\}, \tag{25}$$

where $C(\mathbf{X})$ is calculated in Eq. 18, and $G(\mathbf{X})$ is calculated in Eq. 1. By comparing the differences in C^{*} between the original image and the distorted image, the score of C^{*} can be expressed as

$$Q_{C\star}(\mathbf{X}, \mathbf{X}') = \frac{1}{N} \sum_{i=1}^{N} \left(\frac{min\{C_{\mathbf{X}}(i), C_{\mathbf{X}'}(i)\} + \epsilon}{max\{C_{\mathbf{X}}(i), C_{\mathbf{X}'}(i)\} + \epsilon} \right), \tag{26}$$

where $min\{\cdot, \cdot\}$ is used to find which one is the minimum, and $max\{\cdot, \cdot\}$ is used to find which one is the maximum; the value of N is set to 2; ϵ is a minuscule positive constant, just to make sure that the denominator is not 0. Then we do the above operation with the C^{*} to get the score of $Q_{C*}(\mathbf{X}, \mathbf{X}')$.

Finally we combine $Q_{MASM}(\mathbf{X}, \mathbf{X}')$, $Q_{MISM}(\mathbf{X}, \mathbf{X}')$, $Q_{C\star}(\mathbf{X}, \mathbf{X}')$ and $Q_{C*}(\mathbf{X}, \mathbf{X}')$ together to get the overall quality score $Q(\mathbf{X}, \mathbf{X}')$:

$$Q(\mathbf{X}, \mathbf{X}') = Q_{MASM}(\mathbf{X}, \mathbf{X}') \cdot Q_{MISM}(\mathbf{X}, \mathbf{X}') \cdot Q_{C\star}(\mathbf{X}, \mathbf{X}') \cdot Q_{C*}(\mathbf{X}, \mathbf{X}'). \tag{27}$$

3   Experiment and Analysis

In this section, we introduce the experimental setting and assessment criteria firstly, and then compare the performance of different image quality assessment methods.

3.1   Experimental Setting and Assessment Criteria

We collect 14 image quality assessment methods, which can be divided into six types according to the application scenarios. The first type includes FSIM [11], GSM [12] and VSI [25], these three methods are based on gradient magnitude and full reference information of lossless images. The second type includes VIF-RR [29], WNISM [30], FTQM [31], and SDM [32], which are based on partial reference information of lossless images. The third type includes two non-reference image quality assessment methods, namely BMPRI [33], BPRI [34]. The fourth

Table 1. Comparisons of 14 state-of-the-art QA methods from the five assessment criteria on the SIQAD database.

Metric	Type 1			Type 2				Type 3		Type 4		Type 5		Type 6	
	FSIM	GSM	VSI	**VIF-RR**	WNISM	FTQM	SDM	BMPRI	**BPRI**	SPQA	**GDI**	PBM	**RWQMS**	SCVSR	**Our**
KRC	**0.4253**	0.4054	0.3874	**0.4431**	0.3540	0.3268	0.4322	0.2283	**0.4040**	0.6803	**0.6486**	0.5280	**0.5835**	0.6255	**0.6350**
SRC	**0.5824**	0.5483	0.5381	**0.6082**	0.5188	0.4575	0.6020	0.3325	**0.5612**	0.8416	**0.8436**	0.7168	**0.7815**	0.8213	**0.8307**
PLC	**0.5906**	0.5686	0.5568	**0.5758**	0.5857	0.4691	0.6034	0.3737	**0.5863**	0.8584	**0.8515**	0.7264	**0.8103**	0.8343	**0.8418**
MAE	**9.0066**	9.1660	9.2875	**9.5197**	9.4566	10.132	9.0139	10.693	**8.9966**	5.7890	**5.9744**	7.7632	**6.8200**	6.3196	**6.2028**
RMS	**11.551**	11.775	11.890	**11.703**	11.602	12.641	11.414	13.277	**11.596**	7.3421	**7.5055**	9.8375	**8.3892**	7.8924	**7.7263**

type contains three recent image quality assessment methods SPQA [19], and GDI [24]. The fifth type consists of RWQMS [21] and PBM [23]. The methods of the last type are SCVSR [35] and our method, which are image quality assessment methods based on very sparse reference information.

In order to test the above methods, we use SIQAD [18], which consists of 980 distorted SC images generated with 20 reference images in 7 common distortion types at 7 degradation levels: Gaussian noise (GN), Gaussian blur (GB), motion blur (MB), contrast change (CC), JPEG compression (JP), JPEG2000 compression (J2), and layer segmentation based coding (LC).

The performance of an IQA method is generally evaluated in terms of prediction monotonicity, prediction accuracy, and prediction consistency. To evaluate these properties, we use a logistic regression equation to reduce the nonlinearity of the prediction scores, as shown in Eq. 28.

$$R(\zeta) = \gamma_1 \left(\frac{1}{2} - \frac{1}{1 + e^{\gamma_2(\zeta - \gamma_3)}} \right) + \gamma_4 \zeta + \gamma_5 \qquad (28)$$

where ζ and $R(\zeta)$ represent the input and output scores respectively, $\gamma_i (i = 1, 2, 3, 4, 5)$ are the five parameters determined during the curve fitting process, and the correlation between ζ and $R(\zeta)$ is evaluated. The first and second parameters are Kendall's rank correlation coefficient (KRC) and Spearman rank correlation coefficient (SRC), which are used to evaluate the predictive monotonicity. The third one is Pearson linear correlation coefficient (PLC), which is used to evaluate the prediction accuracy. The fourth and fifth parameters are mean absolute error (MAE) and root mean-squared error (RMS), which are used to evaluate prediction consistency.

3.2   Performance Comparison

By testing our method in SIQAD, KRC, SRC, PLC, MAE and RMS of our method are 0.6350, 0.8307, 0.8418, 6.2028, 7.7263 respectively. We compare the optimal methods in each type with our method, as shown in Table 1. The best performing method of each type is highlighted in bold font. Specifically, compared with the best method FSIM in the first type, the relative performance is improved by 49.31%, 42.63% and 42.53% in terms of KRC, SRC and PLC. Compared with the best performance method VIF-RR in the second type, the relative performance of KRC, SRC and PLC increase by 43.31%, 36.58% and 46.20%. Compared with the best method BPRI in the third type, the relative performance gains of KRC, SRC and PLC are 57.18%, 48.02% and 43.58%. Our method is slightly inferior to the methods in the fourth type, because the methods in the fourth type use the full reference information of lossless SC images, while our method only uses partial reference information of SC images. In addition, compared with the best method RWQMS in the fifth type, the relative performance of KRC, SRC and PLC is improved by 8.83%, 6.30% and 3.89%. Based on the SCVSR in the sixth type, our method improves the relative performance by 1.52%, 1.14% and 0.90% in terms of KRC, SRC and PLC.

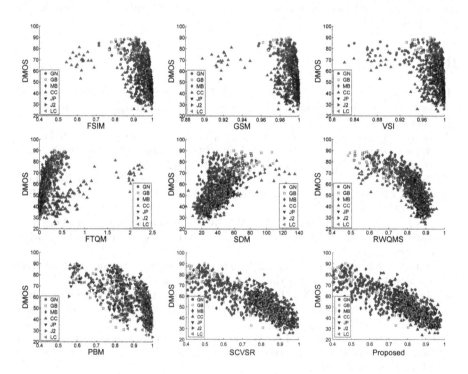

Fig. 3. Scatter plots of the DMOS values and the predictions of FSIM, GSM, VSI, FTQM, SDM, RWQMS, PBM, SCVSR and our proposed model on the SIQAD database.

In order to further illustrate the differences between our method and other eight methods, we provide 9 scatter plots. The ordinates of these scatter plots are the subjective assessment score of SIQAD, and the abscissas are respectively the objective quality score obtained by nine image quality assessment methods, as shown in Fig. 3. It can be seen that the consistency of sample points generated

by our method is significantly better than other eight methods, while the sample
points generated by other eight methods are far apart from each other.

4 Conclusion

In this paper, we propose a new SC image quality assessment method. By integrating the macroscopic structure, microscopic structure and color information
together, the differences between the original image and the distorted image is
compared, and the overall quality score of the SC images is obtained. Then we
compare the method proposed in this paper with the state-of-the-art relevant
methods. By demonstrating the above assessment criteria and scatter plots, it
is proved that our method is superior to other methods in terms of predicting
monotonicity, accuracy and consistency, and the QA model proposed in this
paper is effective.

References

1. Yang, X.K., Ling, W.S., Lu, Z.K., Ong, E.P., Yao, S.S.: Just noticeable distortion
   model and its applications in video coding. Sig. Process. Image Commun. **20**, 742–
   752 (2005)
2. Yang, X.K., Lin, W.S., Lu, Z.K., Ong, E.P., Yao, S.S.: Motion-compensated residue
   preprocessing in video coding based on just-noticeable-distortion profile. IEEE
   Trans. Circ. Syst. Video Technol. **15**(6), 742–752 (2005)
3. Zhai, G., Cai, J., Lin, W., Yang, X., Zhang, W., Etoh, M.: Cross-dimensional perceptual quality assessment for low bit-rate videos. IEEE Trans. Multimed. **10**(7),
   1316–1324 (2008)
4. Zhai, G., Zhang, W., Yang, X., Lin, W., Xu, Y.: Efficient image deblocking based
   on postfiltering in shifted windows. IEEE Trans. Circ. Syst. Video Technol. **18**(1),
   122–126 (2008)
5. Zhu, W., Ding, W., Xu, J., Shi, Y., Yin, B.: Screen content coding based on HEVC
   framework. IEEE Trans. Multimed. **16**(5), 1316–1326 (2014)
6. Gu, K., Zhai, G., Yang, X., Zhang, W., Chen, C.W.: Automatic contrast enhancement technology with saliency preservation. IEEE Trans. Circ. Syst. Video Technol.
   25(9), 1480–1494 (2015)
7. Gu, K., Tao, D., Qiao, J., Lin, W.: Learning a no-reference quality assessment
   model of enhanced images with big data. IEEE Trans. Neural Netw. Learn. Syst.
   29(4), 1301–1313 (2018)
8. Zhu, X., Milanfar, P.: Automatic parameter selection for denoising algorithms using
   a no-reference measure of image content. IEEE Trans. Image Process. **19**(12), 3116–
   3132 (2010)
9. Mittal, A., Moorthy, A.K., Bovik, A.C.: No-reference image quality assessment in
   the spatial domain. IEEE Trans. Image Process. **21**(12), 4695–4708 (2012)
10. Wu, H.R., Reibman, A.R., Lin, W., Pereira, F., Hemami, S.S.: Perceptual visual
    signal compression and transmission. Proc. IEEE **101**(9), 2025–2043 (2013)
11. Zhang, L., Zhang, L., Mou, X., Zhang, D.: FSIM: a feature similarity index for
    image quality assessment. IEEE Trans. Image Process. **20**(8), 2378–2386 (2011)
12. Liu, A., Lin, W., Narwaria, M.: Image quality assessment based on gradient simi-
    larity. IEEE Trans. Image Process. **21**(4), 1500–1512 (2012)

13. Yue, G., Hou, C., Gu, K., Mao, S., Zhang, W.: Biologically inspired blind quality assessment of tone-mapped images. IEEE Trans. Ind. Electron. **65**(3), 2525–2536 (2018)

14. Gu, K., Zhai, G., Yang, X., Zhang, W.: Using free energy principle for blind image quality assessment. IEEE Trans. Multimed. **17**(1), 50–63 (2015)

15. Zhai, G., Wu, X., Yang, X., Lin, W., Zhang, W.: A psychovisual quality metric in free-energy principle. IEEE Trans. Image Process. **21**(1), 41–52 (2012)

16. Gu, K., Zhai, G., Lin, W., Liu, M.: The analysis of image contrast: from quality assessment to automatic enhancement. IEEE Trans. Cybern. **46**(1), 284–297 (2016)

17. Gu, K., Lin, W., Zhai, G., Yang, X., Zhang, W., Chen, C.W.: No-reference quality metric of contrast-distorted images based on information maximization. IEEE Trans. Cybern. **47**(12), 4559–4565 (2017)

18. Gu, K., Zhou, J., Qiao, J., Zhai, G., Lin, W., Bovik, A.C.: No-reference quality assessment of screen content pictures. IEEE Trans. Image Process. **26**(8), 4005–4018 (2017)

19. Yang, H., Fang, Y., Lin, W.: Perceptual quality assessment of screen content images. IEEE Trans. Image Process. **24**(11), 4408–4421 (2015)

20. Gu, K., et al.: Saliency-guided quality assessment of screen content images. IEEE Trans. Multimed. **18**(6), 1–13 (2016)

21. Wang, S., et al.: Subjective and objective quality assessment of compressed screen content images. IEEE J. Emerg. Sel. Top. Circ. Syst. **6**(4), 532–543 (2016)

22. Wang, S., Gu, K., Zhang, X., Lin, W., Ma, S., Gao, W.: Reduced-reference quality assessment of screen content images. IEEE Trans. Circ. Syst. Video Technol. **28**(1), 1–14 (2018)

23. Jakhetiya, V., Gu, K., Lin, W., Li, Q., Jaiswal, S.P.: A prediction backed model for quality assessment of screen content and 3-D synthesized images. IEEE Trans. Ind. Inf. **14**(2), 652–660 (2018)

24. Ni, Z., Ma, L., Zeng, H., Cai, C., Ma, K.: Gradient direction for screen content image quality assessment. IEEE Signal Process. Lett. **23**(10), 1394–1398 (2016)

25. Zhang, L., Shen, Y., Li, H.: VSI: a visual saliency induced index for perceptual image quality assessment. IEEE Trans. Image Process. **23**(10), 4270–4281 (2014)

26. Gu, K., Li, L., Lu, H., Min, X., Lin, W.: A fast reliable image quality predictor by fusing micro- and macro-structures. IEEE Trans. Ind. Electron. **64**(5), 3903–3912 (2017)

27. Min, X., Zhai, G., Gu, K., Yang, X., Guan, X.: Objective quality evaluation of dehazed images. IEEE Trans. Intell. Transp. Syst. **20**, 1–14 (2018)

28. Li, Q., Wang, Z.: Reduced-reference image quality assessment using divisive normalization-based image representation. IEEE J. Sel. Top. Signal Process. **3**(2), 202–211 (2009)

29. Wu, J., Lin, W., Shi, G., Liu, A.: Reduced-reference image quality assessment with visual information fidelity. IEEE Trans. Multimed. **15**(7), 1700–1705 (2013)

30. Wang, Z., Simoncelli, E.P.: Reduced-reference image quality assessment using a wavelet-domain natural image statistic model. In: Proceedings of the SPIE Human Vision and Electronic Imaging X, vol. 5666, pp. 149–159 (2005)

31. Narwaria, M., Lin, W., McLoughlin, I.V., Emmanuel, S., Chia, L.: Fourier transform-based scalable image quality measure. IEEE Trans. Image Process. **21**(8), 3364–3377 (2012)

32. Gu, K., Zhai, G., Yang, X., Zhang, W.: A new reduced-reference image quality assessment using structural degradation model. In: 2013 IEEE International Symposium on Circuits and Systems (ISCAS2013), pp. 1095–1098, May 2013

33. Min, X., Zhai, G., Gu, K., Liu, Y., Yang, X.: Blind image quality estimation via distortion aggravation. IEEE Trans. Broadcast. **64**(2), 508–517 (2018)
34. Min, X., Gu, K., Zhai, G., Liu, J., Yang, X., Chen, C.W.: Blind quality assessment based on pseudo-reference image. IEEE Trans. Multimed. **20**(8), 2049–2062 (2018)
35. Xia, Z., Gu, K., Wang, S., Liu, H., Kwong, S.T.W.: Towards accurate quality estimation of screen content pictures with very sparse reference information. IEEE Trans. Ind. Electron. **67**(3), 2251–2261 (2019)

PMIQD 2019: A Pathological Microscopic Image Quality Database with Nonexpert and Expert Scores

Shuning Xu, Menghan Hu$^{(\boxtimes)}$, Wangyang Yu, Jianlin Feng, and Qingli Li

Shanghai Key Laboratory of Multidimensional Information Processing,
East China Normal University, Shanghai 200241, China
mhhu@ce.ecnu.edu.cn

Abstract. In medical diagnostic analysis, pathological microscopic image is often regarded as a gold standard, and hence the study of pathological microscopic image is of great necessity. High quality microscopic pathological images enable doctors to arrive at correct diagnosis. The pathological microscopic image is an important cornerstone for modernization and computerization of medical procedures. The quality of pathological microscopic images may be degraded due to a variety of reasons. It is difficult to acquire key information, so research for quality assessment of pathological microscopic image is quite necessary. In this paper, we perform a study on subjective quality assessment of pathological microscopic images and investigate whether the existing objective quality measures can be applied to the pathological microscopic images. Concretely, we establish a new pathological microscopic image quality database (PMIQD) which includes 425 pathological microscopic images with different quality degrees. The mean opinion scores rated by nonexperts and experts are calculated afterwards. Besides, we investigate the prediction performance of the existing popular image quality assessment (IQA) algorithms on PMIQD, including 8 no-reference (NR) methods. Experimental results demonstrate that the present objective models do not work well. IQA for pathological microscopic image needs to be developed for predicting the quality rated by nonexperts and experts.

Keywords: Pathological microscopic image · Subjective image quality assessment · No-reference model observer · Database

1 Introduction

Pathological microscopic images are vital tools for medical diagnosis [1]. High quality microscopic images can help doctors or pathologists make the right diagnosis, and be conducive to the further analysis.

In practical applications, the quality of pathological microscopic images will be inevitably degraded due to the following situations: (1) the unstable performance of charge-coupled device under some conditions; (2) the incorrect light

© Springer Nature Singapore Pte Ltd. 2020
G. Zhai et al. (Eds.): IFTC 2019, CCIS 1181, pp. 293–301, 2020.
https://doi.org/10.1007/978-981-15-3341-9_25

source; (3) the dust on the optical path; and (4) the bad focus. The low quality of pathological microscopic images may incur difficulties for doctors or pathologists to diagnose disease. Therefore, high-quality PMIs are essential. Evaluating the quality of pathological microscopic images can greatly reduce the appearance of low-quality pathological microscopic images, thus guaranteeing the diagnostic efficiency and accuracy of medical worker.

In recent years, with the extensive researches in the field of digital image [2], the researches on image quality assessment (IQA) have drawn increasingly attention [3–5]. A variety of image databases such as LIVE [6], CSIQ [7] and TID2013 [8] have been proposed, thus numerous IQA approaches have been developed based on these database.

Unlike traditional images, the content of pathological microscopic image is complex and its distortion is made up of many factors. This paper attempts to explore quality assessment problem of the pathological microscopic images through subjective researches. First, we construct the pathological microscopic image quality database (PMIQD), which consists of 425 pathological microscopic images covering pathological sections of different body parts. Next, according to the ITU-R REC. BT.500-13 [9], 15 subjects are recruited to grade the pathological microscopic images under stringent conditions. The main contributions of this paper are as follows: (1) establish a PMIQD; (2) compare the performance of existing no-reference IQA algorithm [10] on PMIQD; and (3) analyze the differences on MOSs between nonexperts and experts.

In the following of this paper, we introduce the detailed subjective assessment of pathological microscopy images in Sect. 2. Section 3 presents a comparison and assessment of some objective quality indicators of PMIQD and an analysis of algorithmic deficiencies. Finally, Sect. 4 provides some conclusive comments.

2   Subjective Quality Assessment

To study the quality assessment of pathological microscopic images, we establish a special image database that includes pathological microscopic images with different image quality grades. Subsequently, the subjective experiment is designed to obtain the MOS of each image.

2.1   Image Materials

For the sake of the representativeness of our data, we collected pathological microscopic samples from different body parts such as retinal cells, intestinal epithelial cells, and spleen cells. The acquired pathological microscopic images have the same resolution of 2304 × 1728. Our database has two characters: (1) PMIQD consists of 425 pathological microscopic images from different body tissues. Figure 1 shows some samples of pathological microscopic images. The diversity of samples is reflected in morphology of the cells [11], which is distinctly structured, unconspicuous structured, or unstructured. (2) Various types of distortion are included, such as bright spots [12], blur [13–16], and color cast [17]. Figure 2 shows main types of distortion contained in PMIQD.

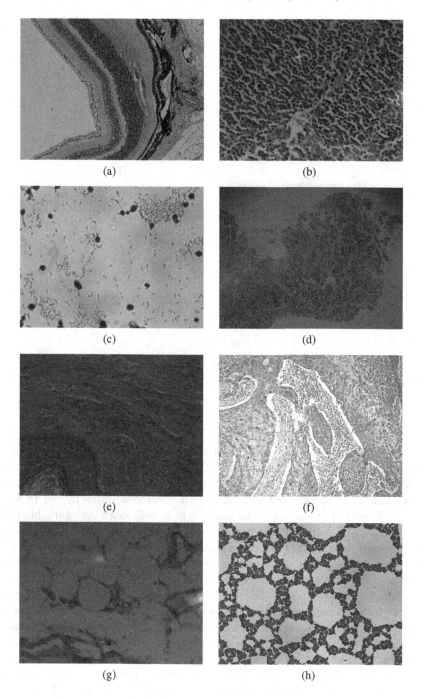

Fig. 1. Some samples of pathological microscopic images.

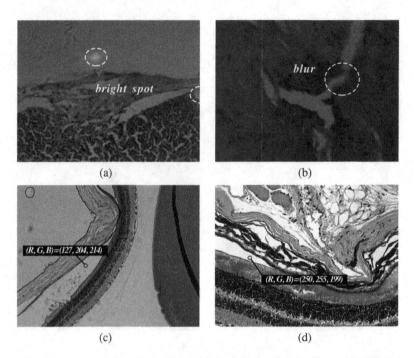

(a)   (b)

(c)   (d)

Fig. 2. Main distortion types for pathological microscopic images: (a) bright spot; (b) blur; (c) cool color cast; (d) warm color cast. (Color figure online)

2.2 Subjective Experiment

In order to collect the quality score of these pathological microscopic images, we conduct subjective experiments on PMIQD. According to the ITU-R REC. BT.500-13 [9], the quality standards are marked as "Bad", "Poor", "Fair", "Good" and "Excellent" corresponding to the numerical value from 1 to 5 [18], listed in Table 1. For instance, the image which is distinct and with high contrast is graded 5 as 'Excellent'. Since the absence of reference images, a single stimulus method is used in our experiment. The viewing distance is set to 2–2.5 times the display height. Illumination of the experimental environment is kept low for a comfortable and clear view [19,20].

Prior to the experiment, the objectives and procedures of this experiment are introduced to each audience. Observers preview 10 pathological microscopic images with 5 different quality grades. For further and more precise assessment, the subject is asked to make a decision based on the overall quality of image. During the experiment, subjects give an appropriate score from 1 to 5 for each pathological microscopic image based on their perception. The order of the images are displayed randomly. According to the guidelines of ITU-R REC BT.500-13 [9], 12 inexperienced subjects and 3 experts are recruited to assess the quality of pathological microscopic images. After the subjective experiment, we obtain the scores given by all observers.

Table 1. THe subjective evaluation criteria employing five-grade scale of quality for pathological microscopic image.

Score	Quality	Standard
5	Excellent	Distinct, high contrast
4	Good	Clear, higher or lower contrast
3	Fair	Fairly clear, a few spots
2	Poor	Fuzzy, some spots
1	Bad	Blurred, bright spots

2.3 Data Analysis

We average 12 nonexperts [10], 3 experts and all 15 subjects respectively, and acquire a mean opinion score (MOS) vector [21] for pathological microscopic images [22]. The MOS of each image is expressed as:

$$MOS_j = \sum_{i=1}^{N}(\frac{P_{ij}}{N_i}) \qquad (1)$$

where N_i is the number of subjects and P_{ij} is the score of image j assigned by i-th subject. Owing to the differences of the subjects, 3 MOSs distributions have some similarities, but slightly different.

3 Comparison of Objective Quality Assessment Models and Discussion

3.1 Comparison of MOSs Given by Nonexperts and Experts

Two histograms of MOS in PMIQD are shown in Fig. 3.

As is depicted in Fig. 3 and Table 2, the MOS of inexperienced subjects are mostly concentrated in the low score region; the MOS of the experts approximately obey normal distribution [23]; and the MOS of all subjects are stable except for the low scores region.

3.2 Performance Comparison of Objective Metrics

We can automatically evaluate the quality of pathological microscopic images through the objective IQA. Although many objective IQA models perform well in assessing the quality of traditional distorted images [24], the performance of these objective IQA models on estimating quality of pathological microscopic images is unknown. Therefore, we will examine the performance of some objective IQA methods for evaluating pathological microscopic images based on PMIQD in this section. We test the following 8 no-reference objective IQA methods on PMIQD.

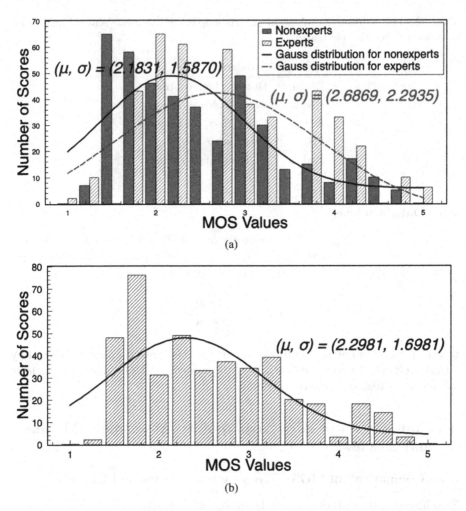

Fig. 3. MOS histogram in PMIQD: (a) MOS of nonexperts and experts; (b) MOS of all subjects.

The selected algorithms are: BLIINDS-II [25], BRISQUE [26], CPBD [14], FIS-BLIM [27], FISH [15], NIQE [28], QAC [29], and SISBLIM [30]. Before performance evaluation, we apply a five-parameter β_1, β_2, β_3, β_4, β_5 monotonic logic function to map the scores predicted by the objective IQA model:

$$Y(x) = \beta_1\left(\frac{1}{2} - \frac{1}{1 + e^{\beta_2(x-\beta_3)}}\right) + \beta_4 x + \beta_5 \qquad (2)$$

Where x and Y are objective scores and map scores [31,32].

Then, we use 3 common indices to evaluate the performance of these objective IQA indicators, namely Pearson linear correlation coefficient (PLCC), Spearman rank correlation coefficient (SROCC) and root mean square error (RMSE) [33].

An excellent IQA method is expected to get values close to 1 in PLCC and SROCC, and values close to 0 in RMSE [34]. Table 2 lists the performance results for these 8 algorithms.

Table 2. Performance of eight IQA models in terms of PLCC, SROCC and RMSE on all subsets of PMIQD.

Model	NONEXPERT			EXPERT			ALL		
	PLCC	SROCC	RMSE	PLCC	SROCC	RMSE	PLCC	SROCC	RMSE
BLIINDS-II [25]	0.4439	0.2150	0.7803	0.4077	0.2267	0.8229	0.4398	0.2180	0.7658
BRISQUE [26]	0.5412	0.5701	0.7323	0.6088	0.6094	0.7149	0.5706	0.5948	0.7003
CPBD [14]	0.0586	0.0049	0.8694	0.2787	0.0108	0.8654	0.0827	0.0070	0.8499
FISBLIM [27]	0.3939	0.0184	0.8005	0.3226	0.0976	0.8530	0.3886	0.0105	0.7857
FISH [15]	**0.6279**	**0.5870**	**0.6778**	**0.7379**	**0.7177**	**0.6082**	**0.6656**	**0.6325**	**0.6364**
NIQE [28]	0.4460	0.3002	0.7795	0.4518	0.4063	0.8039	0.4533	0.3323	0.7601
QAC [29]	0.6387	0.3201	0.6701	0.5665	0.4074	0.7426	0.6390	0.3496	0.6559
SISBLIM [30]	0.3858	0.1197	0.8034	0.2645	0.0238	0.8691	0.3701	0.0982	0.7922

From Table 2, we find that all of IQA methods are not of good performance for evaluating quality of pathological microscopic images. Majority of their PLCC and SROCC values are lower than 0.5, which represents bad performance. The relatively good two metrics are BRISQUE and FISH, whose PLCC values are between 0.5 and 0.8, respectively. Considering RMSEs of BRISQUE and FISH respectively, RMSE of FISH is relatively low, thus FISH performs better than BRISQUE. However, since the RMSEs are higher than 0.6 without exception, the present IQA methods are not very suitable for pathological microscopic images. The possible reasons may be that the distortions of pathological microscopic images are more complicated and multiple [35], comparing to traditional distorted images. For MOSs of 3 groups *viz.* nonexperts, experts and all subjects, almost all algorithms have failed. In terms of professional image, more work need to be done of IQA algorithm design for nonexperts and experts [10].

4 Conclusion

In this paper, we focus on a new quality assessment problem of pathological microscopic images. Firstly, we conduct a new image database named PMIQD, including 425 pathological microscopic images with different degrees of quality, covering pathological sections of different body parts, to investigate the subjective quality of pathological microscopic images. Moreover, we compare 8 mainstream objective no-reference IQA models in PMIQD. In addition, we discuss the differences between MOS of nonexperts and experts of pathological microscopic images.

Acknowledgment. This work is sponsored by the Shanghai Sailing Program (No. 19YF1414100), the National Natural Science Foundation of China (No. 61831015, No. 61901172), the STCSM (No. 18DZ2270700), and the China Postdoctoral Science Foundation funded project (No. 2016M600315).

References

1. Snead, D.R.J., et al.: Validation of digital pathology imaging for primary histopathological diagnosis. Histopathology **68**(7), 1063–1072 (2016)
2. Sheikh, H.R., Bovik, A.C., De Veciana, G.: An information fidelity criterion for image quality assessment using natural scene statistics. IEEE Trans. Image Process. **14**(12), 2117–2128 (2005)
3. Wang, Z., Bovik, A.C., Sheikh, H.R., Simoncelli, E.P., et al.: Image quality assessment: from error visibility to structural similarity. IEEE Trans. Image Process. **13**(4), 600–612 (2004)
4. Zhang, L., Shen, Y., Li, H.: VSI: a visual saliency-induced index for perceptual image quality assessment. IEEE Trans. Image Process. **23**(10), 4270–4281 (2014)
5. Zhang, L., Zhang, L., Mou, X., Zhang, D.: FSIM: a feature similarity index for image quality assessment. IEEE Trans. Image Process. **20**(8), 2378–2386 (2011)
6. Sheikh, H.R.: LIVE image quality assessment database release 2 (2005). http://live.ece.utexas.edu/research/quality
7. Larson, E.C., Chandler, D.M.: Categorical image quality (CSIQ) database (2010)
8. Ponomarenko, N., et al.: Image database TID2013: peculiarities, results and perspectives. Sig. Process. Image Commun. **30**, 57–77 (2015)
9. Recommendation ITU-R BT.500-13: Methodology for the subjective assessment of the quality of television pictures. Technical report, International Telecommunication Union (2012)
10. Zhu, W., Zhai, G., Menghan, H., Liu, J., Yang, X.: Arrow's impossibility theorem inspired subjective image quality assessment approach. Sig. Process. **145**, 193–201 (2018)
11. Karbowski, M., Youle, R.J.: Dynamics of mitochondrial morphology in healthy cells and during apoptosis. Cell Death Differ. **10**(8), 870 (2003)
12. Shrestha, P., Kneepkens, R., Vrijnsen, J., Vossen, D., Abels, E., Hulsken, B.: A quantitative approach to evaluate image quality of whole slide imaging scanners. J. Pathol. Inform. **7**, 56 (2016)
13. Li, L., Lin, W., Wang, X., Yang, G., Bahrami, K., Kot, A.C.: No-reference image blur assessment based on discrete orthogonal moments. IEEE Trans. Cybern. **46**(1), 39–50 (2015)
14. Narvekar, N.D., Karam, L.J.: A no-reference perceptual image sharpness metric based on a cumulative probability of blur detection. In: 2009 International Workshop on Quality of Multimedia Experience, pp. 87–91. IEEE (2009)
15. Vu, P.V., Chandler, D.M.: A fast wavelet-based algorithm for global and local image sharpness estimation. IEEE Sig. Process. Lett. **19**(7), 423–426 (2012)
16. Vu, C.T., Phan, T.D., Chandler, D.M.: A spectral and spatial measure of local perceived sharpness in natural images. IEEE Trans. Image Process. **21**(3), 934–945 (2011)
17. Winkelman, K.-H.: Method and apparatus for the automatic analysis of density range, color cast, and gradation of image originals on the basis of image values transformed from a first color space into a second color space, 16 September 1997. US Patent 5,668,890 (1997)

18. P ITU-T Recommendation: Subjective video quality assessment methods for multimedia applications. International Telecommunication Union (1999)

19. Sheikh, H.R., Sabir, M.F., Bovik, A.C.: A statistical evaluation of recent full reference image quality assessment algorithms. IEEE Trans. Image Process. **15**(11), 3440–3451 (2006)

20. Mantiuk, R.K., Tomaszewska, A., Mantiuk, R.: Comparison of four subjective methods for image quality assessment. In: Computer Graphics Forum, vol. 31, pp. 2478–2491. Wiley Online Library (2012)

21. Kumar, B., Singh, S.P., Mohan, A., Anand, A.: Performance of quality metrics for compressed medical images through mean opinion score prediction. J. Med. Imaging Health Inform. **2**(2), 188–194 (2012)

22. Streijl, R.C., Winkler, S., Hands, D.S.: Mean opinion score (MOS) revisited: methods and applications, limitations and alternatives. Multimed. Syst. **22**(2), 213–227 (2016)

23. Pelgrom, M.J.M., Duinmaijer, A.C.J., Welbers, A.P.G.: Matching properties of MOS transistors. IEEE J. Solid-State Circ. **24**(5), 1433–1439 (1989)

24. Sheikh, H.R., Bovik, A.C.: Image information and visual quality. In: 2004 IEEE International Conference on Acoustics, Speech, and Signal Processing, vol. 3, pp. iii–709. IEEE (2004)

25. Saad, M.A., Bovik, A.C., Charrier, C.: Blind image quality assessment: a natural scene statistics approach in the DCT domain. IEEE Trans. Image Process. **21**(8), 3339–3352 (2012)

26. Mittal, A., Moorthy, A.K., Bovik, A.C.: No-reference image quality assessment in the spatial domain. IEEE Trans. Image Process. **21**(12), 4695–4708 (2012)

27. Gu, K., et al.: FISBLIM: a five-step blind metric for quality assessment of multiply distorted images. In: SiPS 2013 Proceedings, pp. 241–246. IEEE (2013)

28. Mittal, A., Soundararajan, R., Bovik, A.C.: Making a "completely blind" image quality analyzer. IEEE Sig. Process. Lett. **20**(3), 209–212 (2012)

29. Xue, W., Zhang, L., Mou, X.: Learning without human scores for blind image quality assessment. In: Proceedings of the IEEE Conference on Computer Vision and Pattern Recognition, pp. 995–1002 (2013)

30. Ke, G., Zhai, G., Yang, X., Zhang, W.: Hybrid no-reference quality metric for singly and multiply distorted images. IEEE Trans. Broadcast. **60**(3), 555–567 (2014)

31. Moorthy, A.K., Bovik, A.C.: Visual importance pooling for image quality assessment. IEEE J. Sel. Top. Sig. Process. **3**(2), 193–201 (2009)

32. Chen, M.-J., Su, C.-C., Kwon, D.-K., Cormack, L.K., Bovik, A.C.: Full-reference quality assessment of stereopairs accounting for rivalry. Sig. Process. Image Commun. **28**(9), 1143–1155 (2013)

33. Gu, K., Zhai, G., Yang, X., Zhang, W., Liu, M.: Subjective and objective quality assessment for images with contrast change. In: 2013 IEEE International Conference on Image Processing, pp. 383–387. IEEE (2013)

34. Dhanachandra, N., Manglem, K., Chanu, Y.J.: Image segmentation using K-means clustering algorithm and subtractive clustering algorithm. Procedia Comput. Sci. **54**, 764–771 (2015)

35. Ameisen, D., et al.: Automatic image quality assessment in digital pathology: from idea to implementation. In: IWBBIO, pp. 148–157 (2014)

Telecommunication

Communication

A Generalized Cellular Automata Approach to Modelling Contagion and Monitoring for Emergent Events in Sensor Networks

Ru Huang[(✉)], Hongyuan Yang, Haochen Yang, and Lei Ma

School of Information Science and Engineering,
East China University of Science and Technology, Shanghai 200237, China
huangrabbit@ecust.edu.cn

Abstract. In order to improve the invulnerability and adaptability in sensor networks, we propose a cellular automata (CA) based propagation control mechanism (CACM) to inhibit and monitor emergent-event contagion. The cellular evolving rules of CACM are figured in multi-dimension convolution operations and cell state transform, which can be utilized to model the complex behavior of sensor nodes by separating the intrinsic and extrinsic states for each network cell. Furthermore, inspired by burning pain for Wireworld based monitoring model, network entropy theory is introduced into layered states on CACM to construct particle-based information communication process by efficient distribution of event-related messages on network routers, thus an invulnerable and energy-efficient diffusion and monitoring being achieved. Experiment results prove that CACM can outperform traditional propagation models in adaptive invulnerability and self-recovery scalability on sensor networks for propagation control on malicious events.

Keywords: Sensor network · Cellular automata · Invulnerability · Contagion

Sensor networks are modern communication networks consisting of nodes that undertake data acquisition functions. The nodes of the network are deployed in the specified monitoring area. Through cooperative perception and self-organizing interconnection, they can complete the long-term reliable monitoring task of the area under limited manual intervention and limited resource allocation. Network is responsible for transmitting daily data and reporting events with centralization [1]. Event information is enormous and shared with other data links [2]. If events are propagated based on the original forwarding protocol, it will cause a large energy consumption, and conventional methods can not

This work was supported by the Science and Technology Commission of Shanghai Municipality (17511108604), the National Natural Science Foundation of China (61501187 and 61673178).

© Springer Nature Singapore Pte Ltd. 2020
G. Zhai et al. (Eds.): IFTC 2019, CCIS 1181, pp. 305–323, 2020.
https://doi.org/10.1007/978-981-15-3341-9_26

guarantee the safe transmission of reports. The self-repairing ability of sensor networks in incident scene is also the main content of their robustness. Therefore, modeling and analyzing the transmission process of events in the network is an important part of the research on the survivability of sensor networks.

The propagation in the network belongs to the contagion phenomenon. Friedberg et al. [3] built a crowd's emotional infection model based on 2-dimensional cellular automata (CA). The model reflects the factors affecting the infection rate. Shaw et al. [4] discussed that the network structure with small-world effect can promote the transmission process.

Early models focused on infection control based on topology, but lacked consideration of human intervention factors. In some literatures, network propagation control is realized by immunization technology. Tang et al. [5] use the scheduling mechanism of the sensor network to make the nodes enter the sleep mode in turn, and start the adjustment program before the sensor nodes enter the sleep mode to realize the control of propagation. However, such an approach requires a priori knowledge of the control objective function. Xu [6] proposed a discrete SIS model based on cellular automata, which adapts to network virus propagation, and uses network evolution to respond to immune process to realize the control of propagation.

Peng et al. [7] established a smartphone worm propagation model based on CA, and made a beneficial attempt to control the infection of the model by setting clear meaning parameters. However, the evaluation method of the control effect is still based on the traditional complex network propagation measure. CA is also introduced to solve the sensor network optimization problem [8]. Based on CA, Baryshnikov et al. [9] and Athanassopoulos et al. [10] implemented the scheduling and topology control mechanisms of sensor networks respectively. Mansilla and Gutierrez [11] give the constants representing the average growth rate of population size and achieve infection control, but their local area rules lack the description of signaling and nodes bidirectional communication of sensor networks. Starting from the perspective of information dissemination, this paper improves the description of evolution, establishes a CA model based on information transmission, and discusses the mechanism of event dissemination monitoring and control in sensor networks.

1   System Model

1.1   Information and Energy in Sensor Networks

Sensor network node is the main body of sensor network, storage resources are limited [12]. After the node generates or receives data, it is fused and forwarded [1]. When it is redundant, it does not save backup to economize storage. If the symbol U is used to represent the information carried by the node, and there are c independent components, i.e. the rank of the state space of U is c. According to Shannon's expression of the information quantity, the self-information quantity [13] of symbol U is expressed as Eq. (1), and the p_i in equation represents the probability of the source producing the class i symbol.

$$H(U) = -\sum_{i=1}^{c} p_i \log_2 p_i \tag{1}$$

Then, the information quantity H of the node satisfies.

$$H = H^1 + H^2 + \cdots + H^n = \sum_{i=1}^{c} H^i \tag{2}$$

Where H^i represents the i-dimensional component of $\boldsymbol{H}$, then $\boldsymbol{U}$ can be replaced by a symbol $\boldsymbol{V}$ with arbitrary information quantity $\boldsymbol{H}$.

The representation of each dimension information carried by nodes is given based on the finite-state machine. Referring to the infection model, the state transition of FSM is described by the SIS process or SIR process of Kermack-Mckendrick atrioventricular model [6]. The difference is that the contact between all individuals is insufficient. Each finite state machine on a node is called a particle, and the corresponding state transition is shown in Fig. 1. In the network, the nodes with 0 information bits and the nodes with 1 information bits adjacent exchange with each other in probability β, and the nodes with information bit 1 automatically change to 0 in probability γ.

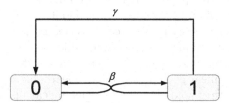

Fig. 1. State transition of particles in the information spreading (SIS) model

For a system described by a set of finite state machines, the relevance between some components and other components is sometimes observed. At this time, it can be considered that some components are in the upper layer of some components, and the information of the system is no longer the simple sum of components. When the adjacent lower layer is binary, there is a correlation similar to binary logic operation [14].

The energy constraint of sensor nodes in unattended environments is particularly critical in defining optimization problems and solving them. Under such restrictions, nodes should ensure that they can work steadily for a long time and provide reliable relay protection for their neighbors. In addition to the collection of environmental data, the energy consumption of nodes is mainly generated when the data is received, sent and processed within the nodes.

Based on such a mode of sensor network node operation, several nodes are waked up synchronously in each round to perform information sensing tasks and complete the in-network flow of data processing. This network slice can finish the round before the energy-exhausted nodes appear, and then collect energy to supplement the nodes [15]. This model is no longer concerned about the residual energy of each node and the balance of energy consumption between nodes, but pays attention to the energy allocation of each operation, so that the energy consumption generated by sending, receiving and processing messages can be confirmed separately.

1.2   Automata Model

In order to investigate the occurrence and propagation of events in sensor networks, a Vertical and horizontal grid cellular automata on a 2-dimensional Euclidean plane is introduced to describe the network and nodes deployed uniformly and intensively in the monitoring plane.

Cellular automata is composed of a cellular space and transformation functions defined in that space [16]. Neighbor type of cellular automata is an important attribute of cellular automata. The neighborhood of cell r is defined by the neighborhood template as Eq. 3, where b is the neighborhood length.

$$N_b \triangleq \{r + c_i | c_i \in N_b, i = 1, 2, \cdots, b\} \tag{3}$$

The CA model in this paper is realized by finite state machine and perceptron. Each cell is a set of finite state machines, and its neighbor discovery and rule definition are implemented by perceptron. Perceptron (perceptron) is a machine learning algorithm for binary classification, which can determine whether the input represented by a digital vector belongs to a specific class [17].

In particular, the linear perceptron defines a learnable separating hyperplane, which compares the threshold value and output by calculating the linear objective function. This paper further discusses the state and transition of FSM. Traditionally, the transition of FSM is expressed by the initial state and termination state when the transition occurs and the necessary conditions to ensure its occurrence. In this paper, the FSM model is modified. The transition occurs randomly and is described by the transition probability matrix, forming the so-called stochastic finite-state automata, SFSA [18].

Figure 2 shows the linear perceptron and the FSM side by side.

This paper defines the layered state. The information carried by the shallow state enters the deep state with the passage of time, and the transmission only occurs at the relatively shallow level, which ensures the safe transmission of messages. In this paper, cellular automata (CA) is used to model the infectious process of networks, and proposed the cellular state of separation of attributes and identities. The cell m is assigned state variables s_m^i and p_m, where $m \in$, $s_m^i \in \{0,1\}$, $p_m \in$, $i \in$, $\subseteq_+$, m is the serial number of nodes, the set of all network nodes is denoted as $\mathbb{M}$, $\mathbb{P}$ is the identity set of nodes, and the monotone mapping from $\{s_m^i | i \in \mathbb{I}, \forall m \in \mathbb{M}\}$ to p_m is defined. $\mathbb{I}$ is the information space of events, and the different dimension attributes of the same layer are independent of each other. Specifically, it is assumed that the node is a cell defined on a $N \times N$

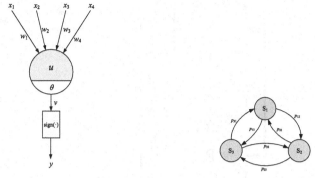

(a) A linear perceptron with 4 inputs   (b) A SFSA with 3 states

Fig. 2. Schematics for a simple perceptron and a simple FSM

rectangle lattice, and s_m^i represents the perception of events on the i-dimension of the node m.

The improved cellular automata can simplify the problem conveniently by marking individuals with discrete states, but at the same time, it conceals the actual attributes of individuals. On the time axis, the past and current attributes of individuals, the type and occurrence time of the next transition state of individuals are transparent to the network. In this paper, the integral of the first attribute value with respect to time is used to assign the second attribute value, and then the identity is output by linear mapping.

The cell of each label S in each round of SIS model contacts the cell of label I, and the state transition occurs with probability $\lambda = \beta e(i)$ and is labeled as I. $e(i)$ is the expectation of the number of cells of the adjacent label I. At this time, each cell of label I is labeled S with probability λ transition state. In SIRS model, R state cells are changed through I state, and the corresponding cells no longer respond to messages carrying the same information; R individuals are restored to the former state with probability λ_1.

Based on hierarchical automata, Brian Silverman's Wireworld model [19] is introduced to characterize SIR and SIRS propagation. Wireworld defines four cellular states: empty, electronic head, electronic tail and conductor.

Figure 3 is a graphical description of Wireworld evolution rules. Identity cond, head and tail represent cell types as conductor, electronic head and tail respectively. Let cells have two layers of state variables. Layer 1 states s_1 contains two independent components s_1^1 and s_1^2, which describe the "conductivity" and "initial charged state" of cells respectively. Layer 2 states s_1 is used to describe the evolution process of conductors, electronic heads and electronic tails. The four states of cell space, electron head, electron tail and conductor represented by tuple $\{s_1^1, s_1^2, s_2\}$ are $\{0, \vartheta, \vartheta\}$, $\{1, 1, \vartheta\}$, $\{1, 0, 1\}$ and $\{1, 0, 0\}$, in which θ represents any value in set $\{0, 1\}$, and the model represents standard SIR when $\theta = 0$ and $Con.3$ and $Con.4$ are not established at all.

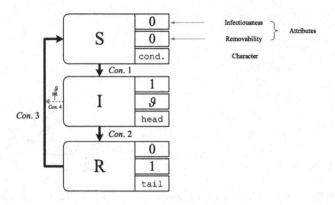

Fig. 3. State transition of a Wireworld-style contagion model

1.3   Template Operation of Cellular Automata

Cellular automata is discrete and finite in time domain, space domain and state. Spatial interaction and temporal causality are both local. Cells have homogeneity, which can be described by using finite size template operations. Based on the state of segmentation of cellular automata by means of template operation, this paper implements a simple modeling of short-term memory of nodes, so that each cell follows the same evolutionary rules and can still present different behaviors, thus describing the heterogeneity of the network. The cell calculates the value of its next state based on its current state and neighbor's current state., so evolution rules can be expressed by Hadamard product with the model. This kind of operation defines several finite size matrices called templates, and considers cell space and transition rules as matrices composed of cell states and their 2-D convolution results.

M is the rule template of the defined network propagation model, as shown in Formula 4. The vector $\lambda = [\lambda_1, \lambda_2, \lambda_3, \ldots, \lambda_8]^T$ varies with the position of convolution, and the elements indicate whether the cell chooses to transmit the information to the corresponding neighbor or not. According to the nature of two-dimensional convolution, information transfer operation can be represented by matrix convolution.

The weight vector indicates that the way of the cell choosing the next hop of information subjects to Multinomial Distribution. Hence, M is a random matrix.

$$\mathbf{M} = \begin{bmatrix} -\lambda_1 & -\lambda_2 & -\lambda_3 \\ -\lambda_8 & 1 & -\lambda_4 \\ -\lambda_7 & -\lambda_6 & -\lambda_5 \end{bmatrix} \tag{4}$$

In view of the same distribution of elements in M, let $\lambda \sim \Lambda$, thus the template D is defined as formula 5.

$$\mathbf{D} = \begin{bmatrix} -\varLambda & -\varLambda & -\varLambda \\ -\varLambda & 1 & -\varLambda \\ -\varLambda & -\varLambda & -\varLambda \end{bmatrix} \tag{5}$$

D is a matrix whose elements obey random distribution, and 1 represents the normalized reference value.

To implement the rules of message movement, when a cell has a message to send, the L-1 norm of the weight vector of its location equals to 1; that is $\|\lambda\|_1 = 1$, when M becomes a random matrix with the sum of all elements being 0, considering $\sum_{j=1}^{8} \lambda_j = 1$. REQ-REP mechanism notifies the sender of the message about the message transmission. The original sender can release the message in time, reduce the occupation of resources, and help the network work reliably. In sensor networks, with the help of REQ-REP mechanism, the transmission under Probabilistic transfer conditions can be realized.

As shown in Fig. 4, let's assume that cells A and B carry particles on $\prod i$ at time t. C and D are neighbors of A, and C is neighbor of B at the same time. Contact propagation may occur synchronously in $A \rightarrow D$ and $B \rightarrow C$. At time $t+1$, the perception of $\prod i$ by cells C and D will be activated. With REQ-REP mechanism, cell A releases the storage of corresponding particles due to the change of the state of neighbors C and D from time t to $t+1$. Therefore, if the propagation of the particle $B \rightarrow C$ is completed normally, even if the communication of $A \rightarrow D$ fails, A will respond to the state transition of C. In fact, $B \rightarrow C$ is equivalent to $A \rightarrow C$; $A \rightarrow D$, $B \rightarrow C$ is equivalent to $A \rightarrow C$, D.

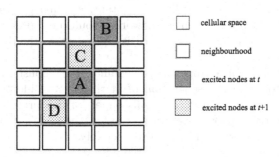

Fig. 4. A delivery rule in CA

1.4   Gridding of Sensor Networks

As mentioned above, the node neighbors of wireless sensor networks are not as fixed as those of wired sensor networks. Under the time-energy model, node degeneration can be realized, cell normalization and neighbor template homogenization can be achieved (Fig. 5 regards the six nodes in the center as the same cell according to the adjacency relationship between node groups and surrounding nodes.

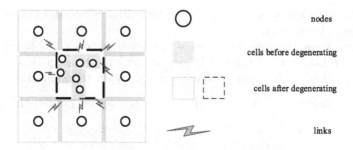

Fig. 5. Cell degenerating based on time-energy models

2 Propagation and Control of Events

The effective energy transmission and control mechanism of events not only hopes that the nodes will respond quickly to events, but also hopes that the nodes will not respond repeatedly to events under the designed mechanism, which has strong dynamic adaptability. The traditional approach is to set a forwarding limit for each message, but this is not appropriate for sensor networks where the network diameter is not explicitly constant. By analogy with combustion, the propagation of events in sensor networks should be represented by a rapid one-time traversal of most nodes, which is equivalent to the duration of event sources.

The process of particle propagation is similar to Combustion [20]. Based on the cellular automata with separation of attributes and identities, a burning pain bionic algorithm is proposed in this paper for SIS propagation in sensor networks. By setting up two levels of attributes, the node's knowledge of data and the node's behavior are described respectively.

2.1 Dissemination of Controllable Event with Bionic Burning Pain

Algorithm 1 depicts the Bio-inspired Burning Pain Event Communication Model; the cell space is denoted as O, the set of cell-$\mathbb{H}$, infection rate-β, recovery rate-γ, information dimension of the first layer-c, output identity threshold of the second layer-Φ, neighborhood template-M.

2.2 Tracking and Monitoring of Events

In Fig. 1, the recovery rate γ of particles carrying information 1 is changed to 0 in each round. If γ is a constant, the lifetime of particles obeys Poisson distribution.

Algorithm 1. Bio-inspired Burning Pain Event Communication Model
Input information: infection rate β, recovery rate γ, information dimension of the first layer c, output identity threshold of the second layer Φ, neighborhood template M.

Input: $\beta \ \gamma \ c \ \Phi \ M$
1: Initialize O and event sources;
2: Initialize information particle tags $\mathbb{P}$ in each cell;
3: $t \leftarrow 0$
4: **while** events exist **do**
5:      $t \leftarrow t + 1$
6:      **for** each cell in$\{i \in \mathbb{H}(t) \,|\, \mathbb{P}_i = 1\}$ **do**
7:          With probability β select a neighbor with tag 0, exchange these two tags,
8:          **otherwise** with probability γ transform into 0;
9:      **end for**
10:     **for** each cell in $\mathbb{H}(t)$ **do**
11:         All particles of this cell boost the change of tag;
12:     **end for**
13:     Update cellular automaton;
14: **end while**

In application scenarios, most sensor networks adopt carrier sense multiple access (CSMA) based media access control protocol, which enables network nodes to know the channel occupancy near their location.

Similarly, CA can count the number of designated state neighbors of the template composed by a finite definition field and a square matrix whose elements are all 1. When the particle's walk is a stationary random process, if the two-dimensional convolution of a specified all-1 square matrix is recorded as M, two pooling operations with overlapping domain centers but large scale differences are defined. M^1 and M^2 are sampled in a relatively small size and a relatively large size area respectively, and M^2 can track the information particles emitted by M^1 at time t. $M^3, M^4, \cdots, M^n$ respectively denotes an arbitrary scale template operation between M^1 and M^2, and the order of the template increases in turn.

In this paper, a monitoring mechanism is designed to deploy the monitoring network formed by monitoring nodes. When the monitoring points detect the increase of information in the initial monitoring range, the monitoring radius will be automatically adjusted to maintain the information in the monitoring range for a period of time without reducing. Because of the transfer nature of particles carrying event-related information, and the radius of the two methods changing differently, hence we can identify the occurrence of isolated events. As shown in Fig. 6, the monitoring range of monitoring points in cell space is described in the form of a template.

When events continue to occur, a template with original radius called M^1 and a template with larger radius which is adjustable called M^2 are set in advance. By continuously adjusting the template radius statistics corresponding to the

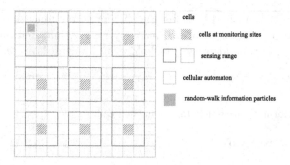

Fig. 6. Event monitoring in CA by placing monitoring sites

monitoring range, the ratio of the statistics obtained with fixed radius is maintained, and the possibility of events is evaluated by step size. Denoting the time when the event happens as t, the order of template M2 will increase over time as shown in Fig. 7. For the case of multi-information sources, although particles may be generated by other regions and then enter the monitoring range, it is necessary to design reasonable monitoring points to complete event recognition.

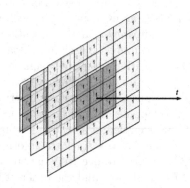

Fig. 7. Scale of mark tracking information particles changes over time

2.3   Event Propagation Control Mechanism

There are two kinds of intervention mechanisms for event propagation based on burning pain algorithm. One is to disseminate instructions from the control center as special events in the network, and to achieve manual intervention in event transmission through antagonism with standard events [21]. Multi-event communication in the field of shared knowledge will produce antagonism. The transmission of signaling in the network follows a distributed or centralized mechanism.

Centralized signaling always keeps communication with base station, and the propagation trend is controllable. Distributed signaling propagates in the network in accordance with the presupposed mode, which is consistent with the law of event propagation. In Fig. 8, the control signaling is propagated antagonistically with events, and the cells are not sensitive to events with known attributes, which reduces the energy consumption caused by transmission events and improves the energy efficiency of the network.

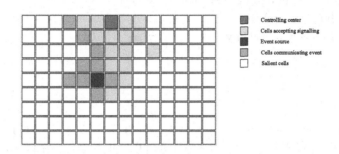

Fig. 8. Antagonistic propagation of control signaling with event in CA

3 Experiments and Analysis

In order to grasp the rule of particle random walk, the validity of information propagation model based on particle swap cellular automata is studied, and the reliability of the algorithm is analyzed. The simulation results show that the algorithm has good propagation simulation ability. The applicability of the algorithm is tested by constrained grid size.

3.1 Single Event Communication Experiment

Three transfer probabilities $\lambda, \gamma_1, \gamma_2$, of the SIRS propagation model are experimented in finite-size cellular space, which improved by Wireworld. The number of I-state cellular is counted. The simulation parameters of the experiment are given in Table 1, in which $\delta(\cdot)$ is a Dirac function. The following is the same.

The matrix corresponding to the cell space is denoted as G shown in formula 6, to determine whether each cell transmits information to its neighbor expression, and symbol * represents the convolution operation. Besides, construction method of matrix M is given by the formula 7.

Table 1. Simulation parameters for contagion (SIRS) model.

Parameters		Settings
Cellular space size	$m \times n$	100×100
Information	H	2 (in a Wireworld)
Neighbor	M	Moore neighbor
Event intensity	η_{event}	$\delta(0)$
Number of repeated experiments	num_rpt	30
Infection probability	λ	0.1
Recovery probability	γ_1, γ_2	Fig. 9
Path loss	Φ	0

$$
\lambda_1 \begin{bmatrix} -1 & 0 & 0 \\ 0 & 1 & 0 \\ 0 & 0 & 0 \end{bmatrix} * \mathbf{G} + \lambda_2 \begin{bmatrix} 0 & -1 & 0 \\ 0 & 1 & 0 \\ 0 & 0 & 0 \end{bmatrix} * \mathbf{G} + \lambda_3 \begin{bmatrix} 0 & 0 & -1 \\ 0 & 1 & 0 \\ 0 & 0 & 0 \end{bmatrix} * \mathbf{G} + \lambda_4 \begin{bmatrix} 0 & 0 & 0 \\ 0 & 1 & -1 \\ 0 & 0 & 0 \end{bmatrix} * \mathbf{G} +
$$

$$
\lambda_5 \begin{bmatrix} 0 & 0 & 0 \\ 0 & 1 & 0 \\ 0 & 0 & -1 \end{bmatrix} * \mathbf{G} + \lambda_6 \begin{bmatrix} 0 & 0 & 0 \\ 0 & 1 & 0 \\ 0 & -1 & 0 \end{bmatrix} * \mathbf{G} + \lambda_7 \begin{bmatrix} 0 & 0 & 0 \\ 0 & 1 & 0 \\ -1 & 0 & 0 \end{bmatrix} * \mathbf{G} + \lambda_8 \begin{bmatrix} 0 & 0 & 0 \\ -1 & 1 & 0 \\ 0 & 0 & 0 \end{bmatrix} * \mathbf{G},
$$

$$(6)$$

$$
\mathbf{M} = \lambda_1 \begin{bmatrix} -1 & 0 & 0 \\ 0 & 1 & 0 \\ 0 & 0 & 0 \end{bmatrix} + \lambda_2 \begin{bmatrix} 0 & -1 & 0 \\ 0 & 1 & 0 \\ 0 & 0 & 0 \end{bmatrix} + \lambda_3 \begin{bmatrix} 0 & 0 & -1 \\ 0 & 1 & 0 \\ 0 & 0 & 0 \end{bmatrix} + \lambda_4 \begin{bmatrix} 0 & 0 & 0 \\ 0 & 1 & -1 \\ 0 & 0 & 0 \end{bmatrix} +
$$

$$
\lambda_6 \begin{bmatrix} 0 & 0 & 0 \\ 0 & 1 & 0 \\ 0 & -1 & 0 \end{bmatrix} + \lambda_7 \begin{bmatrix} 0 & 0 & 0 \\ 0 & 1 & 0 \\ -1 & 0 & 0 \end{bmatrix} + \lambda_8 \begin{bmatrix} 0 & 0 & 0 \\ -1 & 1 & 0 \\ 0 & 0 & 0 \end{bmatrix}.
$$

$$(7)$$

Figure 9 is the superposition result of thirty experimental curves under 4th experimental conditions. It can be seen that the setting of the recovery rate determines the maximum range of infection at the same time in the initial stage of infection. The relapse of infection is affected in the middle stage of infection, and the continuity of infection is determined in the stable stage of infection.

For the results of $\gamma_1 = 0.001$ and $\gamma_2 = 0.02$, an energy-like visualization is shown in Fig. 10. The possibility of network outbreak infection can be intuitively grasped. In the figure, the quantitative width of the number axis is 200, and the brighter the area is, the more frequent the outbreak occurs.

When events occur in a certain time window with persistence, the event propagation model of the Burning Pain Algorithm balances the tendency of network energy over-speed by introducing path loss. Table 2 is the simulation parameters of the event propagation experiment in a burning pain algorithm.

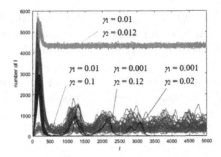

Fig. 9. Results of SIRS-contagion simulation in CACM

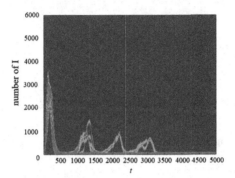

Fig. 10. Seem-periodic outbreak of epidemic in an energy representation

Table 2. Simulation parameters for sensor network event communication.

Parameters		Settings
Cellular space size	$m \times n$	250×250
Information dimension	c	4
Neighbor	$\boldsymbol{M}$	Moore neighbor
Event intensity	η_{event}	$1/d$ step
Number of repeated experiments	num_rpt	250
Infection probability	β	1
Recovery probability	γ	0.01
Path loss	$\varPhi$	0.01

The event source is located in the center of the region. The number of stimulated cellular is maintained by continuously generating events with a certain intensity. The cellular activated state is determined by subtracting the path loss from each step of the attribute value accumulation at the second level. The experiment is repeated 250 times.

The experimental results are processed because the mean values of each experiment are similar after stabilization. So the normalization of the mean values between 1000 and 5000 steps is taken for each statistic. In Fig. 11, the abscissa is the time step, and the corresponding ordinate is the number of normalized stimulated cellular with the mean values. Figure 11 identifies the maximum and minimum values of 250 normalized experiments in the form of error bands. It can be observed that propagation has both stable and fluctuating properties.

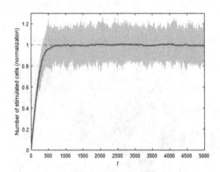

Fig. 11. Propagation experiment for CACM

3.2   Particle Space Sampling Experiment

In this paper, by taking the positive center of the region as the source of particle generation. Only one particle is generated globally, and the particle is transferred to any neighbor at each step. A system constructed by statistical particles appearing in a specified area and the entire cellular automaton. The particles appear in the designed area and in the system which the whole cellular automata build. The number of their occurrences is statistics. That is, the ratio of the times number in the cellular space. The extent to particles appear in a fixed range is discussed. Table 3 gives the simulation parameters of the experiment.

Figure 12 shows the experiment of the relationship between the frequency of particles occurrence and the framed area under the finite size of different checking steps. In a system of finite size m * n, statistics show the frequency of particle occurrence in a square region whose edge length is odd number s $(s < m, s < n)$ centered on the source of particle occurrence. The probability of particle occurrence in this region is estimated. Meanwhile, $l = \min(m, n)$. The system size is $M = n = 3$, 9, 33, 129, 257, and the field size S is repeated at 1, 3, 5, ..., l. In each case, calculating 100, 1000 and 10000 steps respectively. The odd number $2k - 1$ is processed in the kth position and aligned with the same ratio of L in the result. In Fig. 12, the abscissa is the ratio of the area of two regions, and the ordinate is the frequency of particle occurrence. The three subgraphs represent different calculating steps.

Table 3. Simulation parameters for particle probability.

Parameters		Settings
Cellular space size	$m \times n$	$m = n$
Information dimension	c	1
Neighbor	M	Moore neighbor
Visual field size	s	Odd number and less than cell space size
Event intensity	η_{event}	$\delta(0)$
Calculus step	num_step	100, 1000, 10000
Repeated experiments number	num_rpt	100

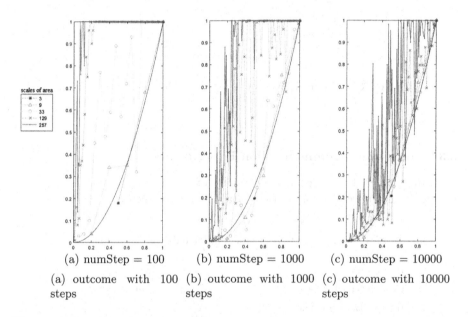

(a) numStep = 100   (b) numStep = 1000   (c) numStep = 10000

(a) outcome with 100 steps   (b) outcome with 1000 steps   (c) outcome with 10000 steps

Fig. 12. Relationship between possibilities of particle being concerned and sizes of areas in CACM

When the walking step size of the particle is comparable to the area of the system, the frequency of the particle being counted is approximately linearly related to the ratio of the two areas (the area ratio curve is drawn with a black solid line in Fig. 12). As the area of the region expands gradually, the frequency instability of particles in the frame becomes more evident.

The probability that random walk particles appear in the region has a large uncertainty. When the region edge length of 1025 * 1025 automaton is taken from 1 to 37 odd numbers respectively, the first time that the particles leave the region is simulated and counted. The situation is repeated 100 times. Under the condition of limited vision field, the frequency of each departure time is counted.

Figure 13 is drawn in double logarithmic coordinates. The scatter points in the figure reflect certain scale-free characteristics.

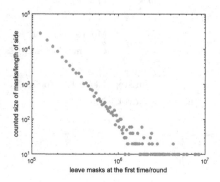

Fig. 13. Constrained counts for time of particles first leaving masks

3.3   Multi-event Communication Experiments

According to Sect. 2.3, the controllability of event propagation is discussed through a multi-event antagonistic propagation experiment.

Table 4. Simulation parameters for particle probability.

Parameters		Settings
Cellular space size	m, n	$m = n = 250$
Information dimension	c_{event}	2
Control signal information dimension	$c_{control}$	1
Neighbor	M	Moore neighbor
Event 1 dissemination rate	λ_{event}	0.1
Event 2 dissemination rate	$\lambda_{control}$	1
Event intensity	η_{event}	$\delta(0)$
Calculus step	num_step	1000
Repeated experiments number	num_rpt	1

Table 4 gives the simulation parameters of the experiment. The reference event point and the manual intervention event point (issued by the controller) are set on both sides of the regional center, coordinates (125, 100) and (125, 150), respectively. The modes of transmission are SI. The case where the signaling bits occupy one-dimensional event information is discussed. Figure 14 compares the message generated by the two types of events under the condition that the bits are consistent and opposite, and the condition without human intervention.

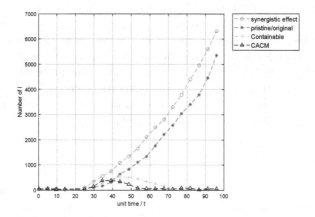

Fig. 14. Antagonistic propagation of events

It can be seen that, compared with the single event propagation (Fig. 14 pristine curve), the opposite antagonism plays a significant role in restraining propagation (Fig. 14 containable curve). And because the dimension of event information is higher than signaling, even if the information bits are consistent, they have a limited role in promoting event propagation (Fig. 14 synergistic curve). In contrast, the proposed CACM algorithm has achieved remarkable results in the simulation process. In the process of transmission of infection events in sensor networks, it produces excellent inhibition. It effectively avoids and terminates the spread of malicious events in the network.

4   Conclusion

Cellular automata can characterize complex systems with simple rules, especially for simulation system contingency and emergencies. In this paper, an event propagation model is established for sensor networks with simple topology and clear tasks. Template operation is introduced to describe them by using cellular evolution rules defined by finite state machine and perceptron. The corresponding monitoring algorithm based on deployment of monitoring points is designed. Meanwhile, control mechanism based on antagonistic propagation is also designed. The event propagation model of the burning pain bionic algorithm is proposed. The modeling and simulation tests are performed on the propagation and delivery of event-related message between network routing nodes. The improved model has strong scalability. Event propagation and monitoring mechanism enables event reports and data to be transmitted in the same message instead of sending event messages separately. It helps sensor networks to maintain energy efficiency and improve the survivability against flooding attacks.

As the convergent node of the whole network data, the monitoring center of the sensor network is often set at the edge of the network. Robust tracking when events occur is very important to the network survivability. How to identify events effectively by cellulars on the edge of event propagation model in sensor networks is still a subject to be studied. With the introduction of wireless technology, the modeling of sensor networks can no longer assume that the number of neighbors of nodes is fixed and the relationship is unchanged. In addition, for the wireless network with sparse connections, the probability of traveling particles returning to the starting point is greatly increased. How to make the event propagation and monitoring algorithm of sensor networks more scalable and higher. It also needs further consideration.

References

1. Ren, F.Y.: Wireless sensor networks. J. Softw. **14**(14), 1513–1525 (2003)
2. Liu, X., Han, J., Ni, G., Zhang, C., Liu, Y.: A multipath redundant transmission algorithm for MANET. In: Liang, Q., Mu, J., Jia, M., Wang, W., Feng, X., Zhang, B. (eds.) CSPS 2017. LNEE, vol. 463, pp. 518–524. Springer, Singapore (2019). https://doi.org/10.1007/978-981-10-6571-2_63
3. Libi, F., Song, W., Wei, L., Lo, S.: Simulation of emotional contagion using modified sir model: a cellular automaton approach. Phys. A Stat. Mech. Appl. **405**, 380–391 (2014)
4. Shaw, A.K., Tsvetkova, M., Daneshvar, R.: The effect of gossip on social networks. Complexity **16**(4), 39–47 (2011)
5. Tang, S., Myers, D., Yuan, J.: Modified SIS epidemic model for analysis of virus spread in wireless sensor networks. Int. J. Wirel. Mob. Comput. **6**(2), 99–108 (2013)
6. Fresnadillo, M.J., García, E., García, J.E., Martín, Á., Rodríguez, G.: A SIS epidemiological model based on cellular automata on graphs. In: Omatu, S., et al. (eds.) IWANN 2009. LNCS, vol. 5518, pp. 1055–1062. Springer, Heidelberg (2009). https://doi.org/10.1007/978-3-642-02481-8_160
7. Peng, S., Wang, G., Shui, Y.: Modeling the dynamics of worm propagation using two-dimensional cellular automata in smartphones. J. Comput. Syst. Sci. **79**(5), 586–595 (2013)
8. Choudhury, S.: Cellular automata and wireless sensor networks. In: Adamatzky, A. (ed.) Emergent Computation. ECC, vol. 24, pp. 321–335. Springer, Cham (2017). https://doi.org/10.1007/978-3-319-46376-6_14
9. Baryshnikov, Y.M., Coffman, E., Kwak, K.J.: High performance sleep-wake sensor systems based on cyclic cellular automata. In: 2008 International Conference on Information Processing in Sensor Networks (IPSN 2008), pp. 517–526. IEEE (2008)
10. Athanassopoulos, S., Kaklamanis, C., Katsikouli, P., Papaioannou, E.: Cellular automata for topology control in wireless sensor networks. In: 2012 16th IEEE Mediterranean Electrotechnical Conference, pp. 212–215. IEEE (2012)
11. Mansilla, R., Gutierrez, J.L.: Deterministic site exchange cellular automata model for the spread of diseases in human settlements (2000)
12. He, Y., Zhang, W., Jiang, N., Luo, X.: The research of scale-free sensor network topology evolution based on the energy efficient. In: 2014 Ninth International Conference on P2P, Parallel, Grid, Cloud and Internet Computing, pp. 221–226. IEEE (2014)

13. Hennebert, C., Hossayni, H., Lauradoux, C.: The entropy of wireless statistics. In: 2014 European Conference on Networks and Communications (EuCNC), pp. 1–5. IEEE (2014)
14. Harris, D., Harris, S.: Digital Design and Computer Architecture. Morgan Kaufmann, Burlington (2010)
15. Wu, T.L., Lai, Y.H., Fung, R.F.: Comparisons of fitness functions in identifying an electromagnetic energy harvester. J. Vib. Eng. Technol. **7**(2), 167–177 (2019)
16. Lopez, L., Burguerner, G., Giovanini, L.: Addressing population heterogeneity and distribution in epidemics models using a cellular automata approach. BMC Res. Notes **7**(1), 1–11 (2014)
17. Panwar, H., Gupta, S.: Optimized large margin classier based on perceptron. In: Wyld, D., Zizka, J., Nagamalai, D. (eds.) Advances in Computer Science, Engineering & Applications. AINSC, vol. 166, pp. 385–392. Springer, Heidelberg (2012). https://doi.org/10.1007/978-3-642-30157-5_38
18. Akram, H., Khalid, S., et al.: Using features of local densities, statistics and HMM toolkit (HTK) for offline Arabic handwriting text recognition. J. Electr. Syst. Inf. Technol. **4**(3), 387–396 (2017)
19. Mata, J., Cohn, M.: Cellular automata-based modeling program: synthetic immune system. Immunol. Rev. **216**(1), 198–212 (2010)
20. Pun-Cheng, L.S.C., Chan, A.W.F.: Optimal route computation for circular public transport routes with differential fare structure. Travel Behav. Soc. **3**(4), 71–77 (2016)
21. Motter, A.E., Timme, M.: Antagonistic phenomena in network dynamics. Annu. Rev. Condens. Matter Phys. **9**(1), 463–484 (2018)

Video Surveillance

Large-Scale Video-Based Person Re-identification via Non-local Attention and Feature Erasing

Zhao Yang, Zhigang Chang, and Shibao Zheng[✉]

Institute of Image Communication and Network Engineering,
Shanghai Key Labs of Digital Media Processing and Transmission,
Shanghai Jiao Tong University, SEIEE Buildings 5-421, Shanghai 200240, China
{10110907,changzig,sbzh}@sjtu.edu.cn

Abstract. Encoding the video tracks of person to an aggregative representation is the key for video-based person re-identification (re-ID), where average pooling or RNN methods are typically used to aggregating frame-level features. However, It is still difficult to deal with the spatial misalignment caused by occlusion, posture changes and camera views. Inspired by the success of non-local block in video analysis, we use a non-local attention block as a spatial-temporal attention mechanism to handle the spatial-temporal misalignment problem. Moreover, partial occlusion is widely occurred in video sequences. We propose a local feature branch to tackle the partial occlusion problem by using feature erasing in the frame-level feature map. Therefore, our network is composed by two-branch, the global branch via non-local attention encoding the global feature and the local feature branch grasping the local feature. In evaluation, the global feature and local feature are concatenated to obtain a more discriminative feature. We conduct extensive experiments on two challenging datasets (MARS and iLIDS-VID). The experimental results demonstrate that our method is comparable with the state-of-the-art methods in these datasets.

Keywords: Video-based person re-ID · Non-local attention · Feature erasing

1 Introduction

Person re-Identification (re-ID) aims to match the same pedestrian across multiple non-overlapping camera at different places or time, which has received more and more attention due to its great significance to applications in recent years, such as video surveillance [25], activity analysis [17], tracking [28] and so on. Deep learning methods have shown to be effective for person re-identification and have made great progress than traditional approaches [1,4,7]. However, it is still a difficult and challenging problem because of the huge variations of pedestrian's

© Springer Nature Singapore Pte Ltd. 2020
G. Zhai et al. (Eds.): IFTC 2019, CCIS 1181, pp. 327–339, 2020.
https://doi.org/10.1007/978-981-15-3341-9_27

appearance caused by person's postures, background clutter, camera viewpoints, blur and occlusion.

In general, with respect to the "probe-to-gallery" pattern, there are four person re-identification strategies: image-to-image, image-to-video, video-to-image, and video-to-video [30]. Image-based person re-identification has been widely studied in the literature where person are matched by only a single still image. It's obviously that video data contain much richer appearance information than a single image which is beneficial to identify a person under complex conditions including occlusion, blur, and the changes of camera viewpoint [31]. Therefore, we will concentrate on the video-based person re-identification in this paper, which associates the trajectories of the pedestrian by comparing his/her video sequences not only one still image.

For the Video-based person re-ID task, the key is to learn a function which encoding the video tracks of person to a single feature in a lower dimensional feature space. A typical pipeline of video-based person re-id methods contains a frame-level feature extractor (Convolutional Neural Network) and a feature aggregation module to aggregate the frame-level features to a single feature, where average pooling or maximum pooling is widely used. However, basic average pooling or maximum pooling is hard to handle spatial misalignment caused by the variation of human poses or viewpoints among sequences. To make full use of the video sequence information to alleviate the noise, many of the recent methods focus on the temporal model. There are two classical temporal modelling methods: Recurrent Neural Network (RNN) based and temporal attention based. In RNN-based methods, McLanghlin et al. [19] proposed to use an RNN to aggregate feature among the temporal frame-level features; Yan et al. [27] also used an RNN to encode sequence features, where the final hidden state is used as the output sequence representation. In temporal attention based methods, Liu et al. [16] proposed a Quality Aware Network (QAN), to measure the quality of every frame among the sequence and score. QAN is actually an attention weighted average of frame-level feature; Zhou et al. [31] proposed to encode the video sequence with temporal RNN and attention to pick out the most discriminative frames. However, these method mentioned above are still difficult to handle temporal attention and spatial misalignment simultaneously. Moreover, these method mainly consider the most discriminative frames in a video sequence by attention weights to learn a global feature representation, which ignored the discriminative body parts in a frame. Local discriminative body parts learning have a significant improvement to occlusion, particularly partial occlusion caused by other pedestrians or blackground cluster. Li et al. [14] proposed a spatial-temporal attention model that automatically discovers a diverse set of distinctive body parts.

In this paper, inspired by the success of non-local block in video classification [26], we propose a spatial-temporal attention mechanism to tackle the spatial misalignment problem by adopting a non-local block in a bottleneck structure. Moreover, We propose a local feature branch to tackle the partial occlusion problem by using unified region feature erasing in the frame-level feature map. Therefore, our network is composed by two-branch, the global branch via non-local attention scheme encoding the global feature and the local feature branch grasping the local feature. In evaluation, the global feature and local feature are concatenated to generate a more discriminative feature. In summary, our contribution of this work are three fold:

1. We propose a non-local block with a bottleneck structure as a spatial-temporal attention module to deal with the misalignment problem intra the video sequences, which could learn a more robust global feature.
2. The positional consistency feature erasing mechanism intra a video sequence in the local feature branch can effectively learn the local distinctive feature, which is beneficial to the partial occlusion problem.
3. A more discriminative feature used to re-identify the person is obtained by fusioning the robust global feature learned by the non-local attention and the local feature learned by feature erasing scheme with positional consistency. We conduct extensive experiments and ablation study to demonstrate the effectiveness of each component in a two challenging datasets: Mars and iLIDS-VID. The experiment results are comparable with the state-of-the-art methods in these datasets.

2   Related Works

In this section, we review the related work including image-based person re-identification, video-based person re-identification and self-attention.

2.1   Image-Based Person Re-identification

Image-based person re-identification has been widely studied in the literature. Recent work focus on discriminative feature learning and metric learning. Hermans et al. [12] proposed a variant of triplet loss to perform end-to-end deep metric learning, which selects the hardest positive and negative example for every anchor example in a batch and outperforms many other published methods by a large margin. Recently many work try to learn fine-grained discriminative features from local part of person. Part-based methods [9,20,23] have achieved state-of-the-art results, which split the feature map horizontally into a fixed number of strips and aggregate features from those strips. However, in evaluation, the high-dimensional feature vector increase the burden of computation and storage. Dai et al. [5] proposed a batch feature erasing method which erasing the feature map of images from different person in the batch at the same position can effectively learn the local feature without much computation. Inspired by this method, we extend it to our video-based person re-identification network to learn local discriminative feature of the video sequence. The feature maps of every frame intra the sequence are synchronously erased in the same position in a batch.

2.2   Video-Based Person Re-identification

Video-based person re-identification is an extension of the image-based person re-identification and has drawing more and more attraction due to the richer information contained in a video sequence. Making full use of the temporal feature of the video sequence is the most important part of video-based person re-identification. Most of the recent methods focus on the design of spatial-temporal attention intra the sequence. Gao et al. [10] revisited many temporal modeling method, including rnn-based, average pooling and proposed a attention scheme by using a fully connected layer to weighted every frame. Distinct with these methods which represent the video sequence in a single feature, Chen et al. [3] devided the video sequence to several overlapped snippets, and then computed the similarity of every snippet of two video sequences with co-attentive embedding and assumed the average scores of the top 20% as the final similarity of the two video sequences, which can implicitly deal with the posture misalignment problem.

2.3   Self-attention

Long-term range dependence modelling makes a great significance to natural language process and computer vision. Non-local means [2] is a classical filtering algorithm in digital image process that computes a weighted average of all pixels in an image. Different from convolution operation which only consider the local neighborhood pixels, non-local means allows distant pixels to contribute to the filtered response at a location based on patch appearance similarity. Vaswani et al. [22] proposed a self-attention module which computes the response at a position (e.g., a word) in a sequence (e.g., a sentence) by taking every position into account and summing weighted responses in an embedding space. Furthermore, Wang et al. [26] proposed a non-local module to bridge self-attention for machine translation to the more general class of non-local filtering operations that are applicable to image and video problems in computer vision. The non-local module has the ability to model the long-term range dependence and could discover the consistent semantice messages of frames in a video sequence. Therefore, we can embed a non-local block to our neural network to capture a long-term range dependence in video both on space and time to deal with the spatial-temporal misalignment problem intra the sequence.

3   The Proposed Method

In this section, we introduce the overall architecture (Fig. 1) of the proposed method, and then describe each of its important part with more details. First the input video is devided into several non-overlapped consecutive clips $\{c_k\}$, each contains T frames. The backbone CNN extracts feature maps for every frame in the clip, and then the non-local block takes these feature maps as input to model the spatial-temporal attention intra the clip, and obtain the weighted feature maps without dimension change of the input feature maps. The rest part have two component: the global feature branch and the local feature branch. The global one uses global average pooling over the weighted feature maps to grasp the global feature, and the local branch adopts feature erasing by a mask intra the clip with position consistency to learn the local discriminative feature. In evaluation, we concatenate the global feature and the local feature to generate a more distinctive and robust clip-level feature. The final video-level feature is the average of all clip-level feature. For the loss function, we combine batch hard triplet loss and softmax loss [12].

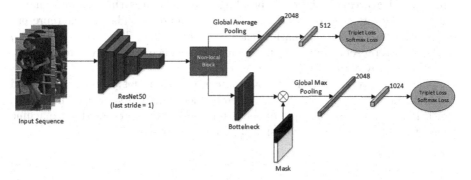

Fig. 1. Overall architecture. The input video clip is fed into the backbone network to be transformed to the feature maps, then the non-local block captures the long range dependence intra the video clip both in space and time, and output weighted feature maps. The global branch adopts global average pooling to the weighted feature maps to generate a global feature representation, and the local branch uses a feature erasing mask to grasp the discriminative local feature. In evaluation, we concatenate the 512 dimension global feature vector (L2 normalized) and the 1024 dimension local feature vector (L2 normalized) to generate the final robust and discriminative representation of the input clip.

3.1   Backbone Network

We adopt ResNet50 [11] as the backbone network. Sun et al. [20] removed the last spatial downsampling in the last residual block in ResNet50 to increase the size of feature map. Luo et al. [18] denoted the last spatial down-sampling operation in the ResNet50 backbone as last stride. Thus, we modify the last stride of ResNet50 from original 2 to 1. For a classical input image size 256×128, the modified ResNet50 outputs a feature map with the spatial size of 16×8 and channel dimension 2048.

3.2   Non-local Block

A non-local attention block has the ability to capture the long-term range dependency in sequence or video both in space and time. Therefore, for video-based person re-identification, a non-local attention block could tackle the spatial mis-alignment problem caused by the viewpoint and distance. In this part, we give an detail definition of non-local operation.

Following the non-local mean operation [2,26], a generic non-local operation in deep neural networks is defined as:

$$y_i = \frac{1}{C(x)} \sum_{\forall j} f(x_i, x_j) g(x_j) \tag{1}$$

Here x_i can be the feature vector of the input signal x (image, sequence, video; often their features) at the corresponding position i. y is the output signal of the same size as x. A pairwise function f calculate a scalar which represent the relationship such as similarity between x_i and all x_j The unary function g computes a representation of x_j in an embedded space. Finally, the response y_j is normalized by a factor $C(x)$.

There are several versions of f and g, e.g. gaussian, dot product, embedded gaussian. In our experiment, we select embedded gaussian version of non-local, which is the essential principle of self-attention module presented for machine translation [26]. Embedded gaussian function f is given by:

$$f(x_i, x_j) = e^{\theta(x_i)^T \phi(x_j)} \tag{2}$$

Here $\theta(x_i) = W_\theta x_i$ and $\phi(x_j) = W_\phi x_j$. If we set $C(x)$ as a softmax function, we would have:

$$y = softmax(x^T W_\theta^T W_\phi x) g(x) \tag{3}$$

A non-local operation is a flexible building block and can be easily inserted to neural network. In the non-local neural network [26], a non-local block is defined as:

$$z_i = W_z y_i + x_i \tag{4}$$

where y_i is defined in Eq. (1) and "$+x_i$" represent a residual connection [26]. The residual connection allows us to insert a new non-local block into any pre-trained model, without breaking its initial behavior (e.g., if W_z is initialized as zero) [11]. An example non-local block is illustrated in Fig. 2. In our experiment, we insert a non-local block within a bottleneck structure [11], which sequentially contains a conv1 × 1, a conv3 × 3, a conv1 × 1, each followed by a batchnorm layer and ReLU function, the non-local block is inserted between the last batchnorm layer and last ReLU function.

3.3 Unified Region Feature Erasing

We extend the batch feature erasing method [5] proposed in image-based person re-id to video-based, which we defined as unified region feature erasing. In a batch of video clips, especially intra every clip, the region erased is the same, which is beneficial to grasp the same pattern local feature in a batch. The region erased should be big enough to include part semantice messages of the original feature map. The detailed algorithm is described below:

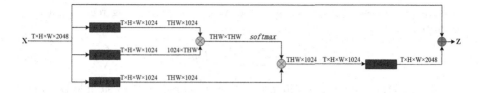

Fig. 2. A spacetime non-local block. The feature maps are shown as the shape of their tensors, e.g., $T \times H \times W \times 2048$ for 2048 channels (proper reshaping is performed when noted). "$\otimes$" denotes matrix multiplication, and "$\oplus$" denotes element-wise sum. The softmax operation is performed on each row. The blue boxes denote $1 \times 1 \times 1$ convolutions. Here we show the embedded Gaussian version, with a bottleneck of 1024 channels.

Algorithm 1. Unified Region Feature Erasing Algorithm.

Require:
    Input tensor of size [N, T, C, H, W], T_{in};
    Ratio of erased height, r_h;
    Ratio of erased width, r_w;

Ensure:
    Erased Tensor, T_e;

1: **if** training **then**
2:    $H_e := H \times r_h, W_e := W \times r_w$
3:    $x_e := Rand(0, H - H_e), y_e := Rand(0, W - W_e)$
4:    $Mask := Ones(H, W)$
5:    $Mask[x_e : x_e + H_e, y_e : y_e + W_e] := 0$
6:    $T_e := T_{in} \times Mask$
7: **else**
8:    $T_e := T_{in}$
9: **end if**
10: **return** T_e;

3.4   Loss Functions

We use a triplet loss function and a softmax cross-entropy loss function with label smoothing regularization [21] to train the network. We use the batch hard triplet loss function which was originally proposed in [12]. To form a batch, we randomly sample P identities and randomly sample K clips for each identity (each clip contains T frames); Totally there are PK clips in a batch. For each sample a in the batch, the hardest positive and the hardest negative samples within the batch are selected when forming the triplets for computing the loss $L_{triplet}$.

$$L_{tripletloss} = \sum_{i=1}^{P} \sum_{a=1}^{K} [m + \overbrace{\max_{p=1...k} D(f_a^i, f_p^i)}^{hardest\ positive} - \underbrace{\min_{\substack{j=1...P \\ n=1...K \\ j!=i}} D(f_a^i, f_n^j)}_{hardest\ negative}] \tag{5}$$

The softmax cross-entropy loss function with label smoothing regularization is given by:

$$L_{softmax}^{smooth} = -\frac{1}{P \times K} \sum_{i=1}^{P} \sum_{a=1}^{K} p_{i,a} log((1-\epsilon)q_{i,a} + \frac{\epsilon}{N}) \tag{6}$$

where $p_{i,a}$ is the ground truth identity, $q_{i,a}$ is the prediction of sample $\{i, a\}$ and N is the number of classes. The final loss function is the combination of the two losses:

$$L_{total} = L_{tripletloss} + L_{softmax}^{smooth} \tag{7}$$

4   Experiments

4.1   Datasets and Evaluation Protocal

Mars Dataset.   [30] is one of the largest video-based person re-identification dataset. It contains 17,503 tracklets from 1,261 identities, and additional 3,248 tracklets serving as distractors. Each tracklet has 59 frames on average. These video tracklets are captured by six cameras in a university campus. The total 1,261 identities are split into 625 identities for training and 636 identities for testing. The ground truth bounding boxes are detected and tracked using the Deformable Part Model (DPM) [8]and GMCP tracker [29].

iLIDS-VID Dataset.   [24] consists of 600 video sequences of 300 persons. Each image sequence has a variable length ranging from 23 to 192 frames, with averaged number of 73. This dataset is challenging due to clothing similarities among people and random occlusion. The ground truth bounding boxes are annotated manually.

Evaluation Protocol.   In our experiments, we report the standard evaluation metrics: mean average precision score (MAP) and the cumulative matching curve (CMC) at rank-1, rank-5, rank-10 and rank-20. For fair comparison, we report MAP and CMC for Mars dataset, CMC for iLIDS-VID dataset.

4.2   Implemantation Details

As mentioned in "The Proposed Method", we use modified ResNet50 which is pre-trained on the ImageNet [6]. The frame number of each clip is set to $T = 4$. we augment the clip data with clip-level random synchronize flipping and cropping. There are $P \times K$ clips in a minibatch, we set $K = 4$ (clips of each person), $P = 32$ (identities) and $K = 4$, $P = 16$ for Mars and iLIDS-VID respectively, where all of P and K are randomly sampled. The input image is resized to 256×128. We set $r_w = 1.0$ and $r_h = 1/3$ to erase the feature map. The margin parameter in triplet loss is equal to 0.3. During training, we use the Adam optimizer [13] with weight decay 5e−4. We use a warm up training scheme, the initial learning rate is linearly increasing from 1e−4 to 1e−3 in first 50 epochs. And the learning rate decays to 1e−4, 1e−5, 1e−6 at 200, 400, 600 epochs respectively. We train the network for 800 epochs in total.

4.3   Ablation Study

We conduct ablation investigation to analyze the effect of each component in our proposed method, including non-local block and the local feature erasing branch.

Effectiveness of Components. In Table 1, we show the results of each component. All these three methods adopt the modified ResNet50 as the backbone network, and use batch hard triplet loss and softmax cross-entropy loss function with label smoothing regularization for a fair comparison. *Baseline* method represents our global branch without the non-local block trained on MARS/iLIDS-VID dataset, where the global average pooling layer encoding the feature maps of clip to a single tensor with dimension 2048. Then the feature dimension is reduced from 2048 to 512 by using a fully connected layer. We choose the final 512 dimension feature vector to represent the clip. *Baseline + NL* method represents our global branch with the non-local block. Compared to *Baseline*, the *Baseline + NL* method improves rank-1 and mAP accuracy by 1.0% and 1.4% on the large-scale Mars dataset, as well as 6.6% and 4.4% on iLIDS-VID dataset. The result demonstrate that the non-local block is effectively to model the spatial-temporal attention to learn a more discriminative feature and deal with the spatial misalignment problem. The visualization results are show in Fig. 3. The *Baseline + NL + FE* method is our whole network with a non-local block to model the long range dependency within a video both space and time, the final feature representation is the concatenation of the feature vector from the global branch and the feature vector from the local branch. Our overall network increase 3.2% and 3.5% in rank-1 and mAP when compared with *Baseline* on Mars dataset, and gain 8.6% and 5.9% improvements on iLIDS-VID dataset. Obviously, combining the global feature and local feature together can effectively obtain a more robust and discriminative feature to re-identify the same person.

Table 1. Comparison of different proposed components.

Model	MARS					ILIDS-VID				
	R1	R5	R10	R20	MAP	R1	R5	R10	R20	MAP
Baseline	83.5	94.3	96.5	97.6	77.5	76.7	94.0	97.0	100.0	84.8
Baseline + NL	84.5	94.9	96.6	97.7	78.9	83.3	96.7	99.3	100.0	89.2
Baseline + NL + FE	86.7	95.4	97.2	97.9	81.0	85.3	98.0	99.3	99.3	90.7

4.4   Comparison with the State-of-the-Arts

Tables 2 and 3 report the comparison of our proposed method with the state-of-the-art methods on Mars and iLIDS-VID respectively.

Fig. 3. Visualization of the behavior of a non-local block to deal with spatial misalignment.

Results on MARS. Mars is one of the largest video-based person re-id dataset and full of challenge. Table 2 show comparisons between our proposed method with most of the state-of-the-art method on Mars dataset. We achieve 81.0% in MAP, which outperforms all these work by nearly 5%. For the rank-1, rank-5, and rank-10, Our approach all keeps ahead than all these methods. Our approach achieve 97.9% in rank-20, which is comparable with the $Snipped + OF^*$. $Snipped + OF^*$ uses optical flow to provide more extra motion information and brings with more computation.

Table 2. Comparison of our proposed method with the state-of-the-art on MARS dataset. '–': no reported results.

Model	Rank1	Rank5	Rank10	Rank20	MAP
Mars [30]	68.3	82.6	–	89.4	49.3
SeeForest [31]	70.6	90.0	–	97.6	50.7
QAN [16]	73.7	84.9	–	91.6	51.7
Non-local + C3D [15]	84.3	94.6	96.2	–	77.0
STAN [14]	82.3	–	–	–	65.8
Snipped [3]	81.2	92.1	–	–	69.4
Snipped + OF^* [3]	86.3	94.7	–	**98.2**	76.1
Our proposed	**86.7**	**95.4**	**97.2**	97.9	**81.0**

Results on iLIDS-VID. iLIDS-VID is a small dataset especially comparied to Mars dataset. Table 3 show comparisons between our proposed method with most of the state-of-the-art method on iLIDS-VID dataset. Our method surpass most of these method, and is comparable with the best method $Snipped + OF^*$. For rank-5 and rank-10, we achieve the best results with some advantages.

Table 3. Comparison of our proposed method with the state-of-the-art on iLIDS-VID dataset. '–': no reported results.

Model	Rank1	Rank5	Rank10	Rank20
Mars [30]	53.0	81.4	–	–
SeeForest [31]	55.2	86.5	–	97.0
QAN [16]	68.0	86.8	95.4	97.4
STAN [14]	80.2	–	–	
Snipped [3]	79.8	91.8	–	–
Snipped + OF^* [3]	**85.4**	96.7	98.8	**99.5**
Our proposed	85.3	**98.0**	**99.3**	99.3

5 Conclusion

This paper concentrates on the large-scale video-based person re-identification. The key of video-based person re-id is to learn a mapping which encoding the clip-level feature to a single feature in a low-dimensional feature space. We adopt the non-local block to capture the long-term range dependence in a video both space and time, which can learn the corresponding consistent semantic information to deal with the spatial misalignment occurred in video sequences. The synchronous feature erasing scheme is beneficial to learn the local discriminative feature, which can alleviate the partial occlusion problem intra the video sequence. A more robust feature is generated by concating the global feature and local feature. Extensive experiments conducted on two challenging datasets (Mars and iLIDS-VID) demonstrate the effect of our method, which is comparable to most of state-of-the-art methods. A valuable direction of person re-identification including image and video is to combine with multi-pedestrian tracking in real world.

References

1. Ahmed, E., Jones, M., Marks, T.K.: An improved deep learning architecture for person re-identification. In: Proceedings of the IEEE Conference on Computer Vision and Pattern Recognition, pp. 3908–3916 (2015)
2. Buades, A., Coll, B., Morel, J.M.: A non-local algorithm for image denoising. In: 2005 IEEE Computer Society Conference on Computer Vision and Pattern Recognition (CVPR 2005), vol. 2, pp. 60–65. IEEE (2005)
3. Chen, D., Li, H., Xiao, T., Yi, S., Wang, X.: Video person re-identification with competitive snippet-similarity aggregation and co-attentive snippet embedding. In: Proceedings of the IEEE Conference on Computer Vision and Pattern Recognition, pp. 1169–1178 (2018)
4. Chen, S.Z., Guo, C.C., Lai, J.H.: Deep ranking for person re-identification via joint representation learning. IEEE Trans. Image Process. **25**(5), 2353–2367 (2016)
5. Dai, Z., Chen, M., Zhu, S., Tan, P.: Batch feature erasing for person re-identification and beyond. arXiv preprint arXiv:1811.07130 (2018)

6. Deng, J., Dong, W., Socher, R., Li, L.J., Li, K., Fei-Fei, L.: ImageNet: a large-scale hierarchical image database. In: 2009 IEEE Conference on Computer Vision and Pattern Recognition, pp. 248–255. IEEE (2009)
7. Ding, S., Lin, L., Wang, G., Chao, H.: Deep feature learning with relative distance comparison for person re-identification. Pattern Recogn. **48**(10), 2993–3003 (2015)
8. Felzenszwalb, P.F., McAllester, D.A., Ramanan, D., et al.: A discriminatively trained, multiscale, deformable part model. In: CVPR, vol. 2, p. 7 (2008)
9. Fu, Y., et al.: Horizontal pyramid matching for person re-identification. In: Proceedings of the AAAI Conference on Artificial Intelligence, vol. 33, pp. 8295–8302 (2019)
10. Gao, J., Nevatia, R.: Revisiting temporal modeling for video-based person ReID. arXiv preprint arXiv:1805.02104 (2018)
11. He, K., Zhang, X., Ren, S., Sun, J.: Deep residual learning for image recognition. In: Proceedings of the IEEE Conference on Computer Vision and Pattern Recognition, pp. 770–778 (2016)
12. Hermans, A., Beyer, L., Leibe, B.: In defense of the triplet loss for person re-identification. arXiv preprint arXiv:1703.07737 (2017)
13. Kingma, D.P., Ba, J.: Adam: a method for stochastic optimization. arXiv preprint arXiv:1412.6980 (2014)
14. Li, S., Bak, S., Carr, P., Wang, X.: Diversity regularized spatiotemporal attention for video-based person re-identification. In: Proceedings of the IEEE Conference on Computer Vision and Pattern Recognition, pp. 369–378 (2018)
15. Liao, X., He, L., Yang, Z., Zhang, C.: Video-based person re-identification via 3d convolutional networks and non-local attention. In: Jawahar, C.V., Li, H., Mori, G., Schindler, K. (eds.) ACCV 2018. LNCS, vol. 11366. Springer, Cham (2019). https://doi.org/10.1007/978-3-030-20876-9_39
16. Liu, Y., Yan, J., Ouyang, W.: Quality aware network for set to set recognition. In: Proceedings of the IEEE Conference on Computer Vision and Pattern Recognition, pp. 5790–5799 (2017)
17. Loy, C.C., Xiang, T., Gong, S.: Multi-camera activity correlation analysis. In: 2009 IEEE Conference on Computer Vision and Pattern Recognition, pp. 1988–1995. IEEE (2009)
18. Luo, H., Gu, Y., Liao, X., Lai, S., Jiang, W.: Bag of tricks and a strong baseline for deep person re-identification. In: Proceedings of the IEEE Conference on Computer Vision and Pattern Recognition Workshops (2019)
19. McLaughlin, N., Martinez del Rincon, J., Miller, P.: Recurrent convolutional network for video-based person re-identification. In: Proceedings of the IEEE Conference on Computer Vision and Pattern Recognition, pp. 1325–1334 (2016)
20. Sun, Y., Zheng, L., Yang, Y., Tian, Q., Wang, S.: Beyond part models: person retrieval with refined part pooling (and a strong convolutional baseline). In: Proceedings of the European Conference on Computer Vision (ECCV), pp. 480–496 (2018)
21. Szegedy, C., Vanhoucke, V., Ioffe, S., Shlens, J., Wojna, Z.: Rethinking the inception architecture for computer vision. In: Proceedings of the IEEE Conference on Computer Vision and Pattern Recognition, pp. 2818–2826 (2016)
22. Vaswani, A., et al.: Attention is all you need. In: Advances in Neural Information Processing Systems, pp. 5998–6008 (2017)
23. Wang, G., Yuan, Y., Chen, X., Li, J., Zhou, X.: Learning discriminative features with multiple granularities for person re-identification. In: 2018 ACM Multimedia Conference on Multimedia Conference, pp. 274–282. ACM (2018)

24. Wang, T., Gong, S., Zhu, X., Wang, S.: Person re-identification by video ranking. In: Fleet, D., Pajdla, T., Schiele, B., Tuytelaars, T. (eds.) ECCV 2014. LNCS, vol. 8692. Springer, Cham (2014). https://doi.org/10.1007/978-3-319-10593-2_45

25. Wang, X.: Intelligent multi-camera video surveillance: a review. Pattern Recogn. Lett. **34**(1), 3–19 (2013)

26. Wang, X., Girshick, R., Gupta, A., He, K.: Non-local neural networks. In: Proceedings of the IEEE Conference on Computer Vision and Pattern Recognition, pp. 7794–7803 (2018)

27. Yan, Y., Ni, B., Song, Z., Ma, C., Yan, Y., Yang, X.: Person re-identification via recurrent feature aggregation. In: Leibe, B., Matas, J., Sebe, N., Welling, M. (eds.) ECCV 2016. LNCS, vol. 9910. Springer, Cham (2016). https://doi.org/10.1007/978-3-319-46466-4_42

28. Yu, S.I., Yang, Y., Hauptmann, A.: Harry Potter's Marauder's map: localizing and tracking multiple persons-of-interest by nonnegative discretization. In: Proceedings of the IEEE Conference on Computer Vision and Pattern Recognition, pp. 3714–3720 (2013)

29. Roshan Zamir, A., Dehghan, A., Shah, M.: GMCP-Tracker: global multi-object tracking using generalized minimum clique graphs. In: Fitzgibbon, A., Lazebnik, S., Perona, P., Sato, Y., Schmid, C. (eds.) ECCV 2012. LNCS, vol. 7573. Springer, Heidelberg (2012). https://doi.org/10.1007/978-3-642-33709-3_25

30. Zheng, L., et al.: MARS: a video benchmark for large-scale person re-identification. In: Leibe, B., Matas, J., Sebe, N., Welling, M. (eds.) ECCV 2016. LNCS, vol. 9910. Springer, Cham (2016). https://doi.org/10.1007/978-3-319-46466-4_52

31. Zhou, Z., Huang, Y., Wang, W., Wang, L., Tan, T.: See the forest for the trees: joint spatial and temporal recurrent neural networks for video-based person re-identification. In: Proceedings of the IEEE Conference on Computer Vision and Pattern Recognition, pp. 4747–4756 (2017)

Design and Optimization of Crowd Behavior Analysis System Based on B/S Software Architecture

Yuanhang He[1], Jing Guo[2], Xiang Ji[2], and Hua Yang[1]([envelope])

[1] Institution of Image Communication and Network Engineering,
Shanghai Jiao Tong University, Shanghai, China
{hyhzxy,hyang}@sjtu.edu.cn
[2] National Engineering Laboratory for Public Security Risk Perception and Control
by Big Data (PSRPC), China Academic of Electronics and Information Technology,
Beijing, China
goodman2000@163.com, 394712680@qq.com

Abstract. With the development of society and economy, the importance of crowd behavior analysis is increasing. However, the system often requires a large amount of computing resources, which is often difficult to meet for personal computers in traditional client/server architecture (C/S architecture). So based on the existing local analysis system [5], we construct a crowd behavior analysis system based on browser/server architecture (B/S architecture). Then we optimize many aspects of this B/S system to improve its communication capability and stability under high load. Finally, the acceleration work of the CGAN-based crowd counting module is carried out. The generator of CGAN (Conditional Generative Adversarial Network) was optimized such as residual layer pruning, upsampling optimization, and instance normalization layer removing, and then deployed and INT8 quantized in TensorRT. After these optimizations, the inferring speed on the NVIDIA platform is increased to 541.6% of the original network with almost no loss of inference accuracy.

Keywords: Crowd behavior analysis · B/S architecture · CGAN · Network acceleration

1 Introduction

With the improvement of informatization, the coverage of camera monitoring has gradually increased. However, it is often difficult for the human to observe too much monitoring information at the same time and is impossible to predict the situation out of control. Automated crowd behavior analysis based on surveillance image data can help to better understand population dynamics and related personnel behavior in a variety of ways, and provide early warning of possible dangerous conditions.

© Springer Nature Singapore Pte Ltd. 2020
G. Zhai et al. (Eds.): IFTC 2019, CCIS 1181, pp. 340–354, 2020.
https://doi.org/10.1007/978-981-15-3341-9_28

However, the system often requires a large amount of computing resources. In traditional C/S (client/server) architecture, the computing power of the user's computer is often difficult to meet. The B/S (Browser/Server) architecture only gives very little computation to the client (ie, the browser). The main computation is running on the server, and the client is only responsible for instruction reception and result presentation, which greatly reduced the user's computing pressure. Moreover, in the traditional C/S software architecture, installation and environment configuration are complex, subsequent maintenance updates are inconvenient, and there are many platform restrictions. But in the B/S architecture, these inconvenience and restrictions have disappeared. Therefore, in recent years, the influence of the B/S architecture is gradually increasing.

Based on the existing local crowd behavior analysis system [5], we built a crowd behavior analysis system based on B/S software architecture. The system consists of 5 major modules and 3 main subsystems, providing user account management, real-time analysis, and historical data query. Then we optimize many aspects of this B/S system.

Furthermore, we carried out the inferring acceleration of the CGAN in the crowd counting module. After many experiments, we performed optimization operations such as residual layer pruning, upsampling optimization, and instance normalization layer removing on the generator, and obtained nearly three times speed improvement and a small amount of memory consumption reduction. Finally, for the NVIDIA platform, we further modified the network to accommodate the API requirements of TensorRT, and then deployed and INT8 quantized in TensorRT. After testing, the above work achieved a speed increase of 541.6% with almost no loss of inference accuracy.

2   Overall System Architecture

The system can be divided into three subsystems according to functions: account system, real-time analysis system and history query system. The functions performed by the three subsystems are shown in Fig. 1.

There are 5 major modules: the web front-end module, the dispatch server module, the instruction forwarding module, the main analysis module, and the crowd counting module. They are shown in Fig. 2.

3   Design and Implementation of Key Modules of the System

3.1   Precise Control of Frame Rate at Normal Speed

In the B/S architecture, the video is played on the web page at normal speed. In the server-side main analysis module, the video is played under another independent time control. When the video frame-to-frame time interval is set to deviate from the normal speed, the deviation will gradually accumulate in one direction, and finally the time difference between the front-end video position and

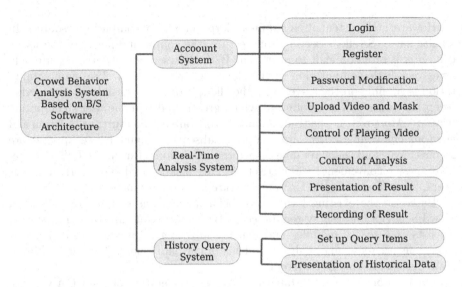

Fig. 1. Overall functions architecture

the server-side analysis position will gradually increase. When the video length is long, the time difference will accumulate, which causes considerable video misalignment. Moreover, the server load level is fluctuating, and the time taken for the frame fetching operation will fluctuate accordingly. Even if the sleep interval in the loop is the same, the actual frame fetch interval will fluctuate and deviate with the server performance fluctuation. When under high load conditions, the time taken for the frame only operation has exceeded the interframe time interval. Even the time interval of sleep in the loop is reduced to zero, the playback speed cannot be restored to normal.

In order to cope with the above situation, the system adopts a control strategy based on time prediction, which uses dynamic sleep time in a cycle combined with frame skipping. Its control logic is shown in Fig. 3.

The core idea of the control strategy is to set and maintain the rhythmic counter of the program. The counter is ideally incremented by the time interval of each video frame, and its rhythm is not affected by the pause of video playback. When the rhythm of the counter deviates from the target value to a certain threshold (in this paper we set it as the rounded down video FPS), the counter adds the value of the threshold once and the video skip the number of frames of the threshold value. During normal playback, the loop capture frame rate is maintained by adjusting the sleep value for each cycle. In the pause state, the counter will not stop counting and the expected time (expectTime) will deviate. After the playback resumes, the current time (nowTime) and the expected time (expectTime) will not be influenced. When under high load conditions, the frame interval will be too large to be disordered, but each time the deviation is greater than FPS * timeInterval (about 1 s), it will be corrected to maintain normal playback speed.

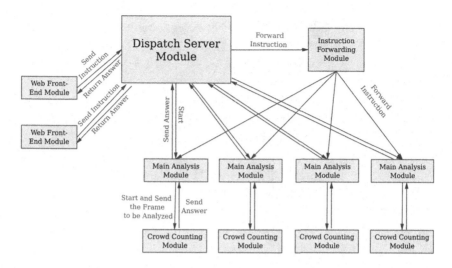

Fig. 2. The 5 major modules

We made a test under high server load conditions. The video playback progress of the web and server under different control strategies is shown in Fig. 4.

As shown in Fig. 4, in the strategy with dynamic sleep time combined with frame skipping, after a certain amount of deviation is accumulated, the frame skipping will be triggered, and the effective control will be taken. The frame rate is consistent with the front end playback speed.

3.2 Optimization of Cross-Language Data Communication

The system uses multiple languages such as JavaScript, Node.js, Python, and C++. So the system has a lot of cross-language communication needs. Common cross-language communication solutions include file communication through the hard disk, socket communication and so on. Because the file communication through the hard disk is limited by the read/write speed and there is synchronization problem, we choose socket communication as the cross-language communication method.

In this system, socket packet information is divided into header and body. The header information records the picture/text type (the 7-byte character of "jpgpict"/"txttext"), the user name, the analysis type, and the length of the packet header of the result contained in the packet. When under higher load pressure, socket communication is prone to fracture and adhesion of long data packets. For the text result data packet in this system, the byte length is short, and this problem does not occur; but the picture result data packet whose length is several times larger is prone to fracture and adhesion. Packet breakage will result in the loss of the packet data after the break, and the packet glue will make the post packet header data hidden and misinterpreted as the pre-packet

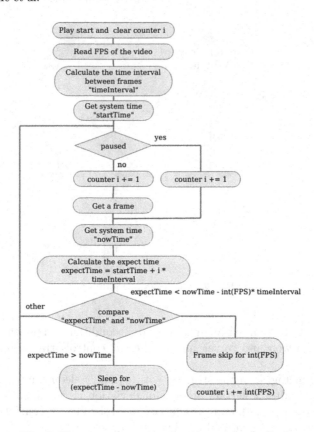

Fig. 3. The control strategy based on time prediction

body, resulting in both packets being damaged. The occurrence of both errors will cause the frame result to not be displayed correctly.

In order to cope with this problem, a solution is to divide the long-byte data packet into a number of short sub-packets and send the packet header information separately. At this time, the sub-packets may be stuck due to the short sending interval, but the receiving end may receive, split and parse the data packets according to the header of each sub-packet, and then flatten together to obtain complete data information. Through implementation and actual testing, this method effectively reduces the probability of damage to the picture results. However, sub-package breaks are still possible, and this method results in a reduction in effective information density.

In order to further improve transmission efficiency and reduce transmission error rate, the following communication scheme is designed. Data is transmitted in large packets, without dividing the data packets. The format of the packet header is the same as the original, and the footer "pictend" is added. When the receiving end receives the data packet, its processing mode is as shown in Fig. 5.

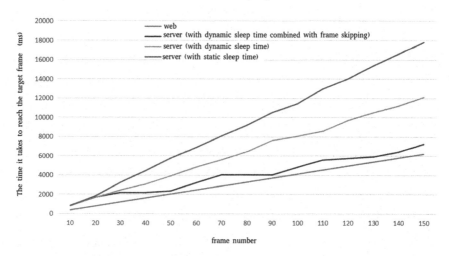

Fig. 4. The video playback progress of the web and server under different control strategies.

The core idea of the communication scheme is to use the binding relationship between the transmission data type and the remote port number of the socket connection, and store various information of the connection, thereby getting rid of the dependency on the header information in the packet parsing. The scheme does not focus on avoiding the packet fracture but in the recovery after the packet is broken. By setting buffers and tail marks, the packets and packages are correctly distinguished, and the data that is not at the end is temporarily stored in order, and finally the entire package is restored.

We compared the number of images that can be correctly restored by sending 100 image analysis results for each socket channel in the method of splitting packets into sub-packets and the recovery method based on the remote port number. The sub-packet splitting method is to transmit a sub-packet whose picture data is divided into 200 lengths (one picture is transmitted using about 2000 sub-packets). The result is shown in the Table 1.

It can be seen that for a relatively large amount of data transmission such as a picture, the socket packet fracture phenomenon is almost inevitable when a single packet is transmitted. This method has almost no communication capability; when the picture is transmitted less than 2 channels, the socket communication with splitting packets into sub-packets has a good transmission capacity, and most of the transmitted pictures can be restored. This transmission mode is insensitive to the sub-package adhesion phenomenon, and the possibility that the sub-package is broken at this time is low. However, when transmitting more than 3 channels at the same time, the communication performance of this mode drops sharply, the recovery rate drops from 97.5% to 34%, and this rate decreases to 20% when the four channels are used. The socket communication with recovery based on the remote port number runs smoothly under high pressure and low pressure, and there is no unresolvable data packet during the test.

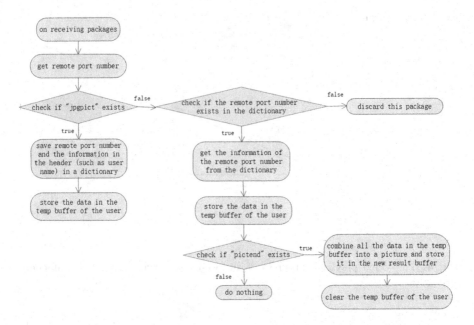

Fig. 5. The processing mode of the receiving end

Due to the existence of read and write conflicts in the file transfer mode, some pictures failed to be restored correctly. As the number of parallel transmissions of pictures increases, the writing pressure of the hard disk increases, so that the writing time of each picture is increased, so that the collision probability of reading and writing is increased, and the image resolution rate is lowered. It can be seen that the socket communication system based on the remote port number recovery mode successfully solves the problem of partial defect of the front-end picture under high load pressure.

Table 1. Performance comparison of various communication methods

	Sent images	Received images			
		(1)	(2)	(3)	(4)
1 channel	100	93	0	98	100
2 channels	200	181	0	195	199
3 channels	300	268	0	102	299
4 channels	400	347	0	80	398

*(1) the file communication through the hard disk (2) the original socket communication (3) the socket communication with splitting packets into sub-packets (4) the socket communication with recovery based on the remote port number

4   Crowd Counting Network Acceleration

4.1   Network Structure

Overall Structure. The crowd counting network is based on CGAN (Conditional Generative Adversarial Nets) in Pix2Pix [4]. It generally contain a two-part structure: Generator and Discriminator. The generator outputs a new picture based on the input picture, and the discriminator is responsible for identifying the difference between the output picture and the reasonable picture [2]. The sum of the pixel values in the standard density grayscale image is the number of people.

The overall structure is shown as Fig. 6.

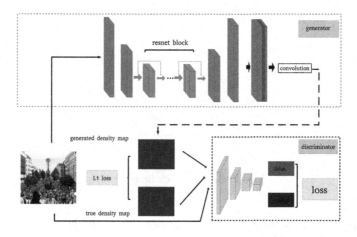

Fig. 6. The overall structure [1]

Generator Structure. The generator structure in this system is based on ResNet (Residual Neural Network) [3]. The residual structure makes the gradient effectively improve when the network is deep, and it resists the phenomenon of gradient disappearance to some extent. This residual structure is also used in the generator of this system. The final generator structure is similar to the generator in CycleGAN [10], and its structure is shown in the Table 2.

Among them, the 1–3 layer and 13–15 layer outputs are connected to the next layer input through a Instance Normalization layer and a Rectified Linear Unit (ReLU). The detailed structure of the 4–12 layer residual structure (ResNet Block) in the network is shown in Fig. 7.

Table 2. The generator structure

Number	Layer
1	$7 \times 7 \times 32$ conv stride = 1
2	$3 \times 3 \times 64$ conv stride = 2
3	$3 \times 3 \times 128$ conv stride = 2
4–12	ResNet Block
13	$3 \times 3 \times 64$ deconv stride = 2
14	$3 \times 3 \times 32$ deconv stride = 2
15	$7 \times 7 \times 3$ conv stride = 1

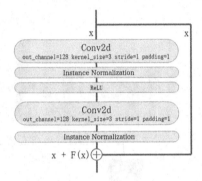

Fig. 7. Resnet block in generator

Performance Measurement. In this paper, Mean Absolute Error (MAE) and Mean Square Error (MSE) are used to measure the inference accuracy.

$$MAE = \frac{1}{n} \sum_{i=1}^{n} |\hat{y}_i - y_i| \tag{1}$$

$$MSE = \sqrt{\frac{1}{n} \sum_{i=1}^{n} (\hat{y}_i - y_i)^2} \tag{2}$$

Where $\hat{y}_i$ represents the inference number for the i-th data, and y_i represents the correct number of people for the i-th data. In this paper, the mean absolute error is taken as the main measure of system accuracy, and the mean square error is used as an auxiliary measure.

The test platform GPU is NVIDIA TITAN Xp, and the system environment is as follows: Ubuntu 16.04, Python 3.6.5, CUDA 9.0, PyTorch 1.0.1.post2, TensorRT 5.0.2.1. The test data set is the Shanghai University of Science and Technology data set [9].

4.2   Residual Structure Pruning

We compare the effects of reducing the different number of channels on the accuracy of the network residual structure. Correspondingly, the output channel of the last layer of downsampling convolution and the input channel of the first layer of upsampling convolution are reduced to achieve channel matching. The experimental results are shown in the Table 3.

Table 3. Inference accuracy with different number of channels in Resnet Block

Number of channels	128	100	64	32
MAE	91.80	91.38	89.95	95.76
MSE	152.02	165.51	166.77	178.95
Inference time (s)	8.343	7.367	6.261	5.231
Memory occupation (MiB)	1932	1884	1896	1904

As can be seen from the data in the table, the network inference time decreases with the number of residual structure channels. When the residual structure retains 100 channels and 64 channels, the network accuracy is still close to the original network; and if it continues to shrink to 32 channels, although the speed is still improved, the inference accuracy begins to show a significant decline. Compared with the original network 128 channels, reducing the channel in the residual structure can reduce the memory usage. Due to some characteristics of the GPU, 64/32 channel memory usage is increased compared to 100 channels.

4.3   Upsampling Optimization

It can be seen from the calculation that the interpolation and convolution method has less calculation amount than the deconvolution method for the input and output data of the same dimension. So we tried to replace the deconvolution with interpolation and convolution, and the results are shown in the Table 4.

Table 4. Inference accuracy with different upsampling method

	Deconvolution	Bilinear interpolation with 1 × 1 convolution	Nearest interpolation with 1 × 1 convolution
MAE	89.95	92.07	93.51
MSE	166.77	175.35	168.26
Inference time (s)	6.261	5.353	5.327
Memory occupation (MiB)	1896	1670	1670

It can be seen from the above table data that there is no significant difference in the inference accuracy between the deconvolution and the interpolation plus convolution, and the interpolation plus convolution method has certain advantages in terms of total time of inferring and memory occupancy. In the interpolation method, the nearest interpolation and bilinear interpolation speed and accuracy performance are basically the same. The method of bilinear interpolation and convolution has a speed increase of about 14.5% compared to the deconvolution speed, and the memory usage is reduced by 11.9%.

4.4   Instance Normalization Layer Removing

Instance Normalization is a method proposed by [8]. And through method proposed by [7], the convolutional layer can be merged with the subsequent batch normalization layer parameters to eliminate the normalization layer during inference.

Instance normalization and batch normalization are similar when batch size is 1. They are the same when the model is in the training mode, but have some difference when in the testing mode. For the mean and variance in the testing mode, instance normalization calculates them basing on the input test data but batch normalization reads the data generated during training. The mean and variance produced by the two methods are different but the values still have similarity.

Based on the similarity between them, we try to remove the instance normalization in the network during the inference process. We first replace the instance normalization in the network with batch normalization whose batch size is 1 and train the model. Then perform parameter fusion to remove the normalization layer according to the method of [7]. And the results are shown in the Tables 5 and 6.

Table 5. Inference accuracy with bilinear interpolation before and after fusing

	Instance normalization	Batch normalization	Batch normalization after fusing
MAE	92.07	133.55	133.57
MSE	175.35	285.11	285.14
Inference time (s)	5.353	5.341	2.961
Memory occupation (MiB)	1670	1670	1638

It can be seen that when the instance normalization is replaced by batch normalization and the parameter fusion is performed, the inference speed is greatly improved, but there is obvious loss of inference precision. Then we continue to train the model on this basis. And the results are shown in the Table 7.

It can be seen that after the second training, the network showed excellent precision performance and speed performance, and the accuracy was similar or even slightly improved compared with the parameters before removing instance

Table 6. Inference accuracy with nearest interpolation before and after fusing

	Instance normalization	Batch normalization	Batch normalization after fusing
MAE	93.51	126.48	126.49
MSE	168.26	215.12	215.14
Inference time (s)	5.327	5.314	2.937
Memory occupation (MiB)	1670	1670	1638

Table 7. Inference accuracy after the second training

	Bilinear interpolation	Nearest interpolation
MAE	85.54	86.94
MSE	151.61	160.31
Inference time (s)	2.961	2.937
Memory occupation (MiB)	1638	1638

normalization layer. For bilinear interpolation, the speed is about 45% higher than before, and the effect is obvious. At this point, compared with the original network, the total time of network inference is only 35.4% of that when the accuracy of the inference does not decrease or even rises slightly.

4.5 Adaptive Modification and Deployment in TensorRT

TensorRT is a high-performance deep learning reasoning platform produced by NVIDIA. It only supports the inference process. It increases throughput and reduce latency during inference when deploying deep learning applications to the GPU. For NVIDIA graphics cards, the model derivation speed deployed in TensorRT will have a certain degree of speed improvement compared to other platforms. In this paper, the above-mentioned accelerated network is deployed in TensorRT, which specifically improves the running speed of the network in NVIDIA graphics cards, and lays a foundation for quantification work.

The bilinear interpolation and nearest interpolation methods are not supported in TensorRT, and the output padding (output_padding) parameter in the deconvolution layer is not supported. In order to adapt to the API in TensorRT, the network was partially modified.

First, we replace the interpolation method with the original deconvolution method, and modify the deconvolution according to the meaning of the output padding parameter (output_padding). The output padding parameter is a parameter set in PyTorch for the output size of the deconvolution process. We replace it by adding an additional padding layer. Subsequently, the deconvolution layer is fused with the subsequent batch normalization layer. Because convolution and deconvolution operations have similarities, both can be written in the form of Eq. 3.

$$y = Wx + B \tag{3}$$

For input channels $i_1, i_2, ..., i_m$, output channels $o_1, o_2, ..., o_n$, the convolution operation maps each input channel value to an output channel, the output channel result can be recorded as Eq. 4.

$$y_{o_k} = W_{o_k i_0} x_{i_0} + W_{o_k i_1} x_{i_1} + \cdots + W_{o_k i_m} x_{i_m} + B_{o_k} \tag{4}$$

The deconvolution operation maps an input channel value to each output channel, and the final value of an output channel is the sum of the mapping results, which can be recorded as Eq. 5.

$$y_{o_k} = W_{i_0 o_k} x_{i_0} + W_{i_1 o_k} x_{i_1} + \cdots + W_{i_m o_k} x_{i_m} + B_{o_k} \tag{5}$$

It can be seen that both the convolution and the deconvolution can be written as a linear form in the Eq. 3. Therefore, the deconvolution can also be fused with the subsequent batch normalization layer. The parameter fusion conclusion is the same except that the specific corresponding position of the W internal parameters is different.

The network of 64-channel residual structure with instance normalization removed and bilinear interpolation plus convolution is used as the base network. And the network is retrained after the adaptability modification. Its performance before and after the adaptive modification tested in the PyTorch platform and TensorRT platform is shown in the Table 8. The test set for this test uses a sizing test set of 1024 * 384 built by the Shanghai University of Science and Technology test set.

Table 8. Inference accuracy in PyTorch and TensorRT

	Original network (PyTorch)	Base network before adaptive modification (PyTorch)	Network after adaptive modification (PyTorch)	Network after adaptive modification (TensorRT)
MAE	83.34	68.80	70.29	70.29
MSE	156.96	140.97	154.17	154.17
Inference time (s)	6.710	2.573	2.949	2.242
Memory occupation (MiB)	1068	1026	978	881

The parameter values in the network are all trained in the original data set. It can be seen that the original network inferring time and the memory usage are the worst, while the generalization in the new data set is also poor, and the accuracy performance has a certain gap with other networks. After the adaptive modification, the network in PyTorch is less efficient than the previous network, but more efficient than the previous network in TensorRT. Compared with the base network in PyTorch, the total network inference time of the modified network deployed in TensorRT is reduced by about 13%, and the memory usage is reduced by 14.1%. Compared with the original network, the total time of inference is reduced by about 66.6%, and the inferring efficiency is the initial network's 299.3%, the memory usage is reduced by about 17.5%.

4.6 INT8 Quantization

The data format in the network described in this article is FLOAT32, which occupies 32 bits, while INT8 only occupies 8 bits, which is 1/4 of FLOAT32. INT8 quantization is to map the FLOAT32 data to the INT8 data type in a certain way, and perform convolution operations according to this type. For GPU platforms that support 8-bit computing, this approach can greatly increase the speed of computing.

We performed INT8 quantization on the model in TensorRT [6]. And the results are shown in the Table 9.

Table 9. Inference accuracy before and after INT8 quantization

	Original network (PyTorch)	Network before INT8 quantization (TensorRT)	Network after INT8 quantization (TensorRT)
MAE	83.34	69.80	70.29
MSE	156.96	141.80	154.17
Inference time (s)	6.710	2.242	1.239
Memory occupation (MiB)	1068	881	1100

It can be seen that the quantization further improves the speed of inference without substantially affecting the accuracy. After quantification, the network is 55.3% of the pre-quantization network inference time, which is 18.5% of the initial network inference time, and the inference efficiency is 541.6% of the initial network. While INT8 quantization is only for the convolution process and not the entire network, and needs to restore the data after convolution, this process takes up extra space. Therefore, the memory usage has increased compared to before quantization, which is 124.9% of the network before quantization and 103.0% of the initial network.

5   Conclusion

Based on the existing local crowd behavior analysis system [5], we built a crowd behavior analysis system based on B/S software architecture. This system is optimized in many points. The use of socket transmission system with recovery based on the remote port number improves the efficiency of cross-language communication and reduces the transmission error rate; the combination of frame skipping and dynamic loop sleep time optimizes the frame-taking performance of the system under high load conditions, and the like. Then in the crowd counting network acceleration, the network generator is optimized, including the residual structure channel reduction, the upsampling method replacement, instance normalization removing, etc., so that the inferring speed of the PyTorch platform upgrade to 282.5% of the original network, and the memory usage is reduced by 15.2%. Finally, for the NVIDIA graphics platform, after adapting to TensorRT API support, it is deployed in TensorRT and INT8 quantized. Relative to the initial network, under the premise that the inference accuracy is not reduced, the inference efficiency is 541.6% of the original network.

Acknowledgement. This work was supported in part by National Natural Science Foundation of China (NSFC, Grant No. 61771303 and 61671289), Science and Technology Commission of Shanghai Municipality (STCSM, Grant Nos. 17DZ1205602, 18DZ1200102, 18DZ2270700), and SJTUYitu/Thinkforce Joint laboratory for visual computing and application. Director is funded by National Engineering Laboratory for Public Safety Risk Perception and Control by Big Data PSRPC.

References

1. Gao, Y.: Intensive crowd counting algorithm based on conditional generative adversarial network (2018)
2. Goodfellow, I., et al.: Generative adversarial nets. In: Advances in Neural Information Processing Systems, pp. 2672–2680 (2014)
3. He, K., Zhang, X., Ren, S., Sun, J.: Deep residual learning for image recognition. In: Proceedings of the IEEE Conference on Computer Vision and Pattern Recognition, pp. 770–778 (2016)
4. Isola, P., Zhu, J.Y., Zhou, T., Efros, A.A.: Image-to-image translation with conditional adversarial networks. In: Proceedings of the IEEE Conference on Computer Vision and Pattern Recognition, pp. 1125–1134 (2017)
5. Li, J.: Crowd analysis research for complex video surveillance scenes (2016)
6. Migacz, S.: 8-bit inference with TensorRT (2017). http://on-demand.gputechconf.com/gtc/2017/presentation/s7310-8-bit-inference-with-tensorrt.pdf
7. TehnoKV: Fusing batch normalization and convolution in runtime (2018). https://tehnokv.com/posts/fusing-batchnorm-and-conv
8. Ulyanov, D., Vedaldi, A., Lempitsky, V.: Instance normalization: the missing ingredient for fast stylization. arXiv preprint arXiv:1607.08022 (2016)
9. Zhang, Y., Zhou, D., Chen, S., Gao, S., Ma, Y.: Single-image crowd counting via multi-column convolutional neural network. In: Proceedings of the IEEE Conference on Computer Vision and Pattern Recognition, pp. 589–597 (2016)
10. Zhu, J.Y., Park, T., Isola, P., Efros, A.A.: Unpaired image-to-image translation using cycle-consistent adversarial networks. In: Proceedings of the IEEE International Conference On Computer Vision, pp. 2223–2232 (2017)

Affective Video Content Analysis Based on Two Compact Audio-Visual Features

Xiaona Guo, Wei Zhong$^{(\boxtimes)}$, Long Ye, Li Fang, and Qin Zhang

Key Laboratory of Media Audio and Video (Communication University of China),
Ministry of Education, Communication University of China, Beijing 100024, China
{guoxiaona,wzhong,yelong,lifang8902,zhangqin}@cuc.edu.cn

Abstract. In this paper, we propose a new framework for affective video content analysis by using two compact audio-visual features. In the proposed framework, the eGeMAPS is first calculated as global audio feature and then the key frames of optical flow images are fed to VGG19 network for implementing the transfer learning and visual feature extraction. Finally for model learning, the logistic regression is employed for affective video content classification. In the experiments, we perform the evaluations of audio and visual features on the dataset of Affective Impact of Movies Task 2015 (AIMT15), and compare our results with those of competition teams participated in AIMT15. The comparison results show that the proposed framework can achieve the comparable classification result with the first place of AIMT15 with a total feature dimension of 344, which is only about one thousandth of feature dimensions used in the first place of AIMT15.

Keywords: Affective video content analysis · eGeMAPS · VGG · Optical flow

1 Introduction

In the generation of massive videos, to analyze the video content automatically has attracted more and more attention in recent years. The audience oriented affective content is an important component for the video content analysis and it can be considered as an objective characteristic of the video. The affective video content analysis aims to automatically recognize the emotions elicited by videos, which has been widely used in video indexing, human-computer interaction, emotion-based personalized content delivery and so on. Due to the subjectivity and complexity of human emotions, the affective video content analysis is always a challenging task.

The affective video content analysis mainly includes two processes: feature extraction and feature mapping. The previous research works [1–3] are to use the

This work is supported by the National Natural Science Foundation of China under Grant Nos. 61801440 and 61631016, and the Fundamental Research Funds for the Central Universities under Grant Nos. 2018XNG1824 and YLSZ180226.

© Springer Nature Singapore Pte Ltd. 2020
G. Zhai et al. (Eds.): IFTC 2019, CCIS 1181, pp. 355–364, 2020.
https://doi.org/10.1007/978-981-15-3341-9_29

traditional statistical or machine learning algorithms to build models based on a dataset. Kang [1] first adopted the classifiers for affective analysis, and mapped the low-level features of video data to high-level emotional events using hidden Markov models (HMMs). Wang and Cheong [2] proposed a complementary approach to understand the affective content of general Hollywood movies using two support vector machines (SVMs). Zhang et al. [3] trained two support vector regression (SVR) based models by using both multimedia features and user profiles for personalized music video affective analysis. Recently with the rapid developments of deep learning, more and more research works [4–6] have been proposed based on convolutional neural networks (CNNs). Acar [4] proposed to use CNNs to learn the mid-level representations from automatically extracted low-level features and then employ SVMs for classification. Baecchi [5] exploited sentiment related features and features extracted from models that exploit deep networks trained on face expressions. Yi [6] utilized four CNNs to extract audio, action, object and scene features to analysis the emotional impact of long movies. Furthermore, a multidisciplinary insight of affective video content analysis can be found in [7].

For the affective video content analysis, MediaEval organized competitions including Affective Impact of Movies Task (AIMT) in 2015 [8], Emotional Impact of Movies Task (EIMT) in 2016 [9], 2017 [10] and 2018 [11]. The AIMT2015 focuses on global affective classification using short video clips, EIMT2017 and EIMT2018 focus on continuous affective regression using long video clips, while the EIMT2016 contains both global and continuous affective regression. And thus AIMT2015 is the largest dataset for video affective content classification to the best of our knowledge.

For the classification task of AIMT2015, there are eight teams participated in [12–19]. The first place in the competition was MIC-TJU [12], they chose MFCC for audio feature and used iDT, DenseSIFT, HSH, CNN_M_2048 as visual features. Finally, the linear SVMs were used for classification and voting was employed for linear late fusion. NII-UIT [13] took the second place in the competition, they also used MFCC as audio feature, iDT and ImageSIFT as visual features, and SVM for classification. The third place was ICL-TUM-PASSAU [14], they chose eGeMAPS for audio feature and extracted visual feature by pretrained CNNs. Finally, Random Forests and AdaBoost were used for classification. Other teams [15–19] also used the traditional descriptors such as ColorSIFT, CENTRIST, GIST and other CNNs to extract audio and video features, and then obtained the classification results. It should be noticed that, although some methods given above can achieve good classification results in the arousal-valence space, the feature vectors involved in the audio and video modalities are always of high dimensions, which leads to the problem of large memory requirement and heavy computational burden.

In this paper, a new framework is proposed to generate two compact audio and visual features for affective video content classification. For the audio modality of the proposed framework, we calculate the eGeMAPS with 88 dimensions as global audio feature. And for the visual modality, we finetune VGG19 network by feeding the key frames of optical flow images as inputs, generating the compact deep feature vectors of 256 dimensions. Finally for model learning, the logistic regression (LR) is employed for affective video content classification. In the experiments, the evaluations of audio and visual features are performed on the AIMT15 dataset, and the comparisons are also made among our results and those of competition teams participated in AIMT15. The comparison results show that the proposed framework can achieve the comparable classification result with the first place [12] of AIMT15. Furthermore by comparing the dimensions of feature vectors, our framework employs two compact feature vectors OFVGG and eGeMAPS with a total dimension of 344, only about one thousandth of feature dimensions used in [12].

The remainder of the paper is organized as follows: Sect. 2 introduces the proposed framework including the audio and visual features and model learning. The experimental results and discussions are given in Sect. 3. Finally, some conclusions are drawn in Sect. 4.

2   The Proposed Framework

In this section, we illustrate the proposed framework for affective video content classification, and its main components are shown in Fig. 1. For the audio modality, the eGeMAPS is calculated as global audio feature. And for the visual modality, we extract optical flow images from video frames and then feed the key frames of optical flow images to VGG19 for transfer learning and feature extraction. Finally for model learning, the LR classifier is employed for affective video content classification.

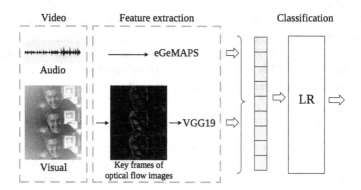

Fig. 1. The proposed framework for affective video content classification.

2.1   Audio Modality

For the audio modality, we utilize eGeMAPS as global audio feature as shown in Fig. 1. In order to facilitate the comparison with other popular features, the

SoundNet is also introduced here and the performance comparison between those two audio features are given in the experiments.

eGeMAPS. eGeMAPS [20] is an extension of the 62-dimensional Geneva Mini-malistic Acoustic Parameter Set (GeMAPS). GeMAPS contains a compact set of 18 low-level descriptors (LLDs). The arithmetic mean and coefficient of variation are applied as functionals to all 18 LLDs, yielding 36 parameters. Additionally, 8 functionals are applied to loudness and pitch, giving a total of 52 parameters. Also, the arithmetic mean of Alpha Ratio and Hammarberg Index, the spectral slopes from 0–500 Hz and 500–1500 Hz over all unvoiced segments, and 6 temporal features are included. In total, 62 parameters are contained in the GeMAPS.

Cepstral parameters have been proven to be highly successful in modeling of affective states and thus in addition to 18 LLDs, 7 LLDs are extended in eGeMAPS. The 7 LLDs include MFCC 1–4, spectral flux and Formant 2–3 bandwidth. The arithmetic mean and coefficient of variation are applied to all 7 additional LLDs to all segments and thus 14 extra descriptors are added. The arithmetic mean of spectral flux in unvoiced regions only, the arithmetic mean and coefficient of variation of spectral flux and MFCC 1–4 voiced regions only are included, resulting in another 11 descriptors. Finally, the equivalent sound level is also included. In total, the eGeMAPS presents a minimalistic set of voice parameters with compact 88-dimensional vectors.

SoundNet. SoundNet [21] is a network which can generate rich natural sound representations and its visualizations suggest some high-level semantics. It proposed a student-teacher training procedure which transfers discriminative visual knowledge from well-established visual models into the sound modality by using unlabeled video as a bridge. It has been shown that the SoundNet representation can yield significant performance improvements on standard benchmarks for acoustic scene/object classification. And thus we also use the features extracted from SoundNet for comparison in the experiments.

2.2 Visual Modality

In the proposed framework shown in Fig. 1, for the visual modality, we first calculate the optical flow images from video frames and then the key frames of optical flow images are extracted by using k-means clustering. With the key frames of optical flow images as input, we implement the transfer learning and feature extraction by finetuning VGG19 model pretrained on the dataset of ImageNet.

Optical Flow Features. The optical flow feature aims to infer the moving speed and direction of an object by detecting the change in the intensity of image pixels with time. Here we calculate the optical flow images from video frames as the motion information.

By defining the brightness of a pixel at (x, y) to be $I(x, y, t)$ with time t, and the object at pixel (x, y) moved to $(x + \Delta x, y + \Delta y)$ in the next frame with time $(t + \Delta t)$, the assumption of constant brightness can be described as:

$$I(x, y, t) = I(x + \Delta x, y + \Delta y, t + \Delta t). \tag{1}$$

By the continuous frame selection and making Taylor series expansion for the right side of Eq. (1), Eq. (1) can be rewritten as

$$I(x, y, t) = I(x, y, t) + \frac{\partial I}{\partial x} \Delta x + \frac{\partial I}{\partial y} \Delta y + \frac{\partial I}{\partial t} \Delta t + H.O.T. \tag{2}$$

And thus it can be concluded that,

$$\frac{\partial I}{\partial x} \Delta x + \frac{\partial I}{\partial y} \Delta y + \frac{\partial I}{\partial t} \Delta t = 0, \tag{3}$$

$$\text{or} \quad \frac{\partial I}{\partial x} \frac{\Delta x}{\Delta t} + \frac{\partial I}{\partial y} \frac{\Delta y}{\Delta t} + \frac{\partial I}{\partial t} = \frac{\partial I}{\partial x} V_x + \frac{\partial I}{\partial y} V_y + \frac{\partial I}{\partial t} = 0. \tag{4}$$

Then we can get V_x and V_y as the optical flow value at $I(x, y, t)$ since they represent the velocity in the directions of x and y separately. In the experiments, the Farneback algorithm is employed to get these vectors and then we transfer them to HSV values and further convert to RGB values to form the final optical flow images.

Key Frame Extraction. It is obviously not feasible to process all the frames in the video as visual information, which results in huge computational burden and excessive information redundancy. Therefore in the proposed framework, we use k-means clustering to extract the key frames of optical flow images to represent the visual information.

For the unlabeled sample set $D = \{x_1, x_2, \ldots, x_m\}$, k-means algorithm aims to seek a reasonable division of the sample set denoted as $C = \{C_1, C_2, \ldots, C_k\}$, where the division C can group similar samples into the same cluster and classify dissimilar objects into different clusters. This process can be formulated as the following optimization problem:

$$\min_C E = \sum_{i=1}^{k} \sum_{x \in C_i} \|x - \mu_i\|_2^2, \tag{5}$$

where μ_i is the center of cluster C_i. And it depicts to some extent the closeness of the samples within the cluster around the cluster mean vector. The smaller the value of E, the higher the sample similarity within the cluster. The closest frames to each of the k cluster center μ_i are selected as the k key frames of a video.

Finetuning Network of VGG19. With the key frames of optical flow images as input, we implement the transfer learning and feature extraction by finetuning VGG19 model. VGG is one of the most frequently used networks in image localization and classification, and it achieves the state-of-the-art results on VOC-2007, VOC-2012, Caltech-101 and Caltech-256 for image classification, action classification and other recognition tasks.

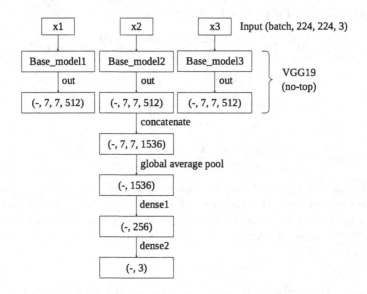

Fig. 2. Finetuning network of VGG19 with three key frames.

Figure 2 shows the finetuning network of VGG19 with three key frames. Since it is very hard to train an entire CNN from scratch with random initialization due to the limited size of data set, and thus we finetune the VGG19 model pretrained on the dataset of ImageNet. It can be seen from Fig. 2 that, three key frames of optical flow images from each of the videos are used to feed the VGG network as inputs, then three outputs of conv5 layer are concatenated and passed through two dense layers to implement the transfer learning. Finally, the outputs of dense1 layer are extracted as deep visual features.

3   Experiments

In the experiments, we first perform the evaluations of audio features and visual features, respectively. Then the evaluation of feature fusion is given and the comparisons with the competition results of AIMT15 are also made. Here all experiments are performed on AIMT15. AIMT15 is the largest dataset for video affective content classification to the best of our knowledge. This dataset is an extension of LIRIS-ACCEDE, including 10900 short video clips extracted from 199 movies. These video clips are ranging from 8 to 12s and are split into a development set of 6144 training/validation videos and a test set of 4756 videos.

3.1 Evaluation of Audio Features

For the evaluation of audio features, we first extract the audio files from AIMT15 dataset using FFMPEG. Then the openSMILE toolkit is used to calculate the eGeMAPS feature vectors by the configuration file 'eGeMAPSv01a.conf'. The SoundNet features are extracted by using the Torch toolbox, and we use the pretrained model and extract the output of conv7 layer as suggested in [21]. Finally, we employ the SVM and LR to perform the classification and the accuracy results are given in Table 1. It can be seen from Table 1, the eGeMAPS outperforms SoundNet features, and the LR modal of eGeMAPS obtains better accuracy than its SVM classification modal.

Table 1. Results of audio features on AIMT15

Audio features	SVM		LR	
	Arousal (%)	Valence (%)	Arousal (%)	Valence (%)
SoundNet	55.551	34.987	55.025	35.072
eGeMAPS	55.803	40.055	**56.981**	**40.517**

3.2 Evaluation of Visual Features

As for visual features, we first extract the images from AIMT15 dataset by using FFMPEG and then the optical flow images are calculated using OpenCV. Here for ease of comparison, the k-means clustering is implemented to extract key frames from both original images and optical flow images by using sklearn.

To finetune the VGG19 model, we use the Keras toolbox with Tensorflow as backend, and the experiments are implemented using a server with 8 Tesla M40 with 24 GB GPU memory for each. We train the dense layers with learning rate of 0.001 and then finetune the conv5 and dense layers with learning rate being 0.0001. The maximum numbers of iterations for both processes are set to be 15360 with batch size of 8. For feature extraction, we extract the vectors from 'dense1' layer, producing a 256-dimensional feature vector for each video. The accuracy results of both SVM and LR classifiers are shown in Table 2.

We can see from Table 2, the deep visual features extracted from VGG19 model finetuned by the key frames of optical flow images, which are denoted as OFVGG in this paper, achieve better results on both SVM and LR than that by the key frames of original images. It suggests that the motion information from optical flow images contributes to the affective video content classification.

3.3 Evaluation of Feature Fusion

Since the videos contain multi-modal contents, we concatenate the audio and visual features together to completely depict the video contents and the results

Table 2. Results of visual features on AIMT15

Visual features	SVM		LR	
	Arousal (%)	Valence (%)	Arousal (%)	Valence (%)
$OVGG$[a]	55.551	37.931	55.467	35.303
$OFVGG$[b]	55.593	**38.025**	**55.593**	36.943

[a]OVGG: feature extracted from VGG19 finetuned with the key frames of original images
[b]OFVGG: feature extracted from VGG19 finetuned with the key frames of optical flow images

of feature fusion are shown in Table 3. It can be seen from Table 3 that, the feature fusion of OFVGG and eGeMAPS together achieves the best result in both arousal and valence dimensions, and their LR modal obtains better accuracy than their SVM classification modal. It indicates the two features OFVGG and eGeMAPS we choose are complementary in this task. Further we can conclude that, the fusion of more features can not necessarily obtain the better result.

3.4   Comparison with the Competition Results of AIMT15

To validate the proposed framework for affective video content classification, we further compare our result with those of competition teams participated in AIMT15 as shown in Table 4. The comparisons are performed in terms of accuracy in arousal and valence dimensions, together with feature dimension.

It can be seen from Table 4 that, the proposed framework can achieve the comparable classification result with the first place [12] of AIMT15. That is to say, compared to the first place [12] of AIMT15, the accuracy of our framework is 1.072% higher in the arousal dimension (57.002 versus 55.930) and 1.117% lower (40.833 versus 41.950) in the valence dimension. Furthermore by comparing the dimensions of feature vectors, the proposed framework employs two compact

Table 3. Results of feature fusion on AIMT15

Multi-model fusion	SVM		LR	
	Arousal (%)	Valence (%)	Arousal (%)	Valence (%)
OVGG+eGeMAPS	55.551	37.931	49.622	34.504
OVGG+SoundNet	55.551	37.447	48.738	34.462
OVGG+eGeMAPS+SoundNet	55.551	37.258	55.299	38.331
OFVGG+eGeMAPS	55.824	39.992	**57.002**	**40.833**
OFVGG+SoundNet	55.551	34.756	55.299	34.273
OFVGG+eGeMAPS+SoundNet	55.509	38.204	55.467	38.856
OVGG+OFVGG+eGeMAPS	55.656	38.604	55.404	37.952
OVGG+OFVGG+SoundNet	55.551	36.922	49.054	34.840
OVGG+OFVGG+eGeMAPS+SoundNet	55.572	37.321	55.635	36.964

Table 4. Comparisons with competition results of AIMT15

AIMT15 results	Arousal (%)	Valence (%)	Feature dimension
TCS-ILAB [19]	48.949	35.660	–
RFA [18]	45.038	36.123	9448,4096
UMons [17]	52.439	37.279	–
KIT [16]	51.892	38.541	113664
Fudan-Huawei [15]	48.844	41.779	23152
ICL-TUM-PASSAU [14]	55.719	41.484	1388
NII-UIT [13]	55.908	41.653	126464
MIC-TJU [12]	55.930	**41.950**	420864
Ours	**57.002**	40.833	**344**

feature vectors OFVGG and eGeMAPS with a total dimension of 344, only about one thousandth of feature dimensions used in [12].

4 Conclusion

In this paper, we propose a new framework utilizing two compact audio-visual features for affective video content classification. In this framework, the eGeMAPS is chosen as the audio feature and the VGG19-based network is fine-tuned by using the key frames of optical flow images for visual feature extraction. Then we employ LR as the affective video content classifier. The experimental results on AIMT15 show that the proposed framework can obtain the comparable accuracy with the first place [12] of AIMT15. Furthermore, our framework employs two compact feature vectors OFVGG and eGeMAPS with a total dimension of 344, only about one thousandth of feature dimensions used in [12].

References

1. Kang, H.-B.: Affective content detection using HMMs. In: Proceedings of the 11th ACM International Conference on Multimedia, pp. 259–262 (2003)
2. Wang, H.L., Cheong, L.-F.: Affective understanding in film. IEEE Trans. Circuits Syst. Video Technol. **16**(6), 689–704 (2006)
3. Zhang, S., Huang, Q., Jiang, S., Gao, W., Tian, Q.: Affective visualization and retrieval for music video. IEEE Trans. Multimed. **12**(6), 510–522 (2010)
4. Acar, E., Hopfgartner, F., Albayrak, S.: Understanding affective content of music videos through learned representations. In: Gurrin, C., Hopfgartner, F., Hurst, W., Johansen, H., Lee, H., O'Connor, N. (eds.) MMM 2014. LNCS, vol. 8325, pp. 303–314. Springer, Cham (2014). https://doi.org/10.1007/978-3-319-04114-8_26
5. Baecchi, C., Uricchio, T., Bertini, M., Bimbo, A.D.: Deep sentiment features of context and faces for affective video analysis. In: Proceedings of the ACM International Conference on Multimedia Retrieval, pp. 72–77 (2017)

6. Yi, Y., Wang, H.-L., Li, Q.Y.: CNN features for emotional impact of movies task. In: MediaEval (2018)
7. Baveye, Y., Chamaret, C., Dellandréa, E., Chen, L.M.: Affective video content analysis: a multidisciplinary insight. IEEE Trans. Affect. Comput. **9**(4), 396–409 (2018)
8. Sjöberg, M., Baveye, Y., Wang, H.L., Quang, V.L., Ionescu, B., et al.: The MediaEval 2015 affective impact of movies task. In: MediaEval (2015)
9. Dellandréa, E., Chen, L. M., Baveye, Y., Sjöberg, M. V. and Chamaret, C.: The MediaEval 2016 emotional impact of movies task. In: MediaEval (2016)
10. Dellandréa, E., Huigsloot, M., Chen, L.M., Baveye, Y., Sjöberg, M.: The MediaEval 2017 emotional impact of movies task. In: MediaEval (2017)
11. Dellandréa, E., Huigsloot, M., Chen, L.M., Baveye, Y., Xiao, Z.Z., Sjöberg, M.: The MediaEval 2018 emotional impact of movies task. In: MediaEval (2018)
12. Yi, Y., Wang, H.L., Zhang, B.W., Yu, J.: MIC-TJU in MediaEval 2015 affective impact of movies task. In: MediaEval (2015)
13. Lam, V., Le, S.P., Le, D.D., Satoh, S., Duong, D.A.: NII-UIT at MediaEval 2015 affective impact of movies task. In: MediaEval (2015)
14. Trigeorgis, G., Coutinho, E., Ringeval, F., Marchi, E., Zafeiriou, S., Schuller, B.W.: The ICL-TUM-PASSAU approach for the MediaEval 2015 affective impact of movies task. In: MediaEval (2015)
15. Dai, Q., et al.: Fudan-Huawei at MediaEval 2015: detecting violent scenes and affective impact in movies with deep learning. In: MediaEval (2015)
16. Vlastelica, P.M., Hayrapetyan, S., Tapaswi, M., Stiefelhagen, R.: KIT at MediaEval 2015-evaluating visual cues for affective impact of movies task. In: MediaEval (2015)
17. Seddati, O., Kulah, E., Pironkov, G., Dupont, S., Mahmoudi, S., Dutoit, T.: UMons at MediaEval 2015 affective impact of movies task including violent scenes detection. In: MediaEval (2015)
18. Mironica, I., Ionescu, B., Sjöberg, M., Schedl, M., Skowron, M.: RFA at MediaEval 2015 affective impact of movies task: a multimodal approach. In: MediaEval (2015)
19. Chakraborty, R., Kumar Maurya, A., Pandharipande, M., Hassan, E., Ghosh, H., Kopparapu, S.K.: TCS-ILAB-MediaEval 2015: affective impact of movies and violent scene detection. In: MediaEval (2015)
20. Eyben, F., Scherer, K.R., Schuller, B.W., Sundberg, J., Andre, E., et al.: The Geneva Minimalistic Acoustic Parameter Set (GeMAPS) for voice research and affective computing. IEEE Trans. Affect. Comput. **7**(2), 190–202 (2016)
21. Aytar, Y., Vondrick, C., Torralba, A.: SoundNet: learning sound representations from unlabeled video. In: Advances in Neural Information Processing Systems, pp. 892–900 (2016)

Virtual Reality

Image Aligning and Stitching Based on Multilayer Mesh Deformation

Mingfu Xie[1,2], Jun Zhou[1,2(✉)], Xiao Gu[1,2], and Hua Yang[1,2]

[1] Shanghai Key Laboratory of Multi Media Processing and Transmissions,
Shanghai Jiaotong University, Shanghai, China
{mingthix,zhoujun,gugu97,hyang}@sjtu.edu.cn
[2] Institute of Image Communication and Network Engineering,
Shanghai Jiaotong University, Shanghai, China

Abstract. This paper aims to solve the problem of strong disparity in image stitching. By studying current mesh deformation methods based on single grid, we propose an image stitching framework based on multilayer mesh deformation for aligning regions in different layer. With development of image depth perception and semantic segmentation technology, we can get layering maps of images or photos expediently. We introduce images representation with layers and get layer corresponding by using depth or disparity information for large parallax scenarios. Registration of each layer is carried out independently. To ensure the integrity of layer synthesis results, we apply deformation with translation and scaling compensation between different layers before blending. The experiment demonstrates that our method can adequately utilize the prior information in layering maps to decouple 2D transformation between different layers, finally achieve outstanding aligning performance in all layers and naturalness in complete stitching result.

Keywords: Image stitching · Image layering · Mesh deformation · Seamless blending

1 Introduction

Immersive experience lately become considerably popular at all ages. We can get more living visual and auditory enjoyment through this new way of generating and presenting content, such as AR/VR application and games, art exhibitions combined with digital technology, panoramic video capturing and playback. Related software algorithms and hardware devices are also developing fast, among which stitching technology used to build panoramic images is crucial for these application scenarios. Image stitching is a process of fusing a group of multi-view images having overlapping areas into a *natural* image having a larger perspective through alignment. In this task, *natural* mainly refers to get a stitched image without ghost, excessive distortion and structure damage after registration and fusion. It is mainly to solve the registration problem of matching all same elements in images captured from different viewpoints. Registration

© Springer Nature Singapore Pte Ltd. 2020
G. Zhai et al. (Eds.): IFTC 2019, CCIS 1181, pp. 367–383, 2020.
https://doi.org/10.1007/978-981-15-3341-9_30

algorithm can be divided into two kinds: parametric and non-parametric. Optical flow is a typical non-parametric registration algorithm, which directly estimates the displacement of each pixel between two frames. By contrast, the parametric algorithm uses a global 2D or 3D motion model to fit the alignment relation of two images. In current works, there are many sticky problems, including: insufficient quantity or poor quality of feature points; excessive parallax; having a large foreground object; too small areas of overlap, etc. Based on the research on the current mainstream image stitching algorithms and their existing problems, we propose a corresponding algorithm with the prospect of large parallax in this paper.

1.1   Related Works

As a milestone of image matching and stitching algorithm, Brown et al. [2] process images by pre-projection, extraction and matching SIFT feature points, then adopt bundle adjustment to solve constraint optimization in parallel to get a global transformation, and finally the stitching result is obtained by multiple band blending.

Global registration often fails to perfectly align the entire overlapping area of input images. It may be caused by many factors, such as unmodeled radial distortion, 3D parallax, small-scale or large-scale motion of object in the scene. Small 3D parallax and small motion can be handled very well with local adjustment. Shum et al. [19] estimated a dense modified displacement field by optimized global registration, and superimposed the correction value on the original image coordinates by bilinear interpolation to register some local elements. Gao et al. [9], first divide the scene into foreground and background planes corresponding to two homography matrices separately, then smooth the registration of the foreground and background boundaries by the weighted combination of the two matrices (DHW). This double homography matrixes approach greatly enhances the global and local registration ability of the parametric approach over the entire image. Inspired by this grouping local homography method, a number of multi-homographies registration methods based on spatial variation have been proposed.

After that, Lin [16] used multi-homographies aligning method, Smoothly Varying Affine (SVA) transformation, to improve the local alignment ability in the entire scene, and it can handle parallax to some extent. Zaragoza et al. [23] proposed the As-Projective-As-Possible image stitching method (APAP), which divides the image into a set of dense grids, which each grid corresponds to a homography matrix. And at the same paper they design an efficient calculation method name Moving DLT. Ever since then, the idea of mesh deformation that commonly used in graphic science begins popular to be used for registering and stitching, and formalizes the image registration as a grid optimization problem. Various optimization terms have been proposed to solve different problems. Zhang [24] and others present similarity constraint term according to classic video debounce methods called Content-Preserving Warps (CPW). And Gao et al. [10] propose optimal geometric transformations using seam finding to improve the stitching effect in large parallax scenes, and they also design optimization terms for curve structure protection to get better image content. These algorithms above can well register most areas in the image. And algorithms based on grid optimization greatly improve the stitching result of complicated scenes. Although their local optimization provides the registration image more freedom, the picture naturalness should be considered.

Recently, for correcting shape perspective, Chang [4] proposed the Shape-Preserving Half-Projective warps (SPHP) to reduce the transformation distortion. However, it is difficult to reduce the distortion of the straight lines in scenes having lots of them. He et al. [3] proposed linear structure constraints to correct the distorted straight line in the stitched image while ensuring the global naturalness, based on the Natural Image Stitching algorithm (NIS) [7]. Lin et al. [14] proposed the Adaptive As-Natural-As-Possible (AANAP) to preserve structure naturalness. Lin et al. [15], inspired by the optical flow method, instead of using the feature point alignment constraint term, realized grid-based optimization mainly by Direct Photometric Aligning (DPA) term [5].

1.2 Motivation

In general, regions with small parallax can't be aligned completely based on methods using single global transformation, usually generating ghosting while applying linear weights to blend. Mesh deformation based methods including GCPW, DHW, SVA, APAP, NIS, DPA and etc. can easily solve this problem. Optimized mainly by a feature point alignment term or photometric error term, with a variety of specific content structure or target protection constraints, a perfect 2D grid to warp images can be obtained for aligning. But these methods will often get in trouble with handling large parallax, especially in the case of too far distance between background and foreground. Because of parallax, the following confusing situation is likely to happen: in order to align foreground, the grid needs get left-shift deformation; in order to aligning background, the grid needs get right-shift deformation. Figure 1 shows a simple example. Even we use seam-cutting for post-processing, the distortion cannot be reshaped and in some situation, aligning errors in the scenario can not be eliminated in final stitching result.

2 Proposed Method

General image stitching algorithm has two stages: image aligning and image blending. Image aligning is the foremost part of the process. It can be divided into two stages basically: rough global registration and refined local registration. Rough global registration includes the following steps: feature points extracting, matching, feature correspondences refining, establishing parametric equations through interior point after purification according to 2D image transformation model, and obtaining precise registration relationship, while imaging conforming to the single planar condition and optical center consistence. For scenario with

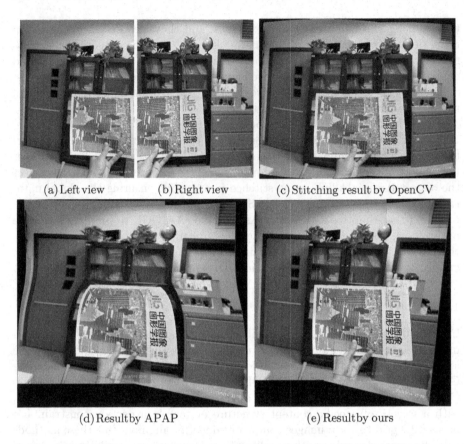

(a) Left view   (b) Right view   (c) Stitching result by OpenCV

(d) Result by APAP   (e) Result by ours

Fig. 1. Stitching result of a sample having large disparity with manual layering

a wide range of DOF or parallax, you need to adopt local registration based on spatial varying deformation to achieve the registration of whole image. The key of local registration refining is restricting the affect on local region by respective homography. With image division by grid, each part of image can be transformed by different homography matrices. In this paper, we propose a method called multilayer mesh deformation, adopting the parallax map or depth map as prior information, to represent the input images by segmenting foreground and background region as different layers, then aligning all layers simultaneously and finally synthesizing each layer to obtain the wide angle result. The whole framework of our method is shown as Fig. 2.

2.1   Disparity-Based Image Layering

To find image segmented layers according to different depth and obtain the corresponding, we can use the existing depth estimation algorithms and devices, existing state-of-the-art semantic segmentation algorithm. We adopt the result

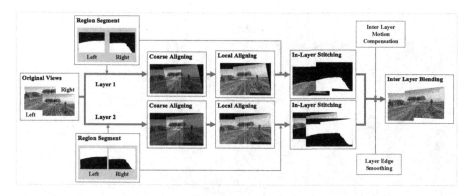

Fig. 2. Our multilayer mesh deformation framework

of these methods as prior information. According methods including optical flow algorithm by using binocular parallax [1] or monocular depth estimation method based on the deep learning [21], we can obtain the depth/disparity map of images with some parallax. In recent applications, there are also many devices that can scan to get the depth map of scenario by using laser radar, infrared, structural light and ect., such as Leap Motion, Kinect [22], RealSense series (Intel), Structure Sensor (Apple iPad), Structural Light Depth Camera [8] and so on. As shown in Fig. 3, through the latest researching, it is relatively easier to obtain high-quality depth information for images or photos in general scenes. We directly take the accurate depth maps obtained by these methods as original input of our method. At the same time, we can also mark different layers manually and assign its corresponding relationship by user, so as to get a more accurate region segmentation boundary. So that we can get a better aligning and blending result as Fig. 1 shows.

For the scene with large foreground, the image can be divided into foreground object and background area. The geometric constraints between two planar regions with different depths are actually very weak, so the constraints between regions with different depth in 2D image can be reduced by segmentation and post-processing. Substantially, getting layering representation is to remove the redundant constraints between different regions, which heavily impact the 2D grid deformation. Images of the view i can be represented as Eq. 1, with layer mask denoted by $M_{i,l}$.

$$I_i = M_{i,1} \times I_{i,1} + M_{i,2} \times I_{i,2} \tag{1}$$

Other segmentation algorithm also works, such as DeepLabv3 for semantic segmentation [6]. After layer segmentation for reference and candidate images, we just need to search and establish the corresponding relationship of each layer in two images determined by feature correspondences and parallax continuity in a same plane. Feature correspondences can roughly estimate corresponding of layers in different image. By merging and re-marking the segmented regions

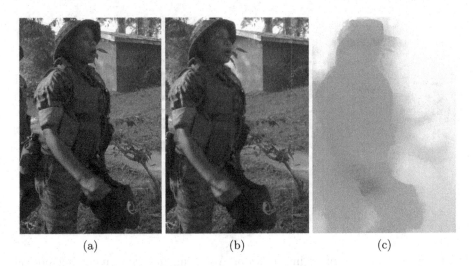

(a)   (b)   (c)

Fig. 3. Sample images (a) (b) and edge-aware flow field (c) from paper [1]

according with parallax continuity condition, we can obtain the accurate layered regions and their refined correspondences.

In order to maintain the integrity of the plane contents with different depths in the final fusion output and keep the smoothness and naturalness of the blending result, we smooth the layers synthesis edges of each layer's mask and padding several pixel inflate along with it. In the experiment of this paper, we take the region segmentation and stratification obtained by depth mask as the base for subsequent processing. Our method has pretty robustness under the condition that the deviation between the segmentation edge and the real depth step boundary conforms to the two-dimensional Gaussian distribution with zero mean and small variance.

2.2   Multilayer Mesh Deformation

For the sake of alleviating the registration error of the subsequent processes, we firstly register the different regions roughly respectively, then use this registration as the initialization for the latter mesh deforming. For each layer of image to be registered, SIFT feature points are firstly extracted, and the KD-tree algorithm is used to get rough matching, and then we adopt the RANSAC algorithm to eliminate the mismatched corresponding pairs [2]. Each image layer's content is basically laying in same plane, so we apply a parametric 2D homography for fitting geometric transformation between images that need to be registered.

To deal with these areas in large parallax scenes, we adopt a grid-based parametric method, which can achieve excellent performance for local content alignment. In particular, for the case with large foreground objects, we achieve local registration of images by the multilayer mesh deformation method. In this paper, we use the vertices position offset between before deforming and after deforming as the optimized variable, so that we can use some iterative algorithms to get a more accurate local aligning results [15]. For all mesh of any layered grid, we set the position coordinates of the four vertices before and after the deformation as V_k and $\hat{V}_k$. Accordingly, the position offset of each vertex can be expressed as:

$$\tau_k = \hat{V}_k - V_k$$

We use one Grid including $M \times N$ Meshes (quads) for each layer's deforming. Mesh deformation is often used to handle certain parallax and register local regions. However, in some cases, such as at the area where the depth gradient is large, even if the feature correspondences is abundant to get enough constraint terms, the general mesh deformation still cannot fully register all regions, and sometimes introduces worse result. As an example shown in Fig. 1, this problem becomes especially serious in cases having foreground occlusion. Essentially, it is all because that at the depth phase step region, no matter how dense the mesh is, the feature alignment term, luminosity error term, and the smoothing constraint term for deformation will all get to optimization conflict. E_{LF}, E_{GF}, E_{LS} denote respectively: local feature aligning term, global feature aligning term, and local similarity constraint term, As Eqs. 2, 3 and 6 show.

$$E_{LF} = \sum_{n=0}^{N-1} \|\hat{p} - \hat{q}\|^2$$

$$= \sum_{n=0}^{N-1} \left\| \sum_{k=0}^{3} \alpha_k \hat{V}_k - \sum_{k=0}^{3} \alpha'_k \hat{V}'_k \right\|_n^2 \tag{2}$$

$$= \sum_{n=0}^{N-1} \left\| \sum_{k=0}^{3} \alpha_k (V_k + \tau_k) - \sum_{k=0}^{3} \alpha'_k (V'_k + \tau'_k) \right\|_n^2$$

$$E_{GF} = \sum_{v=0}^{V-1} \sum_{m=0}^{M-1} \|\hat{\tau}\|_{m,v}^2 \tag{3}$$

In order to ensure that the entire mesh deform smoothly based on the spatial position, it is necessary to introduce a smoothing term [17] to constrain the deformation between adjacent meshes. To achieve this, firstly, we triangulate the mesh [17]. Each quadrilateral mesh can be decomposed into four different triangles. Assume the vertices of each triangle are V_1, V_2 and V_3. Each vertex of a triangle can be represented by a linear combination of the other two vertices. Its linear combination coefficient can be get according to the following equation when triangulated, as Eq. 4 expresses.

$$V_1 = V_2 + u (V_3 - V_2) + vR_{90} (V_3 - V_2), R_{90} = \begin{bmatrix} 0 & 1 \\ -1 & 0 \end{bmatrix} \tag{4}$$

Secondly, it is necessary to ensure that the mesh deformation satisfy the similar transformation, that is to say, the vertices of the four triangles after the quadrilateral mesh is triangulated still have the same linear relationship above represented by u and v, that is

$$\hat{V}_1 = \hat{V}_2 + u\left(\hat{V}_3 - \hat{V}_2\right) + vR_{90}\left(\hat{V}_3 - \hat{V}_2\right) \tag{5}$$

Then convert the Eq. 5 into an error function constraint and take the vertex offset as the optimized variable. The constraint formula can be transferred to Eq. 6.

$$E_{LS} = \sum_{m=0}^{M-1}\sum_{t=0}^{T-1} \|\hat{\tau}_1 - (\hat{\tau}_2 + u(\hat{\tau}_3 - \hat{\tau}_2) + vR_{90}(\hat{\tau}_3 - \hat{\tau}_2))\|_{t,m}^2 \tag{6}$$

Here, we will construct mathematical model for a basic case of problem occurring above. Considering about a single grid initialized by a global transform, when it needs to register the local regions, the depth distribution of the local regions has the following conditions: continuous and relatively smooth (slight fluctuations); continuous but with sudden change in slope; having a discontinuous step (large depth phase step). These situations at these local regions can be illustrated as Fig. 4, with the red dot line denoting the inspected coordinate as d_0.

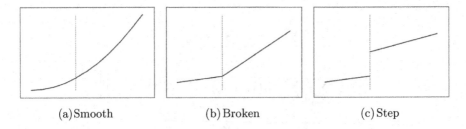

(a) Smooth   (b) Broken   (c) Step

Fig. 4. Disparity distribution categories (Color figure online)

As we know, the first case is easy to be solved by a spatial varying warping. The second case can be optimized by increasing the mesh density, as long as the region feature point pairs is enough and mesh density used for alignment is sufficient. After that the parallax in the mesh near d_0 becomes similar as the first case. For the third case, the parallax step can not be eliminated neither by adding feature point pairs, nor increasing the mesh density.

For the mesh near the depth phase step regions, optimizing vectors (direction and size) of the mesh deformations by feature correspondences at different depth usually contradicts each other, because they are related to different optimization targets. Therefore, it is necessary to perform layer-based mesh deformation into different layers respectively. Compared with DHW, CPW, APAP and other local registration algorithms based on single-grid mesh deformation that can only deal with the first and second local parallax distribution, we can generally handle all these three cases especially the third one which is the most difficult to align well in all regions, as Fig. 5 shows.

For all grids in different layer, we can get the combination of these three optimization terms according to [14,15]:

$$E_\tau = \sum_{l=0}^{L-1} \|w_{LF}E_{LF} + w_{GF}E_{GF} + w_{LS}E_{LS}\|_l^2 \tag{7}$$

Where L is the number of segmented layers, we will optimize the mesh of all layers simultaneously. In this paper, we use parameter $L = 2$ for fundamental demonstration, each coefficient is the corresponding weight of a constraint item. For multilayers cases ($L \geq 3$), we can get deformation and synthesis layer by layer.

(a) Deformed view 1 of layer 1   (b) Deformed view 2 of layer 1

(c) Synthetic intermediates of layer 1.   (d) Synthetic intermediates of layer 2

Fig. 5. Mesh deformation intermediate result of layers and views

There is no significant conflict between the terms, credited to the multilayer mesh deformation of the layered images. Only guaranteeing the priority of the local feature alignment item and the smoothing item is generally enough. Set the Jacobi matrices and residual vectors of all the weighted error as A and b. This is a set of linear equations whose solution can be obtained by least squares method, for $A\tau = b$. All grids for each layer can be optimized simultaneously.

Update the solution of this iteration by the vertex offset τ to the last vertex coordinates to obtain the current vertexes coordinate after mesh deformation.

$$V_{current} = V_{previous} + \tau \tag{8}$$

2.3   Seamless Blending

After multilayer mesh deformation, we get a set of grids that register each layer of images. Apply the corresponding mesh deformation to each image to be registered, then stitch related in-layer content firstly, and get inter-layer blending finally. The following method analysis is proposed according to the different requirements within these two phases.

Intra-layer Synthesis. We get the in-layer stitching result by center-seam cutting. For images with good registration, the parallax step problem has been solved in most region by layered mesh deformation. In general, there will be no repeating large objects caused mismatching in the fusion result. For image after local registration by mesh deformation, most background and foreground content have been registered well. But it is still not fully registered at the layer edge or some fully occluded areas. Optimizing constraints by introducing the photometric error [15] can improve final impression in this situation. In this paper, according to [13], we use a center-seam cutting method to stitch the remaining parts that are not completely fitted.

Inter-layers Synthesis. Firstly, we need introduce global inter-layer motion estimation and compensation. Multilayer mesh can alleviate the single-layer mesh optimization conflict problem, since each layer is independently initialized and applying mesh deformation. For this reason, relative motion may occur between layers, it is necessary to simultaneously solve the relative motion problem while merging layers into final result. We must keep the content of different regional layers remain relatively in the original position after the registration and reduce the black holes after inter-layer synthesis. This can be achieved by estimating and compensating an appropriate inter-layer motion.

We calculate two gravity centers of the foreground region and invert of the background region (corresponding to the foreground region) by layer mask after mesh deformation and use their difference as a rough estimation of the relative inter-layer translation motion. Meanwhile, we use the area ratio as the scaling motion estimation. We use a function $GC(Mask)$ to denote the Gravity Center of a binary Mask. Then we use a warp function $warp(*)$ to denote the final mesh deformation result, where $warp(x)$ are the point coordinate of x after the mesh deformation, and $warp_j(I_i)$ indicate the resulting image of image in ith viewpoint after mesh deformation corresponding to the j-th viewpoint. $A(Mask)$ is used to denote the total non-zero region area of the $Mask$. Therefore, the translational displacement and scaling factor of the layer-1 relative to the layer-2 are expressed as Eqs. 9 and 10.

$$\Delta T = GC\left(\text{warp}_1\left(\text{Mask}_{L1}\right)\right) - GC\left(\text{warp}_2\left(\text{Mask}_{L1}\right)\right) \tag{9}$$

$$s = \sqrt{S_1/S_2} = \sqrt{A\left(\text{Mask}_{L1}\right)/A\left(\text{warp}_2\left(\text{Mask}_{L1}\right)\right)} \tag{10}$$

Then we can obtain inter-layer motion compensation as Eq. 11 shows.

$$H_M = \begin{bmatrix} 1 & 0 & \Delta T_x \\ 0 & 1 & \Delta T_y \\ 0 & 0 & s \end{bmatrix} \tag{11}$$

Because the edges obtained by segmentation algorithm are not enough smooth and precisely attached to the object edge, it is necessary to eliminate the aliasing and flaws of the layer edges using a good fusion algorithm during the inter-layer blending stage. Human eyes are sensitive with the image gradient, especially the edges of subject content, such as brightness changing and color aliasing of which will seriously affect the quality of final result.

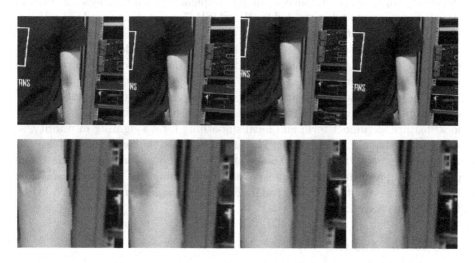

Fig. 6. Top: final blending result with blur kernel size $k = 1$, 3, 5, 7 (*pixels*); Bottom: inset of top respectively

As Fig. 6 shows, the kernel size for layer boundary smoothing is 1 (no smoothing), 3, 5, 7 pixels. We can know from the result that the smaller smooth kernel size, the more jags in blending result, which damage the image naturalness. On the contrary, oversize kernel not means always best, and it may cause blurry blending result, in which boundary between the foreground and background cannot be distinguished. Overall, as shown in Fig. 9, the complex background and foreground regions in the image can be registered accurately. And the final synthetic image is also visually natural without obvious artifacts, ghosting and aliasing at the fusion edge.

3   Experiment and Result

Our algorithm runs on operating system Windows 10 64-bit, with Intel Xeon 3.3 GHz CPU and 16 GB RAM. To demonstrate the ability of our proposed method, as shown below, we implement a basic multilayer mesh deformation framework without complicate pre-processing.

3.1 Implement Details

The experiments in this paper mainly compare the result of each algorithm in large parallax scenario. The following shows and compare the stitching results of algorithm in this paper and following mainstream algorithms: AutoStitcher, a classic global homography estimation algorithm and APAP, a classic single-grid mesh optimization algorithm. The source code is provided by the module in [7]. The classical algorithm of global motion estimation is implemented based on the *Sitcher* component in OpenCV [39], and the relevant parameters are kept in the default options. APAP algorithm gets implementation by adopting codes of APAP-Stitching module provided by NISwGSP [7] open source, and the relevant parameter is set to the default state.

Due to the dense grid, it will take a long time to render the result subsequently. Considering both time consuming and visual effect, we set the mesh density of each layer of grid as 20×20 in our implement, with each size of the grid in accordance with the input image adaptive. The relative weight of different optimization item effects on the final result, but we just need to keep the local feature alignment term and local smoothing term weighting in priority, so as to get a feasible aligning result. Therefore, we share the weight of each layer as $w_{LF} = 0.8$, $w_{GF} = 0.3$, $w_{LS} = 0.6$. Considering the sparse matrix calculating in optimization, and that the complexity of directly using the least-square method is too high, we use the Conjugate Gradient Least Squares method provided by Eigen [11] to solve it.

3.2 Result Analysis

For example shown in Fig. 7, the OpenCV-Stitcher based on single homography fails to align two image planes at the same time, resulting in dislocation of the figure in the foreground. Algorithms similar to APAP cannot deal with the boundary deformation disorder by constraints contradiction. Meanwhile, our algorithm achieves the currently optimal performance under this situation. Our method obtain the accurate registration of different regions, with quite natural transition at the depth phase step area (Fig. 8).

For example as shown in Fig. 1 at the beginning of the article and Fig. 12, large parallax in the image has a serious impact on registration and stitching, due to the presence of foreground subject.

The component of OpenCV can only register one planar region of images, and the other plane will have serious mismatch because of the large parallax. Usually small parallax can be eliminated by using multiple band blending or seam cutting methods in the layer edge as the Fig. 1 shows the results on the left side of the head and body dislocation, the right is the man's head after the sutures, face completely abandoned the content of the image of the final result is terrible. In APAP-class algorithm, although most of the image area have been

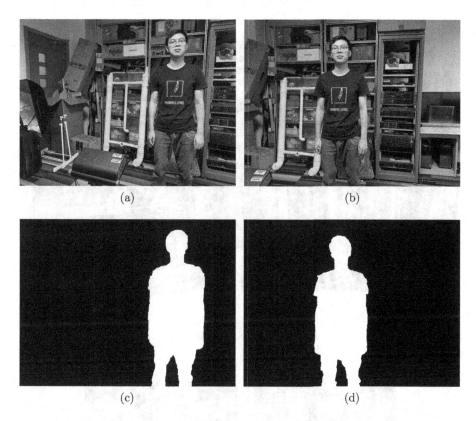

(a)   (b)

(c)   (d)

Fig. 7. Sample with salient object and messy background. (a), (b): input views. (c), (d): layer mask and its corresponding (white to white, black to black). Layering masks are generated by DeepLabv3 [6].

(a)   (b)

Fig. 8. Sample result: (a) by OpenCV, (b) by APAP

Fig. 9. Sample result by ours

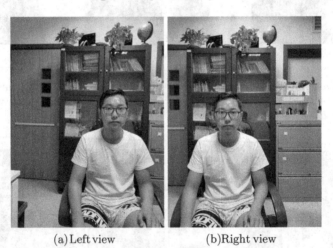

(a) Left view   (b)Right view

Fig. 10. Images with large disparity. There is large relative motion between foreground and background after globally aligning.

registration, but the edge of person appears registration disorder: it would like to register background and foreground object at the same time, but failed at the same time (Figs. 10 and 11).

By linear blending, we can see serious ghosting around the person. Even if the seam cutting method can eliminate ghosting, but content and structure in the distorted image has been destroyed. The seam finding algorithm based on

Fig. 11. Results by OpenCV. It shows two categories of seam-cut failures.

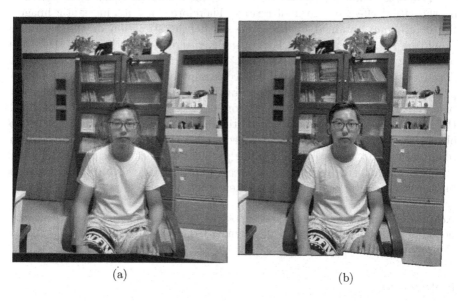

(a)   (b)

Fig. 12. Stitching result. (a): by APAP. (b): by ours

content aware [18] can partly solve this problem, which let the seam bypass the foreground content, however, in the beginning case we can see, it will fail while the prospects of image content (book) across the non-overlap with the overlap area. For such example, after registration background, there is no one available seam to bypass the foreground content for final stitching. Only by decoupling the image in different layers we can get accurate registration and perfect fusion for regions in different depth such as our result in Fig. 12.

4   Conclusion and Future Work

Through the research on disparity distribution of the images that need to be registered for stitching, we propose a framework to decompose different layers of the large parallax image based on the layering by depth map or semantic segmentation, and applying a multilayer mesh deformation algorithm for local registrations, so as to solve the problem of misregistration usually caused by existing algorithms which only use single-layer mesh deformation. In our method, in order to obtain stratified results of the final output, we use a center-seam cut method for intra-layer stitching, and estimate relative motion between layers and then compensate it in the final fusion step. We demonstrate through experiments that our method can achieve better results in the scenario with large foreground or large parallax by comparing with the existing single-layer mesh deformation algorithms and classic global registration algorithm.

As analysis shows above, the accuracy of the layer edge had a great influence on the visual effect of the final results. Therefore, for stitching static images, the depth map obtained by the preposed depth estimation or the hardware should be relatively accurate. For example, the related work in [22], the time redundancy of video sequence will also help improve the edge accuracy of the depth map. Using multi-dimensional constraint edge precision will also be a productive researching direction, combined with the recent work, such as super pixel integral [20], semantic segmentation [6,12] or heat map. At the same time, the introduction of layering in video stitching task will also be helpful [26] to obtain the motion information of objects, and the redundant information in temporal can be used to eliminate the problem of black holes caused by view occlusion while fusing result. We can keep on promoting this algorithm through further research work on prospect as mentioned above.

Acknowledgement. The paper was supported by Science and Technology Commission of Shanghai Municipality (STCSM) under Grant 18DZ1200102 and NSFC under Grant 61471234, 61771303.

References

1. Anderson, R., et al.: Jump: virtual reality video. ACM Trans. Graph. (TOG) **35**(6), 198 (2016)
2. Brown, M., Lowe, D.G.: Automatic panoramic image stitching using invariant features. Int. J. Comput. Vis. **74**(1), 59–73 (2007)
3. He, C., Zhou, J.: Mesh-based image stitching algorithm with linear structure protection (2018)
4. Chang, C.H., Sato, Y., Chuang, Y.Y.: Shape-preserving half-projective warps for image stitching. In: Proceedings of the IEEE Conference on Computer Vision and Pattern Recognition, pp. 3254–3261 (2014)
5. Chen, K., Tu, J., Yao, J., Li, J.: Generalized content-preserving warp: direct photometric alignment beyond color consistency. IEEE Access **6**, 69835–69849 (2018)
6. Chen, L.C., Papandreou, G., Schroff, F., Adam, H.: Rethinking atrous convolution for semantic image segmentation. arXiv preprint arXiv:1706.05587 (2017)

7. Chen, Y.-S., Chuang, Y.-Y.: Natural image stitching with the global similarity prior. In: Leibe, B., Matas, J., Sebe, N., Welling, M. (eds.) ECCV 2016. LNCS, vol. 9909, pp. 186–201. Springer, Cham (2016). https://doi.org/10.1007/978-3-319-46454-1_12
8. Furukawa, R., Sagawa, R., Kawasaki, H.: Depth estimation using structured light flow-analysis of projected pattern flow on an object's surface. In: Proceedings of the IEEE International Conference on Computer Vision, pp. 4640–4648 (2017)
9. Gao, J., Kim, S.J., Brown, M.S.: Constructing image panoramas using dual-homography warping. In: CVPR 2011, pp. 49–56. IEEE (2011)
10. Gao, J., Li, Y., Chin, T.J., Brown, M.S.: Seam-driven image stitching. In: Eurographics (Short Papers), pp. 45–48 (2013)
11. Guennebaud, G., Jacob, B., et al.: Eigen v3 (2010). http://eigen.tuxfamily.org
12. Johnson, J.W.: Adapting mask-RCNN for automatic nucleus segmentation. arXiv preprint arXiv:1805.00500 (2018)
13. Li, N., Liao, T., Wang, C.: Perception-based seam cutting for image stitching. SIViP **12**(5), 967–974 (2018)
14. Lin, C.C., Pankanti, S.U., Natesan Ramamurthy, K., Aravkin, A.Y.: Adaptive as-natural-as-possible image stitching. In: Proceedings of the IEEE Conference on Computer Vision and Pattern Recognition, pp. 1155–1163 (2015)
15. Lin, K., Jiang, N., Liu, S., Cheong, L.F., Do, M., Lu, J.: Direct photometric alignment by mesh deformation. In: Proceedings of the IEEE Conference on Computer Vision and Pattern Recognition, pp. 2405–2413 (2017)
16. Lin, W.Y., Liu, S., Matsushita, Y., Ng, T.T., Cheong, L.F.: Smoothly varying affine stitching. In: CVPR 2011, pp. 345–352. IEEE (2011)
17. Liu, F., Gleicher, M., Jin, H., Agarwala, A.: Content-preserving warps for 3D video stabilization. In: ACM Transactions on Graphics (TOG), vol. 28, p. 44. ACM (2009)
18. Shi, H., Guo, L., Tan, S., Li, G., Sun, J.: Improved parallax image stitching algorithm based on feature block. Symmetry **11**(3), 348 (2019)
19. Shum, H.Y., Szeliski, R.: Construction of panoramic image mosaics with global and local alignment. In: Benosman, R., Kang, S.B. (eds.) Panoramic Vision. MCS, pp. 227–268. Springer, New York (2001). https://doi.org/10.1007/978-1-4757-3482-9_13
20. Tu, W.C., et al.: Learning superpixels with segmentation-aware affinity loss. In: Proceedings of the IEEE Conference on Computer Vision and Pattern Recognition, pp. 568–576 (2018)
21. Wang, C., Miguel Buenaposada, J., Zhu, R., Lucey, S.: Learning depth from monocular videos using direct methods. In: Proceedings of the IEEE Conference on Computer Vision and Pattern Recognition, pp. 2022–2030 (2018)
22. Xu, K., Zhou, J., Wang, Z.: A method of hole-filling for the depth map generated by Kinect with moving objects detection. In: IEEE International Symposium on Broadband Multimedia Systems and Broadcasting, pp. 1–5. IEEE (2012)
23. Zaragoza, J., Chin, T.J., Brown, M.S., Suter, D.: As-projective-as-possible image stitching with moving DLT. In: Proceedings of the IEEE Conference on Computer Vision and Pattern Recognition, pp. 2339–2346 (2013)
24. Zhang, F., Liu, F.: Parallax-tolerant image stitching. In: Proceedings of the IEEE Conference on Computer Vision and Pattern Recognition, pp. 3262–3269 (2014)

Point Cloud Classification via the Views Generated from Coded Streaming Data

Qianqian Li, Long Ye$^{(\boxtimes)}$, Wei Zhong, Li Fang, and Qin Zhang

Key Laboratory of media Audio and Video, Communication University of China,
Ministry of Education, Beijing 100024, China
{liqianqian,yelong,wzhong,lifang8902,zhangqin}@cuc.edu.cn

Abstract. Point cloud has been widely used in various fields such as virtual reality and autonomous driving. As the basis of point cloud processing, the research of point cloud classification draw many attentions. This paper proposes a views-based framework for streaming point cloud classification. We obtain six views from coded stream without fully decoding as the inputs of the neural network, and then a modified ResNet structure is proposed to generate the final classification results. The experimental results show that our framework achieve comparable result, while it could be used when the input is streaming point cloud data.

Keywords: Streaming · Point cloud classification · Views-based · ResNet

1 Introduction

Polygonal mesh and point cloud are two popular ways to represent a three-dimensional model. By comparing them, the dominant one is the mesh models, mostly due to its native support in modern graphics chips. But mesh models have complex topological structures, consisting of a series of vertices together with their connectivity information in terms of edges and faces, which makes it difficult to represent the real world. In real world representation, point cloud is more appropriate, because it consists of discrete 3D points, and each point has position information and attribute information such as color and normal. Meanwhile, it does not need to store topology information.

With the coming of 5G Era, point cloud shows its great potential to be one of the mainstream media forms in the future. A few of the latest applications of point cloud processing and analysis include virtual reality, augmented reality, mixed reality, autonomous navigation system and free-view broadcasting. As the basis of point cloud processing, the research for point cloud classification become a hot topic recently.

This work is supported by the National Natural Science Foundation of China under Grant Nos. 61971383 and 61631016, and the Fundamental Research Funds for the Central Universities under Grant Nos. 2018XNG1824 and YLSZ180226.

© Springer Nature Singapore Pte Ltd. 2020
G. Zhai et al. (Eds.): IFTC 2019, CCIS 1181, pp. 384–392, 2020.
https://doi.org/10.1007/978-981-15-3341-9_31

The existed methods for 3D object classification can be divided into the following three categories: volume-based methods [1–3], pointset-based methods [4–6] and view-based methods [7–11]. Volume-based methods voxelize point cloud and then use 3D convolutional neural network for training. 3D ShapeNets [1] uses a Convolutional Deep Belief Network to convert 2.5D depth map into probability distributions of binary variables on a 3D voxel grid, and classifies the voxelized shape. Pointset-based methods propose a new kind of neural network that directly uses unordered point sets as input, which respect the alignment invariance of input points well. Pointnet [4] is the pioneer of directly using the point cloud as input, which uses spatial transformation network and symmetry function to learn the global representation of each point cloud object. Pointnet++ [5] on the basis of Pointnet [4], learns the local features of the point cloud, and adaptively combines features from multiple scales to overcome the problem that point cloud classification with different densities. View-based methods transform point clouds into a series of depth maps or multiple perspective views for classification. MHBN [9] proposed Multi-view Harmonized Bilinear Network (MHBN), which uses patches-to-patches similarity measurement to classify 3D shape.

On the whole, the view-based approaches achieved higher accuracy. But this kind of method is to render several 2D views from 3D mesh model. In this process, light will be added to create a realistic shading effect. For the classification task of point cloud data in streaming media, we need to get the characteristics of point cloud directly or indirectly from the bit stream. So, rendering views is not an efficient method. According to the projection-based point cloud compression method [12,13], We can easily recover the correspondence between point cloud and image from the data stream, and further recover the six views of the point cloud. The six views are different from the rendered views in two ways: one is that the six views are orthogonal projection rather than perspective projection; the other is that each pixel of the six views represents the color or depth of the point itself in the point cloud, and does not produce information such as illumination and shadow brought by rendering. Such six views can be easily obtained from data stream and can use view-based method to classify point cloud.

This paper focuses on the research of point cloud classification based on the views generated from the coded data stream and organizes as follows. Section 2 introduces the details of our method. Section 3 shows the experiments and experimental results. And the conclusion is given in Sect. 4.

2   Our Method

When we classify streaming point cloud, we need to get the input of the classifier network, that is, views generated from compressed data stream. In this paper, we take projection-based point cloud compression approach to obtain six views of point cloud. The main idea of this approach is projecting point cloud to the images and then compresses images by video encoder.

For point cloud compression, Schwarz et al. [13] proposes point cloud compression of MPEG, in which the MPEG V-PCC divides the point cloud into

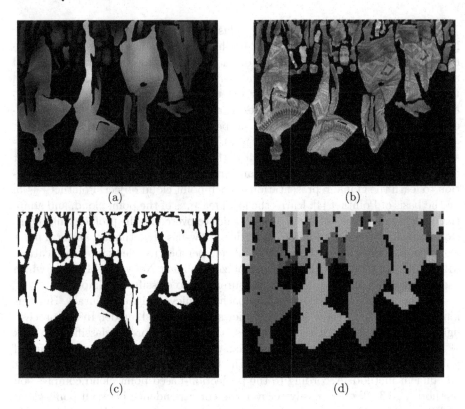

(a)   (b)

(c)   (d)

Fig. 1. (a) is a depth image containing geometric information after packaging all patches, (b) is a color image containing color information, (c) is an occupation map, indicating whether a point is included, (d) is a image indicating which patch each block (16 * 16) belongs to. (Color figure online)

several point cloud patches and projects them to an image, as shown in Fig. 1. Such images cannot be used for classification directly, so these patches need to be further recombined to form six views.

We recombine the patches in Fig. 1(b) to generate six views. This will recover the lost spatial information and be suitable for classification.

We sequentially get each patch's index of the projection plane $index1$, position in the image $(U0, V0)$, size of the patch $(sizeU0, sizeV0)$, and position in 3D space $(U1, V1, D1)$. After, determine whether the block in the image belongs to the patch according to blockToPatchMap (storing patch index information), whether the pixel contains a point according to the occupancyMap, and project the point that belonging to the patch to six views.

If the original projection plane indexed by $index1$ is the same as the target projection plane indexed by $index2$, the patch can be directly placed in the corresponding position. If not, the specific projection method is as follows:

We first get the 3D coordinates of each point in the patch P by Eq. (1):

$$(x, y, z) = f^{-1}(index1, u + U1, v + V1, d + D1) \tag{1}$$

Where $f(index, x, y, z)$ is a function that project the 3D point (x, y, z) onto the image (u, v, d) according to the projection plane index, $f^{-1}(index, u, v, d)$ is a reverse mapping.

Then six views can be generated by a projecting process as Eq. (2):

$$(u', v', d') = f(index2, x, y, z) \tag{2}$$

In this way, six views of the point cloud can be obtained, as shown in Fig. 2.

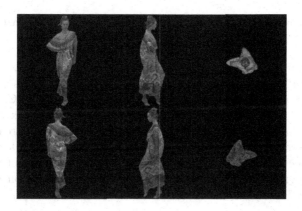

Fig. 2. Six views of the point cloud model shown in Fig. 1.

In our method, six views are recovered from data stream and then fed into the classifier network presented in Fig. 3 for classification. We adopt deep residual network (ResNet) [14] as our base model, because of its superior efficiency and fast convergence. We remove the last full connection layer of ResNet, and each view is then fed to ResNet18, and add a fully connected layer at the end.

3   Experiments

3.1   Implementation Details

The ShapeNetCore dataset contains 51300 3D mesh models, and the models are divided into 55 categories, for example, airplane, bench, cabinet, mobile phone, chair, etc. The number of these samples is unevenly distributed, and there are intersections among some categories. So, in order to facilitate the comparison with other methods, we finally choose 13 categories from them with more than 1000 models per category in our experiments, resulting in 31577 models in total. 80% of the models is used for training, and 20% for testing.

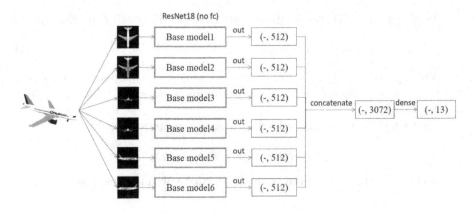

Fig. 3. Streaming media point cloud data classification network based on six views.

We first uniformly sample the original mesh model and voxelize it to generate a $512 * 512 * 512$ colored voxelized point cloud, and get six views of point cloud. Figure 4 shows six views of car and sofa model in the ShapeNetCore dataset.

We can decode the data stream to obtain six RGB views. Six views of point cloud are fed into ResNet [14] separately, and then six features are concatenated and passed through a dense layers. We train the network with learning rate of 0.0001, and we get a 13-dimensional vector as the classification result finally. We use the pytorch toolbox and implement the experiment on one NVIDIA Corporation GP102 [TITAN XP] graphics card.

3.2 The Impact of Views

The direction of the view have a certain impact on the classification performance. We utilized six single views for the classification. And we also combined the single views to explore the impact of numbers of the views on the classification task. The results are shown in Table 1.

We can see, multiple views achieve better results than other single view, indicating that these views contain rich information for the classification, and it meets the common sense. Furthermore, the combination of left, right and front views achieves best performance, and the accuracy is higher than that of the other combinations of three views, which shows the importance of view selection.

But it's not that the more views there are, the better the classification performance will be. Note that the accuracy of six views is not higher than that of three views. This might be attributed to the fact that some views are so similar that the information of three views is saturated, and adding more views will not change much.

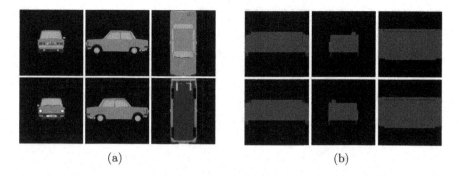

        (a)   (b)

Fig. 4. Six views of car and sofa model in the ShapeNetCore dataset.

Table 1. Accuracy of point cloud classification in different views.

Number of views	Views	Accuracy
1	Top	84.43%
1	Bottom	83.20%
1	Front	90.14%
1	Back	88.04%
1	Left	91.44%
1	Right	91.18%
3	Front, left, top	94.10%
3	Right, back, bottom	93.13%
3	Front, left, right	94.21%
4	Front, back, left, right	93.47%
6	Top, bottom, front, back, left, right	94.01%

3.3 The Impact of Depth

The depth information is essential information for 3D models, and it is very important for the identification of the point cloud models. The view we mentioned above only contains RGB information, which is insufficient. So, we add additional depth information in our experiments. The brightness is used to represent depth here. Higher brightness means lower depth and vice versa. Figure 5 shows an example of the depth information in six views of the car and sofa in the ShapeNetCore dataset.

Table 2 shows the experimental results, and we can see that the upper part of the table is the result of the classification of the view with only RGB information, while the lower part is the result with the depth information added to the RGB views. Compared to the RGB views' results, views with depth information achieve higher accuracy in general. The combination of front, left and right views achieves comparable results in two experiments whether with or without the depth information. While the other combinations achieve higher accuracy

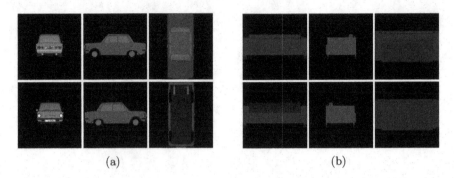

(a)   (b)

Fig. 5. Six views of car and sofa model with depth information in the ShapeNetCore dataset.

with depth information. Among them, six views together with depth information performs best, indicating that the depth information could recover local spatial information from 2D images to 3D objects, which has a positive impact on classification.

Table 2. Accuracy of point cloud classification with depth information.

Modality	Views	Accuracy
RGB	Front, left, top	94.10%
RGB	Front, left, right	94.21%
RGB	Front, back, left, right	93.47%
RGB	Top, bottom, front, back, left, right	94.01%
RGB+depth	Front, left, top	94.32%
RGB+depth	Front, left, right	94.15%
RGB+depth	Front, back, left, right	94.18%
RGB+depth	Top, bottom, front, back, left, right	95.10%

3.4   Comparison with Other Methods

Finally, we compare with other methods. PointNet++ [5] takes discrete 3D points as input; the input views of GVCNN [8] and MHBN [9] is six views generated by rendering every 60° around the mesh model with virtual cameras; we entered the top, bottom, left, right, front and back views generated by projecting.

As shown in Table 3, our method achieves 95.10% average instance accuracy using six views, is lightly higher than that of other methods. And in this situation, our input of the classifier network can be obtained from the data stream without fully decoding, so it is suitable for streaming media point cloud classification.

Table 3. Accuracy of point cloud classification with different method.

Methods	Views	Accuracy
PointNet++ [5]	–	94.75%
GVCNN [8]	6	94.85%
MHBN [9]	6	95.06%
Ours	6	94.09%
Ours	6 (with depth)	95.10%

4 Conclusion

In this paper, we propose a point cloud classification method for streaming media based on six views. We first project point cloud models into six views, and then feed the views as input of modified ResNet for transfer learning and achieve good results in point cloud classification. For point cloud is streaming media, so we can classify it without fully decoding, which makes the unification of point cloud compression and analysis possible. Next, we will further research the unified framework of point cloud compression and classification.

References

1. Wu, Z., et al.: 3D ShapeNets: a deep representation for volumetric shapes. In: Proceedings of the IEEE Conference on Computer Vision and Pattern Recognition, pp. 1912–1920. (2015)
2. Maturana, D., Scherer, S.: VoxNet: a 3D convolutional neural network for real-time object recognition. In: IEEE/RSJ International Conference on Intelligent Robots and Systems (IROS), pp. 922–928 (2015)
3. Qi, C.R., Su, H., Niener, M., Dai, A., Yan, M., Guibas, L.J.: Volumetric and multi-view CNNs for object classification on 3D data. In: IEEE Conference on Computer Vision and Pattern Recognition (CVPR), Las Vegas, NV, pp. 5648–5656 (2016)
4. Charles, R.Q., Su, H., Mo, K., Guibas, L.: PointNet: deep learning on point sets for 3D classification and segmentation. In: IEEE Conference on Computer Vision and Pattern Recognition (CVPR), Honolulu, HI, pp. 77–85 (2017)
5. Charles, R.Q., Yi, L., Su, H., Guibas, L.: PointNet++: deep hierarchical feature learning on point sets in a metric space (2017)
6. Shen, Y., Feng, C., Yang, Y., Tian, D.: Mining point cloud local structures by kernel correlation and graph pooling. In: IEEE/CVF Conference on Computer Vision and Pattern Recognition, Salt Lake City, UT, pp. 4548–4557 (2018)
7. Roveri, R., Rahmann, L., Öztireli, A.C., Gross, M.: A network architecture for point cloud classification via automatic depth images generation. In: IEEE/CVF Conference on Computer Vision and Pattern Recognition, Salt Lake City, UT, pp. 4176–4184 (2018)
8. Feng, Y., Zhang, Z., Zhao, X., Ji, R., Gao, Y.: GVCNN: group-view convolutional neural networks for 3D shape recognition. In: IEEE/CVF Conference on Computer Vision and Pattern Recognition, Salt Lake City, UT, pp. 264–272 (2018)

9. Yu, T., Meng, J., Yuan, J.: Multi-view harmonized bilinear network for 3D object recognition. In: IEEE/CVF Conference on Computer Vision and Pattern Recognition, Salt Lake City, UT, pp. 186–194 (2018)

10. Kanezaki, A., Matsushita, Y., Nishida, Y.: RotationNet-joint object categorization and pose estimation using multiviews from unsupervised viewpoints. In: IEEE/CVF Conference on Computer Vision and Pattern Recognition, Salt Lake City, UT, pp. 5010–5019 (2018)

11. Zhang, Z., Lin, H., Zhao, X., Ji, R., Gao, Y.: Inductive multi-hypergraph learning and its application on view-based 3D object classification. IEEE Trans. Image Process. **27**(12), 5957–5968 (2018)

12. He, L., Zhu, W., Xu, Y.: Best-effort projection based attribute compression for 3D point cloud. In: 23rd Asia-Pacific Conference on Communications (APCC), pp. 1–6 (2017)

13. Schwarz, S., et al.: Emerging MPEG standards for point cloud compression. IEEE J. Emerg. Sel. Top. Circuits Syst. **9**, 133–148 (2019)

14. He, K., Zhang, X., Ren, S., Sun, J.: Deep residual learning for image recognition. In: Proceedings of the IEEE Conference on Computer Vision and Pattern Recognition, pp. 770–778 (2016)

Geometry-Guided View Synthesis with Local Nonuniform Plane-Sweep Volume

Ao Li, Li Fang$^{(\boxtimes)}$, Long Ye, Wei Zhong, and Qin Zhang

Key Laboratory of Media Audio and Video,
Communication University of China, Ministry of Education, Beijing 100024, China
{iamoleo,lifang8902,yelong,wzhong,zhangqin}@cuc.edu.cn

Abstract. In this paper we develop a geometry-guided image generation technology for scene-independent novel view synthesis from a stereo image pair. We employ the successful plane-sweep strategy to tackle the problem of 3D scene structure approximation. But instead of putting on a general configuration, we use depth information to perform a local nonuniform plane spacing. More specifically, we first explicitly estimate a depth map in the reference view and use it to guide the planes spacing in plane-sweep volume, resulting in a geometry-guided manner for scene geometry approximation. Next we learn to predict a multiplane images (MPIs) representation, which can then be used to synthesize a range of novel views of the scene, including views that extrapolate significantly beyond the input baseline, to allow for efficient view synthesis. Our results on massive YouTube video frames dataset indicate that our approach makes it possible to synthesize higher quality images, while keeping the number of depth planes.

Keywords: Image-based rendering · View synthesis · Deep neural networks · Plane sweep volume

1 Introduction

Synthesis of novel views from a given set of captured views of a 3D visual object or scene is usually referred to as image-based rendering (IBR). It is an essential technology of many attractive applications, such as cinematography, virtual reality, free-viewpoint video [1], image stabilization [2], and video replays.

Novel view synthesis is an extremely challenging, under constrained problem. To render a novel viewpoint, the standard image-based rendering or novel-view generation [3, 4] first estimates the underlying 3D structure and synthesize novel views by applying geometric transformation to pixels in the given images. This means that the synthesized image quality depends directly on the accuracy of the estimated depth image. However, depth estimation is a complex non-linear process and sensitive to textureless and occluded regions. This is a substantial problem since small deviations in the estimated depth map might introduce visually annoying artifacts in the rendered views. Handling complex, self-occluding (but commonly seen) objects such as trees is particularly challenging for traditional approaches.

© Springer Nature Singapore Pte Ltd. 2020
G. Zhai et al. (Eds.): IFTC 2019, CCIS 1181, pp. 393–403, 2020.
https://doi.org/10.1007/978-981-15-3341-9_32

An alternative strategy focuses on interpolating rays from dense imagery based on the concept of plenoptic function [5] and its light field (LF) approximation [6, 7]. The scene capture and intermediate view synthesis problem can be formulated as sampling and consecutive reconstruction (interpolation) of the underlying plenoptic function. Buehler et al. [8] presented a generalized model that combines light field rendering and depth image-based rendering using irregularly sampled images. This approach requires little geometric information and can give potentially photorealistic results but requires many more input images. These two approaches can be thought of as opposite extremes of a spectrum where a reduction of one resource, geometric completeness, requires a corresponding increase in another, the number of images, to maintain a consistent quality. Plenoptic sampling theory [9, 36] gives us a theoretical framework to understand this trade-off. Pearson et al. [10] introduced a depth layer-based method for synthesizing an arbitrary novel view. They assigned each pixel to a particular layer and then warped the pixels using a probabilistic interpolation approach.

In recent years, while deep learning approaches have achieved huge successes in image recognition tasks, some studies based on deep learning aimed at maximizing the quality of the synthetic views have been presented. Tatarchenko et al. [11] learn to build a parametric model of the object class, and then use it together with the input image, to generate a novel view. Appearance flow [12] exploits the observation that the visual appearance of different views of the same instance is highly correlated and learns to copy pixels from the input image to the target view. It is further generalized to multiple input views by learning how to optimally combine single-view predictions. Thereafter visibility map and pixel generation [13] are introduced to improve the prediction. These methods are specifically designed to work on objects and do not work well on general scenes. Flynn et al. [14] proposed a view interpolation method called DeepStereo that predicts a plane-sweep volume from a set of input images, and trains a model using images of street scenes. Takeuchi et al. [15] introduced partial plane-sweep volume to DeepStereo and synthesized higher quality images, while keeping the number of depth planes. Liu et al. [16] proposed to exploit the geometry to learn to predict a set of homographies and their corresponding region masks to transform the input image into a novel view. In [17], view synthesis technique has been presented based on learning-based approach using two convolutional neural networks (CNNs) for disparity and color estimation. Four corner views from the light fields captured by a Lytro camera are used to synthesize an intermediate view. These methods predict a representation in the coordinate system of the target view. Therefore, these methods must run the trained network for each desired target view, making real-time rendering a challenge.

A variety of recent works has explored the problem of modeling scenes in view synthesis tasks. Different scene representations are proposed, which can be predicted once and then reused to render multiple views at runtime. Srinivasan et al. [18] proposed to learn a 4D RGBD light field from a single image. In [19], a soft 3D representation is produced for each input view and then interpolate between them. Zhou et al. [20] combined several attractive properties of prior methods, including multiple layers and "softness" of layering, to learn one multiplane image (MPI) representation of a scene. This work focuses on only narrow-baseline stereo pairs.

In this paper, we propose to account for 3D geometry, but instead of applying geometric transformation we use inaccurate depth to guide a local nonuniform plane-sweep volume in the process of novel view synthesis from a stereo image pair. To this end, we estimate a rough depth distribution and approximate the scene by a fixed number of planes spaced nonuniformly according to the depth distribution of the scene. For the depths that occur frequently more layers are placed to reduce the aliasing error in the synthesized view, while for the infrequently occurring depths fewer layers are used. Even though the rendering quality at less often occurring depths decreases, the overall rendered image quality is improved. At the scene representation part, MPI is used to synthesize a range of novel views of the scene. Finally, we train and test our network on the RealEstate10K dataset [20], which consists of frames extracted from YouTube online videos. Experimental result shows that our method achieves better numerical performance and visual experience.

2 Approach

Given a set of images with known camera parameters, our goal is to leverage information about the 3D scene structure to infer a global scene representation suitable for synthesizing novel views of the same scene. In this section, we first describe the idea of local nonuniform plane-sweep, and then present our pipeline and objective for learning to predict MPI representation.

2.1 Local Nonuniform Plane-Sweep Volume

Naively training a deep network to synthesize new views by supplying the input images as inputs doesn't work well. The pose parameters of the target view and of the input views would need to be supplied to the network as well. The relationship between the pose parameters, the input pixels and the output pixels is complex and nonlinear. It is difficult for the network to learn how to interpret rotation angles and to perform image reprojection effectively. To encode the pose parameters, we use plane-sweep volume (PSV) and choose these depth planes to coincide with those of the output MPI following [20]. However, the traditional PSV doesn't involve any 3D scene structure information. To leverage the scene geometry while avoiding the aliasing error caused by inaccurate depth, we combine depth information and PSV to build a so called local nonuniform plane-sweep volume (LNPSV).

The concept of our local nonuniform plane-sweep volume is to determine the position of planes in each grid in an image as shown in Fig. 1. Since the local depth distribution in a grid can be assumed to be smooth and quite different from each other, we can use a plane-sweep volume to approximate the local 3D structure with fewer planes than a global one. This idea was first presented in [10] with evenly spaced planes as usual. However, as objects in a real scene are not uniformly distributed in depth so there are advantages to assigning the planes with uneven spacings. To do so a more detailed knowledge of the scene geometry is needed to assign planes to the best positions.

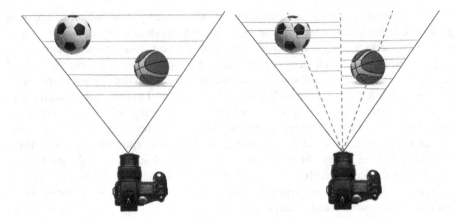

Fig. 1. Schematic diagram of original and our local nonuniform plane-sweep volume. (Left) Planes in original PSV. (Right) Planes in our LNPSV. We individually set a range and sweep planes in each grid.

A. Depth Distribution Estimation

The first stage of the algorithm is to determine the depth distribution for the visible scene. To achieve this aim as efficiently as possible, the pixel-wise depth in the reference image is estimated using the network proposed in [23] for single image depth estimation. Figure 2 gives an example of depth estimation. And then the depth map is converted to a disparity map. In practice, it is done by only take the reciprocal.

Fig. 2. Example of estimated depth maps; left: RGB input, right: depth image output.

To reduce the depth range for a better approximation, both input stereo pair and the estimated depth map are divided into several patches with the size of $H \times W$. The size of each patch is equal to the size defined in the following learning based MPI prediction step.

B. Local Nonuniform Plane Spacing

The Previous works [14, 15] have selected planes that are uniformly spaced in disparity as suggested by Plenoptic theory [9]. For the case of a precisely bandlimited Plenoptic spectrum with an ideal reconstruction filter, the minimum number of planes for alias-free rendering is given. Because the assumptions underlying Plenoptic theory are not fully met in practice, some aliasing is always present and its impact on rendered images can be reduced by increasing the number of planes beyond that indicated by the theory. We employ the scene depth distribution to determine the position of planes. Intuitively, for the depths that occur frequently more planes are placed to reduce the aliasing error in the synthesized view, while for the infrequently occurring depths fewer planes are used. Even though the rendering quality at less often occurring depths decreases, the overall rendered image quality is improved, as will be shown in Sect. 4.

We want to minimize the error from quantizing disparities to these plane positions. To this end, the Lloyd-Max algorithm [24] is used to find the disparity values for each of the planes. Unless specified otherwise, we use $D = 32$ planes set at uneven disparity (inverse depth) with the possible near and far planes at 1 m and 100 m, respectively.

2.2 Learning from Plane-Sweep Volume

To learn to predict a novel view from a stereo pair while exploiting the 3D geometry of the scene, we develop the network shown in Fig. 3.

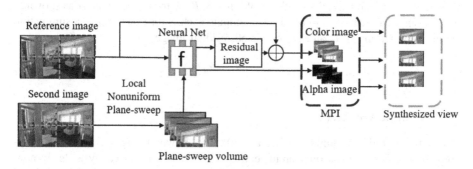

Fig. 3. Our pipeline for view synthesis

A. Multiplane image representation

For global scene representation, we adopt the Multiplane image representation presented in [20]. It consists of a set of fronto-parallel planes at a fixed range of depths with respect to a reference coordinate frame, where each plane d encodes an RGB color image C_d and an alpha/transparency map α_d. An MPI can thus be described as a collection of such RGBA layers $\{(C_1, \alpha_1), \ldots, (C_D, \alpha_D)\}$, where D is the number of depth planes. In [20] the pixels in each layer are fixed at a certain depth, and an alpha channel per layer is used to encode visibility. But in our case, we choose these depth layers to coincide with those of the input PSV, i.e. the layers in MPI have the same nonuniform spacing as the corresponding planes in PSV, making it more efficient to approximate the complex scene structure.

To render from an MPI, the layers are composed from back-to-front order using the standard "over" alpha compositing operation.

B. Predict MPIs from LNPSV

Using plane-sweep volumes as input to the network removes the need to supply the pose parameters since they are now implicit inputs used in the construction of the plane-sweep volume. We compute a nonuniform PSV that reprojects the second input image I_2 into the reference camera at a set of D depth planes and results in a stack of reprojected images $\{\hat{I}_2^1, \ldots, \hat{I}_2^D\}$. Thus, corresponding pixels are now aligned across the plane-sweep volume and so that a given output pixel depends only on a small column of voxels.

The LNPSV $\{\hat{I}_2^1, \ldots, \hat{I}_2^D\}$ is concatenated along the color channels as well as the reference image I_1, resulting in a $H \times W \times 3(D+1)$ tensor. By simply comparing I_1 to each layer of I_2 reprojections, the network can reason about the scene geometry. If a pixel of I_2 is reprojected correctly, it would agree with the corresponding pixel in I_1 and thus the scene depth at this pixel is as same as the depth plane. Just like what we do in traditional plane-sweep stereo [26, 27].

For the d-th depth layer in MPI, the color image C_d captures the scene appearance and the alpha map α_d encodes the visibility and transparency. Since the target MPI is with respect to the reference camera coordinate frame, the color images are highly correlated to the reference image I_1. Therefore, instead of learning the whole color image, we choose to learn the residual color image R_d with respect to the I_1 as shown in Fig. 3. Hence, for each depth plane, we compute each color image C_d as a summation of the reference image I_1 and the predicted residual image R_d:

$$C_d = I_1 + R_d \tag{1}$$

2.3   Network Architecture

The network follows similar design as [20]. As shown in Fig. 4, we use a fully-convolutional encoder-decoder architecture. Each layer is followed by a ReLU non-linearity and layer normalization [31] except for the last layer, where tanh is used and no layer normalization is applied. N branches (in our case $N = 4$) of such networks parallelly process all patches simultaneously. Each branch outputs 32 color images and 32 alpha images. The alpha images are further scaled to a valid range of [0, 1].

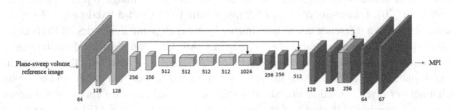

Fig. 4. The network diagram. (Color figure online)

In this work, we use a deep feature matching loss (also referred to as the "perceptual loss" [33, 34]), and specifically use the normalized VGG-19 [28] layer matching from [35]:

$$\mathcal{L}(\hat{I}_t, I_t) = \sum_l \lambda_1 \| \phi_l(\hat{I}_t) - \phi_l(I_t) \|_1 \tag{2}$$

3 Data Setup and Training Details

To train and evaluate our novel view synthesis network, we require triplets of images together with their relative camera poses and intrinsics. We choose the YouTube video frames dataset proposed in [20], which consists of over 7,000 video clips from 1 to 10 s in length, together with the camera position, orientation and field of view for each frame in the sequence. For the comparison with [20] to be fair, we adopt the same data splits as them. The dataset is first divided into different scenes. For each scene, we then randomly choose two different frames and their poses to be the inputs I_1, I_2, c_1 and c_2, and a third frame to be the target I_t, c_t, resulting in a set of tuples $\langle I_1, I_2, I_t, c_1, c_2, c_t \rangle$.

We chose to learn to predict views from a variety of positions relative to the source imagery so as not to overfit to generating images at a particular distance during training. In order to perform a local plane-sweep, we divide I_1, I_2, I_t into four patches evenly and resize them to a same size in a same manner. Slight overlap is employed to avoid blocking effect. These patches are processed parallelly because they belong to the same stereo pair.

At each training iteration, in one branch a reference image patch is fed to the depth estimation module and get a rough depth map in the reference view. The depth map is then inversed to disparity and quantized using Lloyd-Max algorithm with a quadratic cost function to find the positions of planes for both PSV and MPI. Next we use the pose information c_1, c_2 as well as depth of plane positions to reproject a I_2 patch into the reference view. As mentioned in Sect. 2.2, the thus obtained PSV together with the corresponding I_1 patch forms a 3D tensor which is used to predict a local MPI. We use four such branches to process parallelly and get four local MPIs. Finally, four local MPIs synthesize four patches in the target view individually and compose one target image.

As for training details, we implement our system in TensorFlow [30]. We train the network using the ADAM solver [32] for 600K iterations with learning rate 600K iterations with learning rate 0.00025, $\beta_1 = 0.9$, $\beta_2 = 0.999$, and batch size 1. The depth estimation module is initialized by the model provided by [23]. And the MPI prediction module is pre-trained by using the code from [20] with uniformly spaced PSV. Training takes about one day on double NVidia 1080Ti GPUs.

4 Experiment

In this section, we evaluate our approach both quantitatively and qualitatively, and compare it with several view synthesis baselines. Our test set consists of 1,329 sequences that did not overlap with the training set. For each sequence we randomly sample a triplet (two source frames and one target frame) for evaluation. To show the superiority of our method, we first demonstrate the visualization of the MPI representation inferred by our model comparing with the method in [20], and then provide detailed comparison with other recent view synthesis methods with respect to PSNR and SSIM metrics.

We visualize several layers of the alpha map inferred by our network as well as Zhou et al. [20] in Fig. 5. A critical difference compared to our method is that Zhou et al. use a serial of planes set at equidistant disparity in both PSV and MPI for scene geometry approximation. In contrast, we use rough depth information as the guidance of plane spacing, making it more efficient to approximate the 3D scene structure, while avoiding the weakness of performing depth-based warping. From Fig. 5, we observe that the scene structure information distributes more evenly in our method than that in [20]. The Nos. 28–32 layers inferred by [20] almost capture nothing, as shown in the bottom row, while our alpha maps utilize every depth layer to the maximum, as shown in the top row. This observation shows that the proposed local nonuniform plane spacing model can provide a better approximation of the 3D scene structure.

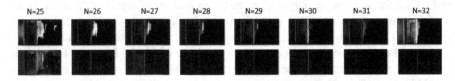

Fig. 5. Alpha image comparison between our model and Zhou et al. [20]

Furthermore, we compare our method with Kalantari et al. [21] and Zhou et al. [20]. Table 1 gives mean SSIM and PSNR similarity metrics for each method across the test set. The result shows that our model outperforms both Kalantari et al. and Zhou et al., indicating the high-quality of novel views rendered from our geometry-guided model.

Table 1. Quantitative comparison between our model and variants of the Zhou et al. [20] and Kalantari et al. [21]. Higher SSIM/PSNR mean are better.

Method	Loss	SSIM	PSNR (dB)
Kalantari et al.	VGG	0.810	32.68
Zhou et al.	VGG	0.823	32.85
Ours	VGG	0.838	33.18

In addition, we also give some qualitative comparison in Fig. 6. It can be seen that our method yields more realistic novel views than Zhou et al. [20] on both indoor and outdoor scenes. Note that we box out some fine details in the synthesized images, including both near scene and far scene. We can see that our model can always place them at the correct depth, while [20] cannot, proving the strength of our geometry-guided model.

Our synthesized view   Zhou et al.   Ours   Ground-Ttruth

Fig. 6. Comparison between Zhou et al. [20] and our approach in synthesized view on RealEaste10K dataset [20].

5 Conclusion

We present a novel geometry-guided view synthesis network. Our method generates realistic images and outperforms existing techniques for novel view synthesis on massive YouTube video frames dataset. The proposed local nonuniform plane-sweep volume provides a more accurate model to approximate the 3D scene structure. The experimental results have demonstrated the benefit of our model. By leveraging geometry information, our method yields predictions that better match the scene structure, while keeping the number of depth planes. In the future, we will improve our model to handle more general cases, such as large baseline and multiple inputs.

References

1. Tanimoto, M.: Overview of FTV (free-viewpoint television). In: Proceedings of the IEEE Conference on Multimedia and Expo (ICME 2009), pp. 1552–1553, June 2009
2. Kopf, J., Cohen, M.F., Szeliski, R.: First-person hyperlapse videos. In: SIGGRAPH (2014)
3. Scharstein, D., Szeliski, R.: A taxonomy and evaluation of dense two-frame stereo correspondence algorithms. Int. J. Comput. Vis. **47**(1–3), 7–42 (2002)
4. Kim, C., Zimmer, H., Pritch, Y., Sorkine-Hornung, A., Gross, M.: Scene reconstruction from high spatio-angular resolution light fields. ACM Trans. Graph. **32**(4), 1–12 (2013)
5. Adelson, E., Bergen, J.: The plenoptic function and the elements of early vision. In: Computational Models of Visual Processing. MIT Press, Cambridge (1991)
6. Levoy, M., Hanrahan, P.: Light field rendering. In: Proceedings of the ACM SIGGRAPH, pp. 31–42 (1996)
7. Gortler, S.J., Grzeszczuk, R., Szeliski, R., Cohen, M.F.: The lumigraph. In: Proceedings of the ACM SIGGRAPH, pp. 43–54 (1996)
8. Buehler, C., Bosse, M., Mcmillan, L., et al.: Unstructured lumigraph rendering. In: Conference on Computer Graphics & Interactive Techniques. ACM (2001)
9. Chai, J., Tong, X., Chan, S., et al.: Plenoptic sampling. In: Proceedings of the ACM SIGGRAPH, pp. 307–318 (2000)
10. Pearson, J., Brookes, M., Dragotti, P.L.: Plenoptic layer-based modeling for image based rendering. IEEE Trans. Image Process. **22**(9), 3405–3419 (2013)
11. Tatarchenko, M., Dosovitskiy, A., Brox, T.: Single-view to multi-view: reconstructing unseen views with a convolutional network. Knowl. Inf. Syst. **38**(1), 231–257 (2015)
12. Zhou, T., Tulsiani, S., Sun, W., Malik, J., Efros, A.A.: View synthesis by appearance flow. In: Leibe, B., Matas, J., Sebe, N., Welling, M. (eds.) ECCV 2016. LNCS, vol. 9908, pp. 286–301. Springer, Cham (2016). https://doi.org/10.1007/978-3-319-46493-0_18
13. Sun, S.-H., Huh, M., Liao, Y.-H., Zhang, N., Lim, Joseph J.: Multi-view to novel view: synthesizing novel views with self-learned confidence. In: Ferrari, V., Hebert, M., Sminchisescu, C., Weiss, Y. (eds.) ECCV 2018. LNCS, vol. 11207, pp. 162–178. Springer, Cham (2018). https://doi.org/10.1007/978-3-030-01219-9_10
14. Flynn, J., Neulander, I., Philbin, J., Snavely, N.: DeepStereo: learning to predict new views from the world's imagery. In: Proceedings of the IEEE Conference on Computer Vision and Pattern Recognition, pp. 5515–5524 (2016)
15. Takeuchi, K., Okami, K., Ochi, D., et al.: Partial plane sweep volume for deep learning based view synthesis. In: ACM SIGGRAPH 2017 Posters. ACM (2017)

16. Liu, M., He, X., Salzmann, M.: Geometry-aware deep network for single-image novel view synthesis. In: 2018 IEEE Conference on Computer Vision and Pattern Recognition (CVPR), pp. 4616–4624 (2018)
17. Kalantari, N.K., Wang, T.-C., Ramamoorthi, R.: Learning-based view synthesis for light field cameras. ACM Trans. Graph. **35**(6), 1–10 (2016)
18. Tao, M.W., Srinivasan, P.P., Malik, J., et al.: Depth from shading, defocus, and correspondence using light-field angular coherence. In: 2015 IEEE Conference on Computer Vision and Pattern Recognition (CVPR). IEEE Computer Society (2015)
19. Penner, E., Zhang, L.: Soft 3D reconstruction for view synthesis. In: Proceedings of the SIGGRAPH Asia (2017)
20. Zhou, T., Tucker, R., Flynn, J., et al.: Stereo magnification: learning view synthesis using multiplane images (2018)
21. Kalantari, N.K., Wang, T.-C., Ramamoorthi, R.: Learning-based view synthesis for light field cameras. In: Proceedings of the SIGGRAPH Asia (2016)
22. Yao, Y., Luo, Z., Li, S., Fang, T., Quan, L.: MVSNet: depth inference for unstructured multi-view stereo. In: Ferrari, V., Hebert, M., Sminchisescu, C., Weiss, Y. (eds.) ECCV 2018. LNCS, vol. 11212, pp. 785–801. Springer, Cham (2018). https://doi.org/10.1007/978-3-030-01237-3_47
23. Hu, J., Ozay, M., Zhang, Y., et al.: Revisiting single image depth estimation: toward higher resolution maps with accurate object boundaries (2018)
24. Lloyd, S.: Least squares quantization in PCM. IEEE Trans. Inf. Theory **28**(2), 129–137 (1982)
25. Shade, J., Gortler, S., He, L., Szeliski, R.: Layered depth images. In: Proceedings of the SIGGRAPH (1998)
26. Collins, R.T.: A space-sweep approach to true multi-image matching. In: CVPR (1996)
27. Szeliski, R., Golland, P.: Stereo matching with transparency and matting. IJCV **32**(1), 45–61 (1999)
28. Jaderberg, M., Simonyan, K., Zisserman, A., et al.: Spatial transformer networks. In: NIPS (2015)
29. Ba, J.L., Kiros, J.R., Hinton, G.E.: Layer normalization. arXiv preprint arXiv:1607.06450 (2016)
30. Abadi, M., et al.: TensorFlow: a system for large-scale machine learning. In: OSDI (2016)
31. Agarwal, S., Mierle, K., et al.: Ceres Solver (2016). http://ceres-solver.org
32. Hasinoff, S.W., et al.: Burst photography for high dynamic range and low-light imaging on mobile cameras. In: Proceedings of the SIGGRAPH Asia (2016)
33. Dosovitskiy, A., Brox, T.: Generating images with perceptual similarity metrics based on deep networks. In: NIPS (2016)
34. Johnson, J., Alahi, A., Fei-Fei, L.: Perceptual losses for real-time style transfer and super-resolution. In: Leibe, B., Matas, J., Sebe, N., Welling, M. (eds.) ECCV 2016. LNCS, vol. 9906, pp. 694–711. Springer, Cham (2016). https://doi.org/10.1007/978-3-319-46475-6_43
35. Engel, J., Koltun, V., Cremers, D.: Direct sparse odometry. IEEE Trans. Pattern Anal. Mach. Intell. **40**(3), 611–625 (2018)
36. Lin, Z., Shum, H.-Y.: A geometric analysis of light field rendering. Int. J. Comput. Vis. **58**(2), 121–138 (2004)

Single View 3D Reconstruction with Category Information Learning

Weihong Cao, Fei Hu, Long Ye$^{(\boxtimes)}$, and Qin Zhang

Key Laboratory of Media Audio and Video, Communication University
of China, Ministry of Education, Beijing 100024, China
{caoweihong,hufei,yelong,zhangqin}@cuc.edu.cn

Abstract. 3D reconstruction from single image is a classical problem in computer vision. Due to the fact that the information contained in one single image is not sufficient for 3D shape reconstruction, the existing model cannot reconstruct 3D models very well. To tackle this problem, we propose a novel model which effectively utilizes the category information of objects to improve the performance of network on single view 3D reconstruction. Our model consists of two parts: rough shape generation network (RSGN) and category comparison network (CCN). RSGN can learn the characteristics of objects in the same category through the comparison part CCN. In the experiments, we verify the feasibility of our model on the ShapeNet dataset, and the results confirm our framework.

Keywords: Single view · 3D reconstruction · Adversarial learning

1 Introduction

3D reconstruction is a technique used in computer vision which has a wide range of applications in areas like object recognition, virtual reality, video games etc. With the establishment of larger 3D model dataset, many 3D reconstruction networks have achieved excellent performance. However, there are still many difficulties when neural networks learn to infer the 3D structure from a single-view image.

When predicting 3D objects, human beings can learn the prior geometric information which network hardly learn from image, which makes neural networks reconstruct 3D object not well. In order to make full use of the prior geometric information, existing methods on single-view 3D reconstruction always need additional annotations, such as perspective and depth, etc. Training the network with additional-annotated data undoubtedly greatly increases the workload of dataset construction and the cost of computation.

According to daily experience, intra-category similarity is an important characteristic of natural objects, and the shapes of objects in a certain category are always roughly the same. Therefore, we consider that the category information of objects is useful. Current work [26] also shows that models based on classification methods (such as clustering or retrieval) can achieve better results. This reflects that the category information of objects is very helpful to the reconstruction work. Furthermore, category

G. Zhai et al. (Eds.): IFTC 2019, CCIS 1181, pp. 404–415, 2020.
https://doi.org/10.1007/978-981-15-3341-9_33

as semantic information is easier to obtain than geometric information of 3D structure, and it doesn't require expensive computational cost.

By the way, the neural network usually tackles image-based 3D reconstruction task with encoder-decoder structure. The encoder maps the input image into a latent variable space, and the decoder is supposed to perform an optimal 3D structure in the output space. This is a difficult problem with the extremely high dimensional solution space. In order to tackle this problem, the category information can be adopted as the restricted condition to shrink the solution space.

According to these ideas, we propose a novel model which effectively utilizes the category information of objects to improve the performance of network on single-view 3D reconstruction. Inspired by adversarial learning, we add two classifiers to a common 3D shape generation network. Our model consists of two parts. The first part is rough shape generation network, which reconstruct the rough shape. The second one is category comparison part; we hope that the generation network can learn the characteristics of objects in the same category through the comparison.

To prove the idea of our work, we compare our model with the pure encoder-decoder network and the results prove that our model can indeed achieve better performance. The main contributions of this paper can be summarized as follows:

1. Based on the existing encoder-decoder method, we add the constraint of category information and propose a novel framework.
2. We designed two classification networks for image and voxel models respectively.
3. We prove that category information is indeed helpful in predicting 3D structures.

2 Related Work

2.1 Datasets of 3D Shapes

In recent years, some large-scale 3D model repositories on the Internet have reached the trillion level, such as Trimbla3D and Yobi3D.ect. As a large 3D dataset, ShapeNet [1] containing 55 common object categories with approximately 50k different 3D models. Xiang et al. constructed PASCAL3D+ [2] and ObjectNet3D [3]. PASCAL3D+ is a dataset for 3D object detection and pose estimation, including 3D annotations of 12 categories of objects in PASCAL VOC 2012, with an average over 3,000 instances in each category. ObjectNet3D is a large database designed for 3D object recognition, witch identifying 3D objects from 2D images. There are also smaller datasets suitable for a particular scenario. 3D models of typical indoor scenes constitute IKEA [4], containing approximately 759 images and 219 3D models. More recently, Sun et al. present Pix3D [5], a new large-scale dataset of diverse image-shape pairs, and they annotated each 3D model with an accurate 3D pose annotation.

2.2 3D Reconstruction with Neural Network

Due to the contribution of 3D datasets, researchers have made significant progress in 3D reconstruction with deep convolutional networks. However, reconstructing 3D model from a single image is still a challenging problem because restoring a 3D structure requires not only color, texture information, but also geometric information.

The depth map contains rich geometric information. Therefore, some RGB-D image based 3D reconstruction models have been proposed. One of the masterpieces is 3Dshapenet [6], which proposes a network model using Convolutional Deep Belief Network [7] to recover the voxel grid representation of the entire 3D object from a single-view 2.5D depth map. A simple encoder-decoder network is proposed [18], which can generate RGB images and depth maps of objects from an arbitrary perspective. MarrNet [20] also try to learn more information to get strong prior knowledge of object shapes via 2.5D Sketches.

There are also some works which use perspective information to assist single-view reconstruction. Perspective Transformer Nets [11] propose an encoder-decoder network with a novel projection loss defined by the projective transformation, which enables the unsupervised learning using 2D observation without explicit 3D supervision. Yang et al. [12] uses camera poses annotations to enforce pose-invariance and view-point consistency. Pose-aware shape reconstruction from a single image [15] design architectures of pose-aware shape reconstruction which reproduce the predicted shape back on to the image using the predicted pose.

Different from the above works, some researchers have implemented more efficient frameworks to achieve the challenge of single-view reconstruction, rather than obtaining more prior information. Hierarchical Surface Prediction (HSP) [17], propose a general framework for which is designed for high resolution 3D object reconstruction. Rezende et al. [16] proposed an unsupervised generation model based on variational autoencoder, which proved for the first time the feasibility of using neural network to infer 3D shape. Differentiable ray consistency (DRC) [19] can compute the gradient of a predicted 3D shape of an object by different types of observations from an arbitrary view. Wang et al. [27] propose an end-to-end deep learning architecture that produces a 3D shape in triangular mesh from a single color image. Groueix et al. [28] represent a 3D shape as a collection of parametric surface elements. Point Set Generation Network [21] studied the framework that can directly output point cloud coordinates. Generative adversarial networks (GAN) [22] are also be used on single-view 3D reconstruction, such as 3D-GAN [23], 3D-VAE-GAN [23] and PrGAN [24].

2.3 Our Work

In this paper, considering of the expensive computational cost, we use category information to assist single-view 3D reconstruction instead of depth maps and perspective information. We design an effective novel network which can take advantage of category information without any additional annotations.

3 The Architectures

In this section, we introduce our network architecture that combines low-level image information with high-level category information. The purpose is to recover 3D structure of objects from 2D images with any perspective. As shown in Fig. 1, our model consists of category comparison network and rough shape generation network. The main idea is to divide the 2d-to-3d reconstruction task to inter-category reconstruction and intra-category optimization. We will introduce our network in the following.

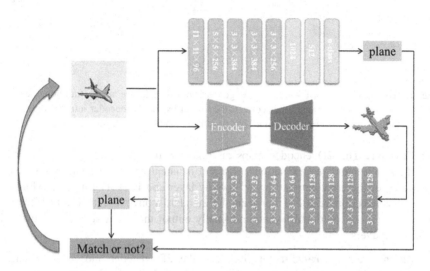

Fig. 1. The architecture of our model. The encoder-decoder (RSGN) produces a rough shape of the input image. The CCN Contains two networks: the 2D classification (top) identify the category of the input image, the 3D classification network (bottom) identify the category of predicted shape.

3.1 Rough Shape Generation Network

Rough shape generation network (RSGN) tries to learn rough shape. As show in Fig. 2, our RSGN is based on the 3D-R2N2 [9] network with some modifications. The 3D-R2N2 consists of 2D encoder, 3D LSTM and 3D decoder. Firstly, we adopt the same encoder as 3D-R2N2. Secondly, LSTM is mainly used to fuse the features of multiple images. Considering our purpose is to reconstruct the object's 3D structure from a single image, we replace the 3D LSTM component with convolutional layers. Finally, we improved the decoder of 3D-R2N2 by replacing all upsampling layers with 3D deconvolutional layers. What's more, the last layer of our network is no longer a classification task but a generation task, therefore we use sigmoid layer instead of the last SoftMax classifier. The output dimensional of the network is no longer $R^{32 \times 32 \times 32 \times 2}$, but $R^{32 \times 32 \times 32 \times 1}$.

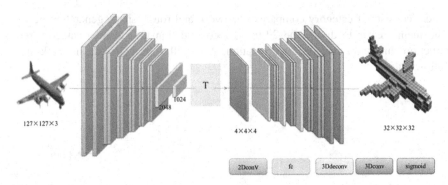

Fig. 2. The architecture of rough shape generation network, we adopt traditional encoder-decoder network, T is a transformer convert the output by the 2D encoder into a 3-dimensional feature, and we replace the up-pooling layers with deconvolutional layers.

2D Encoder. The 2D encoder maps an image x from any perspective in dataset $X = \{x_i\}$ to a latent representation $z \sim p(z|x)$ in latent space. The encoder is mainly composed of the convolutional layer, pooling layer and LeakyReLu layer. Moreover, residual connection is added between the standard convolutional layers. There are altogether 12 convolutional layers, and finally output the latent vector z by two fully connected layers.

2D Latent Vector Transformer. Between the 2D encoder and 3D decoder, we convert the 1-dimensional latent vector z output by the 2D encoder into a 3-dimensional feature z' with of size $4 \times 4 \times 4$ using a convolutional layer. In order to learn the spatial structure of the solid part and the blank part in the 3D voxel model simultaneously, we performed the following operations on z':

$$z_v = [1 - (z')]\tanh(z'), \tag{1}$$

where σ is sigmoid. Then the z_v is fed into the 3D decoder.

3D Decoder. The 3D decoder receives the latent vector z_v from the transformer and reasons it to the voxel representation $v_p \sim p(v|z_v)$ in the 3D data space $R^{32 \times 32 \times 32}$. The decoder mainly consists of 3D convolutional layer. The residual connection is also added between the convolutional layers. It is worth noting that our up-convolution operation uses 'deconvolution+convolution' rather than 'uppooling+convolution'. There are 3 deconvolutional layers and 10 convolution layers in total, and final layer limits the value of each individual predicted element of voxel $v_p(i,j,k) \in (0,1)$ with sigmoid activation function.

All activation layers of the network use LeakyReLu function:

$$y_i = \begin{cases} x_i, x_i > 0 \\ \alpha x_i, x_i \leq 0, \end{cases} \tag{2}$$

where $\alpha = 0.01$. In the residual connections, we all use 1×1 convolutions to expand the size of the feature maps if the output of two convolution layers is unmatched in size.

3.2   Category Comparison Network

The purpose of the category comparison network is to compare the category information of the input image and reconstructed voxel model. The input is an image with its corresponding reconstructed voxel (x, v), and the output is their respective categories information (C_x, C_v).

2D Classification Network. Mainly refer to AlexNet [25], the network consists of five convolution layers and three fully-connected layers. In order to suit our task, we made some changes to AlexNet. In the first convolutional layer, our input is a 127×127 RGB image, and the stride size is reduced to 2. We only train on a single GPU adopting non-parallel computing, so the channels of the kernels for all convolutional layers has doubled ([48, 128, 192, 192, 128] channels to [96, 256, 384, 384, 256] channels). The first two fully-connected layers have 1204 and 512 neurons respectively. The output of the last fully-connected layer is n-way which produces the category distribution C_x over the n-category objects.

3D Classification Network. The network classifies 3D voxel models with nine 3D convolution layers and three fully-connected layers designed by ourselves. The input of our 3D objects classification network is a $32 \times 32 \times 32$ voxel predicted by the RSGN. At the stack of convolutional layers, we use filters with filed $3 \times 3 \times 3$, and the convolution kernels have [128, 128, 128, 64, 32, 1] channels from top to bottom. Pooling operation is carried out by two max-pooling layers, which follow the fourth convolutional layer and the last convolutional layer. The first two fully-connected layers behind the convolutional layer have 1024 and 512 channels respectively. The last fully-connected layer output the category information C_v of voxel model by performing an n-way classification.

The slope of all LeakyReLu at $(-\infty, 0)$ was 0.01 and adopt max-pooling in all pooling layers. In this paper we have selected 8 categories of objects, so $n = 8$.

3.3   Loss Function

In order to transform the 2D-to-3D reconstruction work to inter-category reconstruction and intra-category optimization, two parts of our model, rough shape generation network and category comparison network, respectively correspond to two independent loss functions.

Rough Shape Generation Network: In this part, we construct a Cross Entropy with Logits loss [31] with extra parameters at the element level. The loss function is expressed as:

$$L_v(\omega, v_p, v_g) = \sum_{i,j,k} -\big[\omega v_g(i,j,k) \log\big(v_p(i,j,k)\big) + \big(1 - v_g(i,j,k)\big) \log\big(1 - v_p(i,j,k)\big)\big],$$

(3)

where $v_g(i,j,k)$ represents the ground truth value of the voxel block at (i,j,k), and $v_p(i,j,k)$ represents the value of the predicted shape at (i,j,k), ω is a hyper parameter. In this paper, L_v promotes the network to learn a rough shape.

Category Comparison Network: In this part, we try to restrict the reconstructed voxel model v_p to be the same category as the input image x_i. The loss function is:

$$L_c(c_x, c_v) = \|c_x - c_v\|_2, \tag{4}$$

where c_x the category information of input image x_i and c_v is the category information of reconstructed voxel model v_p. L_c encourages the network to learn more details of a 3D shape which is belong to the same category as the input image.

We combine the loss of rough shape generation network (RSGN) L_v and the loss of category comparison network (CCN) L_c to train our model, and the complete loss function is expressed as follows:

$$L = \alpha L_c(c_x, c_v) + L_v(\omega, v_p, v_g), \tag{5}$$

where α is also a hyper parameter. With some experiments, we finally choose $\alpha = 10$, $\omega = 20$.

4 Experiments

This section introduces the dataset and the experiments details. And the results of several experiments demonstrate the capability of our approach.

4.1 Dataset

ShapeNet. The ShapeNet dataset is an ongoing effort to establish a richly-annotated, large-scale dataset of 3D shapes. It contains 55 common categories object with approximately 50k different 3D models. We used a subset of the ShapeNet which contains about 24,000 3D models from 8 common categories. For each model we rendered a 224×224 RGB image and cropped to 127×127 size by center-clip during training. Within each category, the shapes are randomly split into training and test sets, with 80% for training and 20% for testing.

Metric. We adopt a standard metric Intersection over Union (IoU) in evaluating the reconstruction quality. The voxelized grid value $v(i, j, k)$ can be getting from the predicted value $v_p(i, j, k)$ as follow:

$$v(i, j, k) = \begin{cases} 0, v_p(i, j, k) < t \\ 1, v_p(i, j, k) \geq t, \end{cases} \tag{6}$$

where $t = 0.9$. Then we calculate IoU between a predicted voxel grid v and its ground truth v_g. Higher IoU values indicate better reconstructions.

4.2 Classification Network

We firstly trained the two classification networks independently. The results are shown in Table 1. We find that the accuracy of image classification network is 5.2% lower than

Table 1. Accuracy on ShapeNet test set, the accuracy of image classification network is nearly 5.2% lower than that of voxel. The image classification network with full-connected layer 1024-to-512 and 2048-to-1024 showing comparable results.

Model	Accuracy (%)
ImgClassifyNet (1024 to 512)	89.33
ImgClassifyNet (2048 to 1024)	89.44
VoxClassifyNet	94.60

that of voxel classification network, that is a considerable disparity. We consider that training data contribute to this disparity. The input images rendered from ShapeNet 3D model are simple in shading and lack of rich texture information. Therefore, it is difficult for the network to learn enough priori information to distinguish each category of objects perfectly. For voxel classification network, 3D shapes contain abundant geometric information, so even without coloring, the network can learn enough information to complete the classification task excellently. We converted the fully-connected layers of the image classification network from 1024-to-512 to 2048-to-1024, but the performance of the network has hardly improved, which confirms our statement before.

4.3   Evaluation

To validate our network structure, we now compare our model with 3D-R2N2 on single-image 3D shape reconstruction. The results are shown in Table 2. We achieved better performance in most categories. The IoU of our network is promoted in all categories except cabinet.

Table 2.   3D reconstruction IoU on the ShapeNet test dataset.

	3D-R2N2	Our-CRNet
Bench	0.421	**0.438**
Cabinet	0.716	0.704
Plane	0.513	**0.544**
Couch	0.628	**0.636**
Firearm	0.544	**0.560**
Car	0.798	**0.811**
Cellphone	0.661	**0.671**
Watercraft	0.513	**0.536**
Mean	0.599	0.613

We consider that is on account of the cabinets are generally regular cuboid which contains less characteristic than object in other categories. Therefore, the network can easily learn its features, and may even incorrectly reconstruct other categories into cabinets. So when we append the category information, the performance of our network getting worse on cabinet, because the robustness of network model to cabinets descend. In the other categories, our performance in bench, plane, firearm and watercraft improved considerably. Most of these objects have hollows or prominent components, theirs detailed information is abundant and intra-class characteristics are obvious, which can be easily distinguished from other categories. Therefore, assisted with category information, the reconstruction performance of our network on these categories has been improved. This reveals that the category information can definitely assist the reconstruction task, and multi-task learning helps to boost its performance.

We randomly selected some images and visualized voxelization models reconstructed by our network in the Fig. 3. The background color and perspective of the image are random rendered. Our network can produce precise prediction even from some extremely difficult perspectives (cellphone in row 7 column 4). It is also possible to distinguish objects which have a similar appearance (cabinet in row 1 column 6, couch in row 3 column 6). Figure 4 shows some reconstruction failure cases. Some are incorrect reconstructed about details, such as the hollows on the bench. Others are missing components whose color is different from the body (columns 1, 4, 6). We consider that this is relating to the simple background color of the picture.

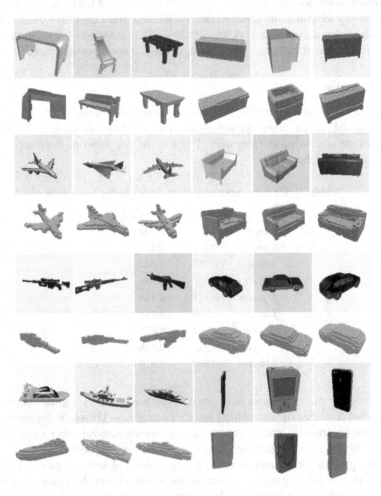

Fig. 3. Sample reconstructions in 8 categories.

Fig. 4. Failure cases. The red grids representative missing components. (Color figure online)

5 Conclusions

In this paper, we design a novel model which can take advantage of category information without any additional annotations. Our model consists of two parts. The first part is rough shape generation network, which reconstruct the rough shape. The second one is category comparison part, with which the generation network can learn the characteristics of objects in the same category of input image. In the experiment, we verify the feasibility of our model on the ShapeNet dataset. The results show that our model achieved better performance on most categories.

Acknowledgment. The work is supported by the National Natural Science Foundation of China under Grant Nos.61971383 and 61631016 and the Fundamental Research Funds for the Central Universities under Grant Nos. 2018XNG1824, YLSZ180226 and 2018XNG1825.

References

1. Chang, A.X., et al.: ShapeNet: an information-rich 3D model repository. Comput. Sci. (2015)
2. Xiang, Y., Roozbeh, M., Savarese, S.: Beyond PASCAL: a benchmark for 3D object detection in the wild. In: Workshop on Applications of Computer Vision, pp. 75–82 (2014)
3. Xiang, Y., et al.: ObjectNet3D: a large scale database for 3D object recognition. In: Leibe, B., Matas, J., Sebe, N., Welling, M. (eds.) ECCV 2016. LNCS, vol. 9912, pp. 160–176. Springer, Cham (2016). https://doi.org/10.1007/978-3-319-46484-8_10
4. Lim, J.J., Pirsiavash, H., Torralba, A.: Parsing IKEA objects: fine pose estimation. In: 2013 IEEE International Conference on Computer Vision (ICCV). IEEE Computer Society (2013)
5. Sun, X., et al.: Pix3D: dataset and methods for single-image 3D shape modeling. In: Computer Vision and Pattern Recognition, pp. 2974–2983 (2018)
6. Wu, N.Z., et al.: 3D ShapeNets: a deep representation for volumetric shape modeling. In: 2015 IEEE Conference on Computer Vision and Pattern Recognition (CVPR). IEEE Computer Society (2015)
7. Lee, H., et al.: Convolutional deep belief networks for scalable unsupervised learning of hierarchical representations. In: International Conference on Machine Learning, pp. 609–616 (2009)
8. Girdhar, R., Fouhey, D.F., Rodriguez, M., Gupta, A.: Learning a predictable and generative vector representation for objects. In: Leibe, B., Matas, J., Sebe, N., Welling, M. (eds.) ECCV 2016. LNCS, vol. 9910, pp. 484–499. Springer, Cham (2016). https://doi.org/10.1007/978-3-319-46466-4_29

9. Choy, C.B., Xu, D., Gwak, J., Chen, K., Savarese, S.: 3D-R2N2: a unified approach for single and multi-view 3D object reconstruction. In: Leibe, B., Matas, J., Sebe, N., Welling, M. (eds.) ECCV 2016. LNCS, vol. 9912, pp. 628–644. Springer, Cham (2016). https://doi.org/10.1007/978-3-319-46484-8_38

10. Hochreiter, S., Schmidhuber, J.: Long short-term memory. Neural Comput. **9**(8), 1735–1780 (1997)

11. Yan, X., et al.: Perspective transformer nets: learning single-view 3D object reconstruction without 3D supervision. In: Neural Information Processing Systems, pp. 1696–1704 (2016)

12. Yang, G., Cui, Y., Belongie, S., Hariharan, B.: Learning single-view 3D reconstruction with limited pose supervision. In: Ferrari, V., Hebert, M., Sminchisescu, C., Weiss, Y. (eds.) ECCV 2018. LNCS, vol. 11219, pp. 90–105. Springer, Cham (2018). https://doi.org/10.1007/978-3-030-01267-0_6

13. Wu, J., et al.: Single image 3D interpreter network. In: Leibe, B., Matas, J., Sebe, N., Welling, M. (eds.) ECCV 2016. LNCS, vol. 9910, pp. 365–382. Springer, Cham (2016). https://doi.org/10.1007/978-3-319-46466-4_22

14. Novotny, D., Larlus, D., Vedaldi, A.: Learning 3D object categories by looking around them. In: International Conference on Computer Vision, pp. 5228–5237 (2017)

15. Zhu, R., et al.: Rethinking reprojection: closing the loop for pose-aware shape reconstruction from a single image. In: International Conference on Computer Vision, pp. 57–65 (2017)

16. Rezende, D.J., et al.: Unsupervised learning of 3D structure from images. arXiv: Computer Vision and Pattern Recognition (2016)

17. Hane, C., Tulsiani, S., Malik, J.: Hierarchical surface prediction for 3D object reconstruction. In: International Conference on 3D Vision, pp. 412–420 (2017)

18. Tatarchenko, M., Dosovitskiy, A., Brox, T.: Multi-view 3D models from single images with a convolutional network. In: Leibe, B., Matas, J., Sebe, N., Welling, M. (eds.) ECCV 2016. LNCS, vol. 9911, pp. 322–337. Springer, Cham (2016). https://doi.org/10.1007/978-3-319-46478-7_20

19. Tulsiani, S., et al.: Multi-view supervision for single-view reconstruction via differentiable ray consistency. arXiv: Computer Vision and Pattern Recognition (2017)

20. Wu, J., et al.: MarrNet: 3D shape reconstruction via 2.5D sketches. In: Neural Information Processing Systems, pp. 540–550 (2017)

21. Fan, H., Su, H., Guibas, L.J.: A point set generation network for 3D object reconstruction from a single image. In: Computer Vision and Pattern Recognition, pp. 2463–2471 (2017)

22. Goodfellow, I.J., et al.: Generative adversarial nets. In: Neural Information Processing Systems, pp. 2672–2680 (2014)

23. Wu, J., et al.: Learning a probabilistic latent space of object shapes via 3D generative-adversarial modeling. In: Computer Vision and Pattern Recognition (2016)

24. Gadelha, M., Maji, S., Wang, R.: 3D shape induction from 2D views of multiple objects. In: International Conference on 3D Vision, pp. 402–411 (2017)

25. Krizhevsky, A., Sutskever, I., Hinton, G.: ImageNet classification with deep convolutional neural networks. NIPS Curran Associates Inc. (2012)

26. Tatarchenko, M., et al.: What do single-view 3D reconstruction networks learn? In: Computer Vision and Pattern Recognition (2019)

27. Wang, N., Zhang, Y., Li, Z., Fu, Y., Liu, W., Jiang, Y.-G.: Pixel2Mesh: generating 3D mesh models from single RGB images. In: Ferrari, V., Hebert, M., Sminchisescu, C., Weiss, Y. (eds.) ECCV 2018. LNCS, vol. 11215, pp. 55–71. Springer, Cham (2018). https://doi.org/10.1007/978-3-030-01252-6_4

28. Groueix, T., et al.: AtlasNet: a Papier-Mâché approach to learning 3D surface generation. In: Computer Vision and Pattern Recognition (2018)
29. Zhao, Y., et al.: 3D point-capsule networks. In: Computer Vision and Pattern Recognition, pp. 1009–1018 (2018)
30. Hu, F., et al.: 3D VAE-attention network: a parallel system for single-view 3D reconstruction. In: Pacific Graphics (2018)

Three-Dimensional Reconstruction of Intravascular Ultrasound Images Based on Deep Learning

Yankun Cao[1], Zhi Liu[1](✉), Xiaoyan Xiao[2], Yushuo Zheng[3], Lizhen Cui[4], Yixian Du[5], and Pengfei Zhang[2]

[1] Intelligent Medical Information Processing, School of Information Science and Engineering, Shandong University, Qingdao 266237, China
liuzhi@sdu.edu.cn
[2] Qilu Hospital, Shandong University, Jinan, China
[3] High School Attached to Shandong Normal University, Jinan 250100, China
[4] School of Software, Shandong University, Jinan, China
[5] School of Computer Science, Fudan University, Shanghai, China

Abstract. Coronary artery disease (CAD), CAD is a common atherosclerotic disease and one of the leading diseases that endanger human health. Acute cardiovascular events are catastrophic, the main cause of which is atherosclerosis (AS) plaque rupture and secondary thrombosis. In order to measure the important parameters such as the diameter, cross-sectional area, volume, wall thickness of the vessel and the size of the AS plaque, it is necessary to first extract the inner and outer membrane edges of the vessel wall in each frame intravascular ultrasound (IVUS) and the plaque edges that may exist. IVUS-based three-dimensional intravascular reconstruction can accurately assess and diagnose the tissue characterization of various cardiovascular diseases to obtain the best treatment options. However, due to the presence of vascular bifurcation in the blood vessels, the presence of bifurcated blood vessels creates great difficulties for the segmentation and reconstruction of the inner and outer membranes. In order to solve this problem, this paper is based on the deep learning method, which first classifies the intravascular bifurcation vessels and normal blood vessels, and then segmentation of the inner and outer membrane, separately. Finally, the three-dimensional reconstruction of the segmented blood vessels is of great significance for the auxiliary diagnosis and treatment of coronary heart disease.

Keywords: Deep learning · Atherosclerosis · Intravascular ultrasound images · Image classification · Three-dimensional reconstruction

Supported in part by the National Key Research and Development Program of China under Grant 2018YFC0831006-3 and in part by the Key Research and Development Plan of Shandong Province under Grant 2017CXGC 1503 and Grant 2018GSF118228, and the fundamental research funds of Shandong University 2018JC009.

G. Zhai et al. (Eds.): IFTC 2019, CCIS 1181, pp. 416–427, 2020.
https://doi.org/10.1007/978-981-15-3341-9_34

1 Introduction

In recent years, the incidence and mortality of coronary heart disease have increased year by year in the world. How to diagnose and prevent coronary heart disease has become an important issue of concern. Acute cardiovascular events are catastrophic, the main cause of which is atherosclerosis (AS) plaque rupture and secondary thrombosis [1]. In the lumen of the coronary arteries, due to the accumulation of fibrosis and calcification, the lumen is narrowed to produce coronary heart disease. Because the images collected by different medical instruments have different characteristics, the application range is also different. At present, the wide application of ultrasound imaging is in the diagnosis of obstetrics and gynecology, ophthalmology and cardiovascular, so the intravenous Ultrasound (IVUS) are selected in this paper [2]. IVUS is a new imaging technology developed in recent years. It combines non-invasive ultrasound technology with invasive catheter technology, which is, connecting the end of the catheter to the ultrasonic probe to emit ultrasonic waves for imaging. It has been commonly used clinically to diagnose vascular lesions. The cross-sectional information of the lumen can be displayed in the IVUS image, from which not only the lumen length can be accurately measured to calculate the volume, but also the size of the plaque. More importantly, it can also provide plaque tissue information to assist in the diagnosis of coronary heart disease and effective interventional therapy. For clinical diagnosis, important parameters such as the diameter, cross-sectional area, volume, thickness of the vessel wall, and the size of the plaque tissue need to be measured. For this purpose, it is necessary to accurately extract and divide the intimal and media-adventitia of the blood vessel wall in each frame IVUS. In the three-dimensional reconstruction of IVUS images, the boundary between the intimal and media-adventitia is also required to be segmented, and the accuracy of the two-dimensional segmentation controls the degree of quantitative analysis of various parameters of the blood vessel and the precision of the three-dimensional reconstruction. IVUS-based three-dimensional intravascular reconstruction can accurately assess and diagnose the tissue characterization of various cardiovascular diseases to obtain the best treatment options. This three-dimensional reconstruction approach monitors and studies the dynamic development and progression of atherosclerotic plaque, thereby minimizing or omitting angiography for clinical navigation and surgery during clinical treatment, reducing surgical time and contrast agent use. However, due to the presence of vascular bifurcation in the blood vessels, the presence of bifurcated blood vessels creates great difficulties for the segmentation and reconstruction of the intimal and media-adventitia. In order to solve this problem, this paper is based on the deep learning method, first classifies the intravascular bifurcation vessels and normal blood vessels, and then divides them separately, and finally reconstructs the segmented blood vessels in three dimensions, which is helpful for coronary heart disease diagnosis and treatment.

Figure 1 below is an IVUS image. It can be seen from the figure that the yellow line indicates the adventitia, the red line refers to the endocardium, and the widened part of the intimal and media-adventitia refers to the plaque tissue.

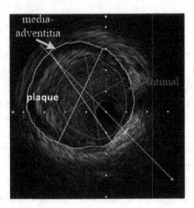

Fig. 1. IVUS image schematic.

In recent years, deep learning has been widely used in medical image processing [3–7]. This thesis is mainly applied to the classification and segmentation methods in medical image analysis. The deep learning method is used to classify and segment the IVUS images, and finally the 3D reconstruction. In the classification algorithm, there is no literature on the classification of IVUS bifurcation vessels and normal vessels. In the segmentation algorithms, there have been many studies on the segmentation of the intimal and media-adventitias. In 2004, Olszewski et al. began to use the application of machine learning to detect plaque plaques [8]. In 2007, Giannoglou et al. used active contour models to detect vascular lumen walls [9]. In 2008, Gerardo et al. applied a probabilistic segmentation method to identify lumen walls [10]. In 2011, Gozde et al. applied a snake-based method to segment the segmentation of each frame of IVUS intimal and media-adventitias [11]. In 2013, Shahed suggested using a texture method for lumen detection to classify between soft and hard plaques [12]. An automated framework for detecting lumen and mediaCadventitia borders in intravascular ultrasound images was developed in 2015 [13]. Gao et al. used the adaptive region growing method and the unsupervised clustering method to perform the detect task. In 2017, Su et al. use the artificial neural network method as the feature learning algorithm for the detection of the lumen and MA borders in IVUS images. Su et al. applies the basic neural network to the segmentation of MediaCAdventitia Borders, and the performance has been improved [14]. In 2018, the latest article used the FCN segmentation network in deep learning to segment the endovascular and extravascular membranes and achieved excellent results [15].

2   Method

Deep learning actually includes a variety of artificial intelligence methods that use a large number of simple interconnected units to perform complex tasks. Deep learning can learn features from a large amount of data rather than simply entering some pre-programmed instructions. These algorithms can solve many tasks, such as positioning and classifying objects in images, natural language processing, playing games, and so on. In deep neural networks, the most basic one is the Artificial Neural Network (ANN). A typical artificial neural network mainly has three network layers, an input layer, a hidden layer, and an output layer. The input layer mainly receives the input information and delivers the information to the hidden layer; the hidden layer functions to pass the received input layer information to the output layer after the operation; the output layer is simpler, that is, the entire calculation result is output.

2.1   Convolutional Neural Network

Currently, the Convolutional Neural Network (CNN) is considered to be the most widely used machine learning technology with good feature extraction and recognition capabilities. CNN is actually a deep artificial neural network. Its biggest feature is weight sharing, which greatly reduces the complexity of the network model, and the number of weights is greatly reduced. A typical CNN network structure is shown in Fig. 2 below:

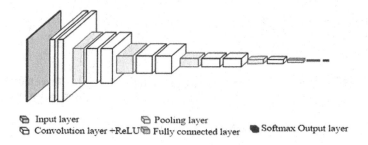

🖻 Input layer   🖻 Pooling layer
🖻 Convolution layer +ReLU🖻 Fully connected layer   🔴 Softmax Output layer

Fig. 2. CNN network structure diagram.

As shown, a typical CNN network structure typically includes alternating convolutional and pooling layers, with one or more fully connected layers at the end. Let's introduce each layer separately: A convolutional layer consists of a set of convolution kernels, each act as a core, which are associated with a small portion of the image. The small block into which the image is divided is called the receptive field [16], and the receptive field is convolved with a specific set of weights, that is, the elements of the filter (weight) are multiplied by the elements of the corresponding receptive field. The calculation formula for the convolution operation is as follows:

$$F_l^k = \left(I_{x,y} * K_l^k\right) \tag{1}$$

In Eq. 1, I represents the input image, x, y represents the spatial position, and K is the convolution kernel. Dividing an image into small pieces helps extract locally correlated pixel values, and this locally aggregated information is also referred to as feature extraction. Different features are extracted by sliding the convolution kernel over an image with the same set of weights. The types of filters are different, the size is different, the filling type and the convolution direction are different, and so the CNN network can be extended to a variety of networks with different characteristics [17].

2.2 Classification and Segmentation

Although LeNet [18], previously proposed by Licun, began the history of deep CNN, this network is limited to handwritten digit recognition tasks and cannot be extended to all image classifications. AlexNet was proposed by Krizhevesky et al. [19] and is the first deep CNN structure that has made a breakthrough in image classification and recognition tasks. This network enhances CNN's learning ability by deepening the number of layers and applying multiple optimization strategies. The basic architecture design of AlexNet is shown in Fig. 3.

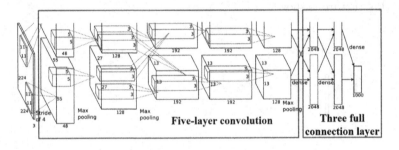

Fig. 3. AlexNet network structure diagram.

As can be seen from Fig. 3, in AlexNet, there are five convolutional layers, making CNN suitable for images of different sizes. Filters using large convolution kernels in the first and second convolutional layers are 11×11 and 7×7, respectively. The convolution kernel used in the third, fourth, and fifth convolutional layers is reduced to 3×3. In addition, AlexNet uses three pooling layers, and the pooling operation is the maximum pooling used. In the last three layers, a fully connected layer is used. The first two fully connected layers are used for feature extraction. The last layer of the fully connected layer is a softmax classifier that outputs the classification result.

Long et al. [20] proposed the first Fully Convolutional Network (FCN), and since then, a major breakthrough has been made in semantic segmentation based on deep learning. At present, the FCN network has become the standard of semantic segmentation, and most of the segmentation networks adopt this architecture. By transforming the last layer of the traditional classification network into a convolutional layer, the FCN becomes a full convolutional network, which can generate an input generation probability map of any size. The FCN recovers spatial information from the downsampling layer by adding an upsampling layer to a standard convolutional network. A leaping architecture (shallow layer) is defined in the FCN network, which combines semantic information and appearance information in the deep layer to produce accurate and in-depth segmentation. The basic idea of image classification is to reconstruct and fine tune the classification model to effectively learn the annotation data of the entire image input and the entire image. This extends these classification models to segmentation and improves the overall architecture through a combination of multiple resolution layers. Figure 4 below shows the network structure of the FCN:

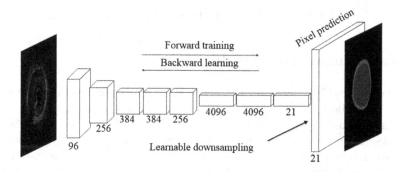

Fig. 4. FCN segmentation network structure.

As shown in the above figure, in the traditional CNN structure, the first five layers are convolutional layers, the sixth and seventh layers are fully connected layers, the output is a one-dimensional vector with a length of 4096, and the eighth layer is a softmax layer, and the output is It is a one-dimensional vector with a length of 1000, which corresponds to a probability of 1000 categories. The last three layers are represented in the FCN as a convolutional layer having a convolution kernel size of 1×1. After multiple convolutions and pooling operations, the resulting image is getting smaller and smaller and the resolution is getting lower and lower. In order to recover the resolution of the original image from this coarse resolution, the FCN uses the upsampling method. After 5 times of convolution and downsampling of the FCN, the resolution of the picture is reduced by 32 times, then the last layer of the output image requires 32 times upsampling to restore the original image size. The specific upsampling is the deconvolution operation, but the result of this recovery is not accurate, and some details cannot be recovered. Therefore, the output of the fourth and third layers of the convolution layer needs to be 16 times and 8 times. Upsampling, which makes the results more detailed, Fig. 5 below shows the process of this convolution kernel deconvolution downsampling:

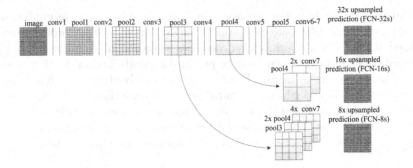

Fig. 5. FCN convolution and deconvolution upsampling.

3   Experiments and Results

This paper is mainly based on IVUS image, which is analyzed and processed. Firstly, the collected IVUS image bifurcation vessels and normal blood vessels are tagged into data sets, which are classified and evaluated with classification networks. Then, the intimal and media-adventitia were manually labeled into bifurcation vessels and normal blood vessels, and the data were collected by classical segmentation networks. The segmentation accuracy was evaluated and the optimal segmentation model was selected again. Finally, the results of the two models are integrated, and the final intact intimal and media-adventitia are reconstructed in three dimensions for further analysis. The data sets used in this paper are all from the clinical data of the hospital and have more clinical application effects. The flow chart of the whole process is shown in Fig. 6:

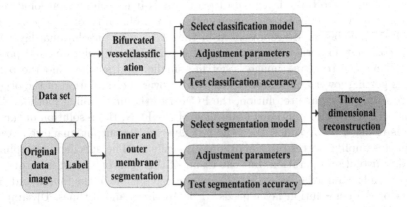

Fig. 6. IVUS image morphology processing flow.

3.1 Data Processing

The data in this article is clinical data collected from the hospital and is of great clinical significance. The first is the data set of human normal blood vessels and bifurcated blood vessels. This data set includes 2288 images, including 1144 normal blood vessel images and 1144 images of bifurcated blood vessels. The bifurcation vessels and normal blood vessels are labeled under the guidance of a professional doctor. Then, the intimal and media-adventitia segmentation data sets of bifurcated vessels and normal vessels were constructed respectively, including 1144 bifurcation blood vessel data and 5216 normal blood vessel data. The labeling of the intimal and media-adventitia in the data set is marked by professional medical students and has certain medical credibility. The images in the data set used herein were obtained using an intravascular ultrasound system from Boston Scientific. The probe frequency is 40 MHz and the rotation frequency is 1800 rpm. A schematic diagram of normal blood vessels and bifurcated vessels is shown in Fig. 7 below:

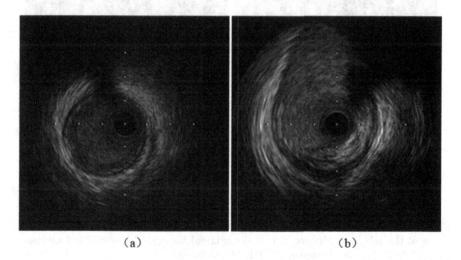

(a)   (b)

Fig. 7. (*a*): IVUS image of normal blood vessels (*b*): IVUS image of bifurcated vessels.

The collected data is tagged, assuming a normal blood vessel label of 1, and a bifurcated blood vessel label of zero. Make a data set after labeling it. After the IVUS image is collected, the inner and outer membranes need to be manually labeled. The labeling software used in this paper is LabelMe [21], assuming that the label of the inner membrane is 0 and the label of the outer membrane is 1. The original data and label data are shown in Fig. 8.

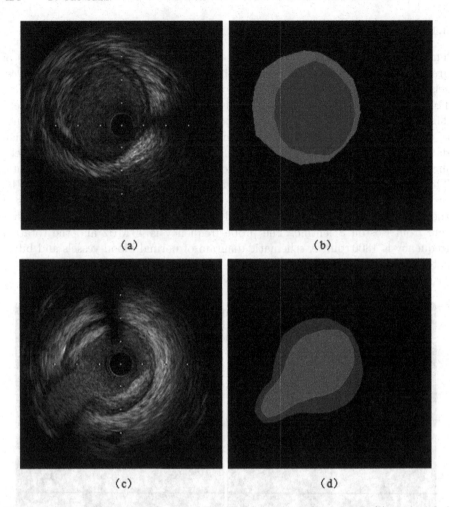

(a)   (b)

(c)   (d)

Fig. 8. (a): is the ultrasound image of the original normal blood vessel, (b): is the label image of the labelme software, (c): is the original bifurcated blood vessel ultrasound image, (d): is the label image of the labelme software.

3.2 Experimental Parameter Setting and Result Analysis

After constructing the data set, this paper uses four basic convolutional network structures to classify blood vessels. The parameters of the classification AlexNet network structures are shown in the following: Maximum number of iterations is 20000, Base learning rate is 0.001, Learning rate adjustment strategy is Step, Weight attenuation rate is 0.0005, and Batch size is 256. The classification accuracy of the classification of the bifurcation vessels and normal blood vessels by this network is 97.67 The parameters of the segmentation FCN network structures are shown in the following: Maximum number of iterations is 100000, Base learning rate is 1e−10, Learning rate adjustment strategy is Fixed,

Weight attenuation rate is 0.0005. The mean IOU of the segmentation of the bifurcation vessels and normal blood vessels by this network is 0.8523. The final 3D reconstruction results are shown as Fig. 9 below:

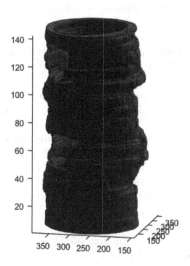

Fig. 9. Three-dimensional reconstructions of blood vessel images.

As shown in the figure, the hollow part of the figure is a bifurcated blood vessel. As can be seen from the figure, this reconstruction effect diagram can well see the bifurcated blood vessel and the normal blood vessel, as well as the wall of the blood vessel. This is of great significance for the next diagnosis and treatment.

4   Conclusion

Coronary artery disease is a common atherosclerotic disease and one of the leading diseases that endanger human health. Acute cardiovascular events are catastrophic, the main cause of which is atherosclerosis plaque rupture and secondary thrombosis. In this paper, the collected IVUS image bifurcation vessels and normal blood vessels are first tagged into a data set. Based on the classical classification network AlexNet, the bifurcated vessels and normal vessels are classified. Then, the intimal and media-adventitia were manually labeled for bifurcation and normal vessels. The layer of FCN is used to segment the intimal and media-adventitia respectively. Finally, the obtained endometrial segmentation results are reconstructed in three dimensions for further analysis and assisting.

References

1. Gao, Z., et al.: Robust estimation of carotid artery wall motion using the elasticity-based state-space approach. Med. Image Anal. **37**, 1–21 (2017)
2. Okada, K., Fitzgerald, P.J., Honda, Y.: Intravascular ultrasound. In: Lanzer, P. (ed.) Textbook of Catheter-Based Cardiovascular Interventions, pp. 329–363. Springer, Cham (2018). https://doi.org/10.1007/978-3-319-55994-0_19
3. Han, G., et al.: Hybrid resampling and multi-feature fusion for automatic recognition of cavity imaging sign in lung CT. Future Gener. Comput. Syst. **99**, 558–570 (2019)
4. Xu, C., et al.: Direct delineation of myocardial infarction without contrast agents using a joint motion feature learning architecture. Med. Image Anal. **50**, 82–94 (2018)
5. Dhungel, N., Carneiro, G., Bradley, A.P.: The automated learning of deep features for breast mass classification from mammograms. In: Ourselin, S., Joskowicz, L., Sabuncu, M.R., Unal, G., Wells, W. (eds.) MICCAI 2016. LNCS, vol. 9901, pp. 106–114. Springer, Cham (2016). https://doi.org/10.1007/978-3-319-46723-8_13
6. Havaei, M., et al.: Brain tumor segmentation with deep neural networks. Med. Image Anal. **35**, 18–31 (2017)
7. Miao, S., Wang, Z.J., Liao, R.: A CNN regression approach for real-time 2D/3D registration. IEEE Trans. Med. Imaging **35**(5), 1352–1363 (2016)
8. Olszewski, M.E., Wahle, A., Mitchell, S.C., Sonka, M.: Segmentation of intravascular ultrasound images: a machine learning approach mimicking human vision. In: International Congress Series, vol. 1268, pp. 1045–1049. Elsevier (2004)
9. Giannoglou, G.D., et al.: A novel active contour model for fully automated segmentation of intravascular ultrasound images: in vivo validation in human coronary arteries. Comput. Biol. Med. **37**(9), 1292–1302 (2007)
10. Mendizabal-Ruiz, G., Rivera, M., Kakadiaris, I.A.: A probabilistic segmentation method for the identification of luminal borders in intravascular ultrasound images. In: 2008 IEEE Conference on Computer Vision and Pattern Recognition, pp. 1–8. IEEE (2008)
11. Zhu, X., Zhang, P., Shao, J., Cheng, Y., Zhang, Y., Bai, J.: A snake-based method for segmentation of intravascular ultrasound images and its in vivo validation. Ultrasonics **51**(2), 181–189 (2011)
12. Dehnavi, S.M., Babu, M.P., Yazchi, M., Basij, M.: Automatic soft and hard plaque detection in IVUS images: a textural approach. In: 2013 IEEE Conference on Information and Communication Technologies, pp. 214–219. IEEE (2013)
13. Gao, Z., et al.: Automated framework for detecting lumen and mediacadventitia borders in intravascular ultrasound images. Ultrasound Med. Biol. **41**(7), 2001–2021 (2015)
14. Su, S., Hu, Z., Lin, Q., Hau, W.K., Gao, Z., Zhang, H.: An artificial neural network method for lumen and media-adventitia border detection in ivus. Comput. Med. Imaging Graph. **57**, 29–39 (2017)
15. Yang, J., Tong, L., Faraji, M., Basu, A.: IVUS-Net: an intravascular ultrasound segmentation network. In: Basu, A., Berretti, S. (eds.) ICSM 2018. LNCS, vol. 11010, pp. 367–377. Springer, Cham (2018). https://doi.org/10.1007/978-3-030-04375-9_31
16. Bouvrie, J.: Notes on convolutional neural networks (2006)
17. LeCun, Y., Bengio, Y., Hinton, G.: Deep learning. Nature **521**(7553), 436 (2015)

18. LeCun, Y., et al.: Learning algorithms for classification: a comparison on handwritten digit recognition. Neural Netw.: Stat. Mech. Perspect. **261**, 276 (1995)
19. Krizhevsky, A., Sutskever, I., Hinton, G.E.: ImageNet classification with deep convolutional neural networks. In: Advances in Neural Information Processing Systems, pp. 1097–1105 (2012)
20. Long, J., Shelhamer, E., Darrell, T.: Fully convolutional networks for semantic segmentation. In: Proceedings of the IEEE Conference on Computer Vision and Pattern Recognition, pp. 3431–3440 (2015)
21. Russell, B.C., Torralba, A., Murphy, K.P., Freeman, W.T.: LabelME: a database and web-based tool for image annotation. Int. J. Comput. Vis. **77**(1–3), 157–173 (2008)

Author Index

Printed in the United States
By Bookmasters